Alternatives to Mineral Nitrogen Fertilizers in Agriculture: State of the Art, Challenges and Future Prospects

Alternatives to Mineral Nitrogen Fertilizers in Agriculture: State of the Art, Challenges and Future Prospects

Guest Editor

Germán Tortosa

MDPI

Basel • Beijing • Wuhan • Barcelona • Belgrade • Novi Sad • Cluj • Manchester

Guest Editor
Germán Tortosa
Department of Soil and Plant
Microbiology
Spanish National Research
Council (CSIC)
Granada
Spain

Editorial Office
MDPI AG
Grosspeteranlage 5
4052 Basel, Switzerland

This is a reprint of the Special Issue, published open access by the journal *Nitrogen* (ISSN 2504-3129), freely accessible at: https://www.mdpi.com/journal/nitrogen/special_issues/Altern_N_Fertil.

For citation purposes, cite each article independently as indicated on the article page online and as indicated below:

Lastname, A.A.; Lastname, B.B. Article Title. *Journal Name* **Year**, *Volume Number*, Page Range.

ISBN 978-3-7258-8079-9 (Hbk)
ISBN 978-3-7258-8080-5 (PDF)
https://doi.org/10.3390/books978-3-7258-8080-5

Contents

About the Editor

Germán Tortosa

Germán Tortosa obtained his degree in Chemistry (2003) and PhD in Agricultural Sciences (2011) from the University of Murcia (Spain). He is currently an R&D Technologist in the Department of Soil and Plant Microbiology at the Estación Experimental del Zaidín (EEZ-CSIC), Spanish National Research Council (CSIC), Granada, Spain. He has published more than 45 scientific publications focused on biological valorisation of waste, composting, organic and biological fertilizers, and biostimulants. He has participated in 17 research projects and has led 13 R&D transfer projects in collaboration with agricultural companies.

His research is focused on agricultural sustainability, microbial biotechnology, and environmental protection and is structured around four main areas. The first focuses on waste valorization and the circular economy, transforming agro-industrial residues into high-value agricultural inputs through composting and related processes. The second addresses biological nitrogen fixation and abiotic stress, examining legume–rhizobia interactions and the impact of environmental stressors on nitrogen fixation efficiency and greenhouse gas emissions. The third area explores microbial ecology and biomarker identification, applying omics methods to characterize active microbial communities and identify indicators of compost maturity and environmental change. Finally, his research evaluates biostimulants and their role in enhancing crop performance, resilience, and nutritional quality, including demonstrated improvements in yield and functional compounds.

nitrogen

Editorial

Alternatives to Mineral Nitrogen Fertilizers in Agriculture: State of the Art, Challenges and Future Prospects

Germán Tortosa

Department of Plant and Soil Microbiology, Estación Experimental del Zaidín (EEZ), CSIC, c/ Profesor Albareda, 1, 18008 Granada, Spain; german.tortosa@eez.csic.es

Despite being one of the most abundant elements in the biosphere, nitrogen remains a limiting factor in agricultural systems [1]. The use of nitrogen fertilizers produced via the Haber–Bosch process has significantly increased crop yields and global food production over the past century. However, the prolonged and excessive application of mineral fertilizers has disrupted the natural nitrogen cycle in soils [2]. Each year, large quantities of reactive nitrogen compounds—including nitrogen oxides (NO_x), ammonia (NH_3), nitrous oxide (N_2O), and nitrate (NO_3^-)—are released into water bodies, soils, and the atmosphere [3,4]. This has led to increased environmental degradation, such as water eutrophication and greenhouse gas (GHG) emissions [5,6]. To develop a more environmentally sustainable agriculture, it is essential to investigate and implement alternatives that reduce reliance on mineral nitrogen fertilizers without compromising crop productivity [7].

In this Special Issue, eleven research teams have published studies addressing the minimization of environmental impacts associated with nitrogen fertilization and the optimization of nutrient management strategies. Hagedorn et al. (Contribution 1) evaluated drainage water management (DWM) as a controlled drainage strategy to reduce nitrogen losses and GHG emissions in a corn–soybean rotation system over three years. They found that while DWM effectively reduced NO_3^- export, it had no statistically significant effect on N_2O or CH_4 emissions, suggesting its potential as a low-risk mitigation strategy. Rebello et al. (Contribution 2) conducted a systematic review on nitrogen fertilization in tea (*Camellia sinensis* L.) cultivation. They assessed the impact of excessive nitrogen inputs on soil health and evaluated alternative practices such as organic and mixed fertilizers, controlled-release formulations, nitrification inhibitors, and biochar amendments. While many of these alternatives reduced nitrogen loading, some trade-offs were observed, including reduced yields or increased N_2O emissions. The authors emphasized the importance of site-specific nutrient management tailored to local conditions. Sanyal and Graham (Contribution 3) investigated the use of the Haney Soil Test to predict optimal nitrogen requirements for winter wheat under semi-arid conditions. Their results highlighted the role of in-season precipitation in determining nitrogen needs and underscored the utility of soil respiration and soil health metrics in refining N recommendations. However, they also identified the need for further research to improve estimates of nitrogen mineralization. Varga et al. (Contribution 4) reviewed nitrogen use efficiency in sugar beet production, where increasing root yield does not necessarily correspond to higher sucrose concentration. They stressed the importance of optimizing nitrogen availability to achieve both high yield and desirable root quality. In viticulture, González-Lázaro et al. (Contribution 5) assessed the foliar application of methyl jasmonate and urea on grape and wine composition. Their findings revealed season-dependent effects on polysaccharide and monosaccharide profiles, highlighting the need for further research into the underlying mechanisms. Using life cycle assessment (LCA), Litskas (Contribution 6) compared the environmental impacts of various nitrogen fertilizers. Results showed that organo-mineral fertilizers (compost enriched with synthetic N) had the highest environmental footprint, while cattle manure emerged as the most environmentally favorable option in terms of water use, fossil energy consumption, and global warming potential. Massey and Davis (Contribution 7) explored the use of nitrogen-fixing cyanobacteria as a sustainable alternative to synthetic fertilizers. Their review covered current applications and challenges, including economic, infrastructural, and geopolitical barriers to widespread adoption. Tortosa et al. (Contribution 8) examined the synergistic effects of *Bradyrhizobium diazoefficiens* inoculation and "alperujo" compost on soybean productivity. Their results demonstrated improved plant biomass, nitrogen content, and symbiotic efficiency, supporting the integration of compost to enhance legume-based fertilization strategies. Da Costa-Cezar et al. (Contribution 9) investigated the use of potassium iodate as a urease inhibitor in urea-based fertilizers. Their findings showed reduced ammonia volatilization and improved nitrogen uptake in maize, although effectiveness varied with soil type and temperature. Ibiapina de Jesus et al. (Contribution 10) evaluated nitrogen mineralization and availability in two different soils amended with three commercial organic fertilizers. Their study emphasized the influence of soil properties and temperature on nitrogen release dynamics. Lastly, Pérez-Álvarez et al. (Contribution 11) assessed the effects of algal extracts, *Rhizobium* sp., and *Trichoderma asperellum* on nitrogen assimilation and yield in hybrid maize. Among the biofertilizers tested, *Rhizobium* sp. had the most significant impact on crop performance, confirming its potential role in sustainable nitrogen management.

The studies presented in this Special Issue provide a comprehensive overview of current strategies aimed at reducing the environmental footprint of nitrogen fertilization in agriculture. Collectively, they demonstrate that while alternatives to mineral nitrogen fertilizers—such as organic amendments, biological nitrogen fixation, and improved management tools—hold significant potential, their effectiveness is often context-dependent and influenced by soil, climate, and crop type. Advancing sustainable nitrogen management will require integrative approaches supported by robust agronomic research, site-specific recommendations, and interdisciplinary collaboration. Future efforts must focus on optimizing these alternatives to ensure environmental protection without compromising agricultural productivity or food security.

I sincerely thank the contributing authors for advancing knowledge toward sustainable nitrogen use and more environmentally responsible agricultural systems.

Conflicts of Interest: The author declares no conflicts of interest.

List of Contributions:

1. Hagedorn, J.G.; Davidson, E.A.; Fisher, T.R.; Fox, R.J.; Zhu, Q.; Gustafson, A.B.; Koontz, E.; Castro, M.S.; Lewis, J. Effects of Drainage Water Management in a Corn–Soy Rotation on Soil N_2O and CH_4 Fluxes. *Nitrogen* **2022**, *3*, 128–148. https://doi.org/10.3390/nitrogen3010010.
2. Rebello, R.; Burgess, P.J.; Girkin, N.T. Identifying Sustainable Nitrogen Management Practices for Tea Plantations. *Nitrogen* **2022**, *3*, 43–57. https://doi.org/10.3390/nitrogen3010003.
3. Sanyal, D.; Graham, C. Using the Haney Soil Test to Predict Nitrogen Requirements in Winter Wheat (*Triticum aestivum* L.). *Nitrogen* **2022**, *3*, 376–386. https://doi.org/10.3390/nitrogen3020024.
4. Varga, I.; Jović, J.; Rastija, M.; Markulj Kulundžić, A.; Zebec, V.; Lončarić, Z.; Iljkić, D.; Antunović, M. Efficiency and Management of Nitrogen Fertilization in Sugar Beet as Spring Crop: A Review. *Nitrogen* **2022**, *3*, 170–185. https://doi.org/10.3390/nitrogen3020013.

5. González-Lázaro, M.; Martínez-Lapuente, L.; Garde-Cerdán, T.; Landín Ross-Magahy, M.; Torres-Díaz, L.L.; Pérez-Álvarez, E.P.; Guadalupe, Z.; Ayestarán, B. Effect of Methyl Jasmonate Plus Urea Foliar Application on the Polysaccharide and Monosaccharide Composition of Tempranillo Grapes and Wines and on the Wine's Quality. *Nitrogen* **2023**, *4*, 263–278. https://doi.org/10.3390/nitrogen4030019.
6. Litskas, V.D. Environmental Impact Assessment for Animal Waste, Organic and Synthetic Fertilizers. *Nitrogen* **2023**, *4*, 16–25. https://doi.org/10.3390/nitrogen4010002.
7. Massey, M.S.; Davis, J.G. Beyond Soil Inoculation: Cyanobacteria as a Fertilizer Replacement. *Nitrogen* **2023**, *4*, 253–262. https://doi.org/10.3390/nitrogen4030018.
8. Tortosa, G.; Mesa, S.; Delgado, M.J.; Amaya-Gómez, C.V. "Alperujo" Compost Improves Nodulation and Symbiotic Nitrogen Fixation of Soybean Inoculated with *Bradyrhizobium diazoefficiens*. *Nitrogen* **2023**, *4*, 223–230. https://doi.org/10.3390/nitrogen4020015.
9. Cezar, J.V.d.C.; Morais, E.G.d.; Lima, J.d.S.; Benevenute, P.A.N.; Guilherme, L.R.G. Iodine-Enriched Urea Reduces Volatilization and Improves Nitrogen Uptake in Maize Plants. *Nitrogen* **2024**, *5*, 891–902. https://doi.org/10.3390/nitrogen5040057.
10. de Jesus, H.I.; Cassity-Duffey, K.; Dutta, B.; da Silva, A.L.B.R.; Coolong, T. Influence of Soil Type and Temperature on Nitrogen Mineralization from Organic Fertilizers. *Nitrogen* **2024**, *5*, 47–61. https://doi.org/10.3390/nitrogen5010004.
11. Pérez-Álvarez, S.; Ochoa-Chaparro, E.H.; Anchondo-Páez, J.C.; Escobedo-Bonilla, C.M.; Rascón-Solano, J.; Magallanes-Tapia, M.A.; Uranga-Valencia, L.P.; Hernández-Campos, R.; Sánchez, E. Nitrogen Assimilation, Biomass, and Yield in Response to Application of Algal Extracts, *Rhizobium* sp., and *Trichoderma asperellum* as Biofertilizers in Hybrid Maize. *Nitrogen* **2024**, *5*, 1031–1047. https://doi.org/10.3390/nitrogen5040066.

References

1. Bhatt, R.; Kunal; Moulick, D.; Bárek, V.; Brestic, M.; Gaber, A.; Skalicky, M.; Hossain, A. Sustainable strategies to limit nitrogen loss in agriculture through improving its use efficiency—Aiming to reduce environmental pollution. *J. Agric. Food Res.* **2025**, *22*, 101957. [CrossRef]
2. Galloway, J.N.; Townsend, A.R.; Erisman, J.W.; Bekunda, M.; Cai, Z.; Freney, J.R.; Martinelli, L.A.; Seitzinger, S.P.; Sutton, M.A. Transformation of the nitrogen cycle: Recent trends, questions, and potential solutions. *Science* **2008**, *320*, 889–892. [CrossRef] [PubMed]
3. Vitousek, P.M.; Aber, J.D.; Howarth, R.W.; Likens, G.E.; Matson, P.A.; Schindler, D.W.; Schlesinger, W.H.; Tilman, D.G. Human alteration of the global nitrogen cycle: Sources and consequences. *Ecol. Appl.* **1997**, *7*, 737–750. [CrossRef]
4. Tortosa, G.; Correa, D.; Sánchez-Raya, A.J.; Delgado, A.; Sánchez-Monedero, M.A.; Bedmar, E.J. Effects of nitrate contamination and seasonal variation on the denitrification and greenhouse gas production in La Rocina stream (Doñana National Park, SW Spain). *Ecol. Eng.* **2011**, *37*, 539–548. [CrossRef]
5. Morrissy, J.G.; Currell, M.J.; Reichman, S.M.; Surapaneni, A.; Megharaj, M.; Crosbie, N.D.; Hirth, D.; Aquilina, S.; Rajendram, W.; Ball, A.S. Nitrogen contamination and bioremediation in groundwater and the environment: A review. *Earth Sci. Rev.* **2021**, *222*, 103816. [CrossRef]
6. Viancelli, A.; Michelon, W. Climate Change and Nitrogen Dynamics: Challenges and Strategies for a Sustainable Future. *Nitrogen* **2024**, *5*, 688–701. [CrossRef]
7. Amaya-Gómez, C.V.; Flórez-Martínez, D.H.; Cayuela, M.L.; Tortosa, G. Compost and vermicompost improve symbiotic nitrogen fixation, physiology and yield of the Rhizobium-legume symbiosis: A systematic review. *Appl. Soil Ecol.* **2025**, *210*, 106051. [CrossRef]

nitrogen

MDPI

Article

Nitrogen Assimilation, Biomass, and Yield in Response to Application of Algal Extracts, *Rhizobium* sp., and *Trichoderma asperellum* as Biofertilizers in Hybrid Maize

Sandra Pérez-Álvarez [1], Erick H. Ochoa-Chaparro [2], Julio César Anchondo-Páez [2], César M. Escobedo-Bonilla [3], Joel Rascón-Solano [1], Marco A. Magallanes-Tapia [3], Luisa Patricia Uranga-Valencia [1], Reinier Hernández-Campos [4] and Esteban Sánchez [2,*]

1 Autonomous University of Chihuahua, Km 2.5 carretera a Rosales, Campus Delicias, Delicias C.P. 33000, Chihuahua, Mexico; spalvarez@uach.mx (S.P.-Á.); jsolano@uach.mx (J.R.-S.); luranga@uach.mx (L.P.U.-V.)

2 Food and Development Research Center A.C. Unidad Delicias, Av. Cuarta Sur 3828, Fracc. Vencedores del Desierto, Delicias C.P. 33089, Chihuahua, Mexico; eriicktronik@hotmail.com (E.H.O.-C.); anchondo_456@hotmail.com (J.C.A.-P.)

3 National Polytechnic Institute-CIIDIR Unidad Sinaloa, Juan de Dios Batiz Paredes No. 250, Guasave C.P. 81101, Sinaloa, Mexico; cesar_escobedomx@yahoo.com (C.M.E.-B.); mmagallanes@ipn.mx (M.A.M.-T.)

4 Unidad Xochimilco, Department of Agricultural and Animal Production, Autonomous Metropolitan University, Calzada del Hueso 1100, Colonia Villa Quietud, Coyoacan, Ciudad de Mexico C.P. 04510, Mexico; reinierhc86@hotmail.com

* Correspondence: esteban@ciad.mx; Tel.: +52-639-549-4681

Abstract: Nitrogen is essential for plants' growth, yield, and crop quality, and its deficiency limits food production worldwide. In addition, excessive fertilization and inefficient use of N can increase production costs and cause environmental problems. A possible solution to this problem is the application of biofertilizers, which improve N assimilation and increase biomass and yield. Therefore, the objective of this research was to evaluate the impact of the application of a combination of green and red algae (*Ulva lactuca* and *Solieria* spp.), *Rhizobium* sp., *Trichoderma asperellum*, and the combination of the above three biofertilizers on N assimilation. A completely randomized design was performed, with 10 plants per treatment and five treatments: T1 = control; T2 = algal extracts; T3 = *Rhizobium* sp.; T4 = *T. asperellum*; T5 = T2 + T3 + T4. Our analyses showed that the biofertilizers' application was better than the control. The application of *Rhizobium* sp. had the best performance amongst all of the biofertilizers, with the highest nitrate reductase activity in maize leaves, which enhanced photosynthesis, increasing biomass and yield. The use of *Rhizobium* sp. showed increases in biomass (13.4%) and yield (11.82%) compared to the control. This research shows that biofertilizers can be a key component for sustainable agricultural practices.

Keywords: algal extracts; biofertilizers; nitrogen assimilation; *Rhizobium* sp.; *Trichoderma asperellum*; *Zea mays* L.

1. Introduction

Maize (*Zea mays* L.) is the most used cereal for human consumption in Mexico, being a fundamental food item in the daily diet [1]. The largest maize producer in the world is the USA, with 348.75 million tons, while Mexico ranks sixth with 26.63 million tons [2]. To achieve high yields, chemical nitrogen (N) fertilizers are widely used in intensive maize production systems, due to the crop's high N demand [3,4].

Nitrogen (N) is essential for plants' growth, yield, and crop quality, and N deficiency constrains food production worldwide [5]. N is crucial for the synthesis of amino acids, proteins, and nucleic acids and is a key element in many physiological processes [6].

N assimilation is a complex process that begins with the uptake of nitrate (NO_3^-) and ammonium (NH_4^+), the main inorganic forms that plants can absorb. Once accumulated inside the roots, the nitrate reductase enzyme (NR) reduces NO_3^- to NO_2^-, and then the nitrite reductase enzyme (NiR) reduces NO_2^- to NH_4^+, which is then converted into organic N compounds [7,8]. Among all of the enzymes that participate in N assimilation, NR is the key player in regulating N's incorporation in plants [9]. Thus, improving N uptake and assimilation can boost yields and crop quality. However, excessive N fertilization and low N-use efficiency can lead to increased production costs and environmental concerns [10].

A potential solution to this issue is the application of biofertilizers, which improve soil fertility [11] and enhance plants' growth and yield [12]. Biofertilizers contain microorganisms or natural substances [13] that enhance soil properties, influence plant growth, and increase productivity both under optimal conditions and under stress, without harming the environment [14]. As examples of these biofertilizers, Plant Growth-Promoting Fungi (PGPF) and Plant Growth-Promoting Bacteria (PGPB) are beneficial microorganisms that are often used as biofertilizers to aid in plant growth [15] and can work as biological control agents [16–18]. Previous studies have demonstrated the effective use of biofertilizers as a sustainable strategy to improve plants' N uptake [19–21]. PGPB can facilitate the absorption of crucial nutrients such as N, phosphorous (P), and iron (Fe) or regulate plant hormones including auxin and cytokinin to promote plant growth [22]. Some of the most important genera of PGPB are *Rhizobium*, *Azotobacter*, *Bacillus*, and *Azospirillum* [23,24].

Algae, essential photosynthetic microorganisms abundant in N and potassium, are present in all terrestrial environments. They are used in agriculture as bio-based fertilizers to enhance soil nutrients, thereby promoting plant growth and increasing yields [25,26]. The levels (2, 4, and 6 kg L^{-1} of water) of seaweed extract (brown seaweed *Ecklonia maxima*) significantly affected all growth traits (plant height, stem diameter, number of cobs per plant, cob seeds' weight) and the yield of maize in both the spring and autumn seasons [27]. Several authors have reported the use of seaweed extracts (*Gracilaria edulis*, *Laminaria* sp., *A. nodosum*, and *K. alvarezii*) in maize crops, with increases in seed germination percentage and rate, seedling vigor, shoot and root growth, net carbon assimilation, and total grain yield [28–30]. When applied to drought-stressed green bean plants, *Ulva rigida* extracts improved vegetative growth by increasing chlorophyll contents, enhancing antioxidant systems, and reducing lipid peroxidation damage [31]. In a similar way, *Solieria chordalis* extracts promoted the height, weight, and chlorophyll contents of in vitro radish seedlings [32]. Different micronutrients (Ca, Mg, Mn, B, Zn, Cu) and plant regulators (abscisic acid, auxins, cytokinins, gibberellins, jasmonates) have been found in these extracts, which stimulate plants' metabolism and growth [33,34].

Several species of the genus *Rhizobium* produce nodules on the roots of legumes, as well as endophytes that can reside within plant tissues [35]. Under certain conditions, these bacteria act as phosphate solubilizers, fix N, and produce hormones in association with non-legume crops [36]. According to Widodo et al. [37], *Rhizobium* can enhance plants' growth directly by regulating the aforementioned plant hormones, which modulate root architecture and growth. Indeed, N absorption and utilization are dependent on root length, root surface area, and root biomass [36]. The inoculation of maize crops with *Rhizobium* sp. from various rhizospheres (rice rhizosphere isolates, maize rhizosphere isolates, soybean rhizosphere isolates, peanut rhizosphere isolates, and edamame rhizosphere isolates) showed a significant effect on plant height, stem diameter, husked ear weight, and dry ear weight compared to controls [37]. Additionally, significant increases in maize grain yield were reported when the crop was inoculated with several species of PGPB (*Azospirillum brasilense*, *Rhizobium tropici*, *Bradyrhizobium* sp., and *Rhizobium* sp.) [38,39].

Species of the genus *Trichoderma* are commonly found across the globe [40]. These microorganisms inhabit nearly all soil types and frequently function as plant mycoparasites, saprotrophs, and symbionts [41,42]. In a similar way to PGPF, *Trichoderma* inoculation improves N accessibility [43] by releasing organic acids in the soil. It has been observed that certain *Trichoderma* species provide benefits to crops, including enhanced plant growth, better seedling emergence, and improved seed germination [44]. As plant growth regulators, *Trichoderma* spp. can produce phytohormones [45,46] such as gibberellic acid [47] and indole-3-acetic acid [48], which are essential for plant growth. These phytohormones increase flowering, photosynthesis efficiency, and yield quality and improve root condition and structure, enhancing seed germination and viability [49]. Earlier studies on maize utilized *T. asperellum* and *Bacillus amyloliquefaciens* to treat seeds and plants in greenhouses, resulting in increased plant growth (shoot, root, and seedling length) [50]. In another experiment with maize, the authors used *T. harzianum* and *B. subtilis* to treat seeds and plants, observing increased plant growth and yield [25].

Incorporating biofertilizers into agricultural systems can address the growing food demand while minimizing environmental concerns, promoting environmental sustainability. These benefits make them a key component of sustainable agricultural initiatives. However, there is scarce literature on the use of biofertilizers based on algal extracts (*Ulva lactuca* and *Solieria* spp.), *Rhizobium* sp., and *Trichoderma asperellum* in hybrid maize. Therefore, the objective of the present study was to evaluate the impact of the application of a green and red algal mixture (*Ulva lactuca* and *Solieria* spp.), *Rhizobium* sp., *Trichoderma asperellum*, and the combination of the above three biofertilizers on N assimilation, biomass, and yield in hybrid maize.

2. Materials and Methods

2.1. Crop Management

The maize crop was grown in an open field. The location was in Lomas del Consuelo, Meoqui, Chihuahua, Mexico, from August to December 2022. Hybrid maize seeds (AS-GROU A 75-73) were sown in loam soil, in 80 cm furrows, with 3 seeds per linear meter (33.3 cm between plants). Three fertilizations were applied, the first two a month apart and the third 22 days later. The first fertilization was 300 kg ha^{-1} of urea and 50 kg ha^{-1} of commercial MAP® 11-52-00, the second was 20 L ha^{-1} of commercial Nitro-Sul®, and the third was 150 kg ha^{-1} of urea dissolved in water and applied in irrigation. The plants were irrigated every 15 days using a rolling irrigation system.

2.2. Experimental Design and Treatments

A completely randomized design was implemented, with five treatments—T1 = control; T2 = algal extracts; T3 = *Rhizobium* sp.; T4 = *Trichoderma asperellum*; T5 = T2 + T3 + T4—using 10 replications per treatment (Figure 1).

Treatment 1 was the control; it was not under the effects of any biofertilizer; it only received the commercial fertilizers.

Treatment 2 was a combination of three commercial green and red algal products: Algomel Push® (10 mL L^{-1} of water), Algomel Proact® (5 mL L^{-1} of water), and Oceamax® (10 mL L^{-1} of water), which were applied to the leaves every 15 days, after the appearance of the first pair of true leaves.

For Treatment 3, the bacterium (*Rhizobium* sp.) was cultured in potato extract and 1% glucose [51] for 72 h to six days, and it was used at 10^9 colony-forming units (CFU) mL^{-1}. These products were provided by Dr. Micah Royan Isaac of the Olmix Company. *Rhizobium* sp. strains were isolated from the root nodules of bean plants (*Phaseolus vulgaris* L.) collected in Delicias, Chihuahua, México. The disinfection of nodules and the bacterium culture were carried out following the methodology of [52]. The bacterium isolate was grown in liquid medium for 72 h at 30 °C, and it was applied at a density of 10^6–10^7 CFU mL^{-1}.

For Treatment 4, *T. asperellum* biofertilizer was used as described by [53], and the fungus concentration used was 1×10^{10} UFM mL^{-1}. The *T. asperellum* strain used in this work is part of the Faculty of Agricultural and Forestry Sciences of Chihuahua, Mexico strain collection and is registered in GenBank under the accession number MN950427.

Treatment 5 was a combination of T2 + T3 + T4; the *Trichoderma asperellum* and *Rhizobium* sp. were applied to the soil near the rhizosphere, immediately followed by foliar application of the algal extracts.

In all treatments except the control, *Trichoderma asperellum* and *Rhizobium* sp. were applied to the soil near the roots of the plants at a rate of 20 L ha^{-1} every 15 days.

Figure 1. Application of biofertilizers in maize plants. T1 = control: plants without biofertilizers applied. T2 = algae extracts: foliar application of algae extracts to maize plants. T3 = *Rhizobium* sp.: root inoculation of maize plants with bacteria from the genus *Rhizobium* sp. T4 = *Trichoderma asperellum*: root inoculation of maize plants with the fungus *Trichoderma asperellum*. T5 = combination of algae extracts, *Rhizobium* sp., and *Trichoderma asperellum*: combined application of the three aforementioned biofertilizers.

2.3. Plant Sampling

At physiological maturity of the plants, 100 days after sowing, plant samples were collected and the roots, aerial parts, and cobs were separated from the 10 plants of each treatment and then carefully washed with distilled water. The fresh plant material was used for analysis of biomass, yield, nitrate reductase enzyme activity, and photosynthetic pigments.

2.4. Plant Analysis

2.4.1. Total Biomass and Yield

After 24 h of washing with distilled water, the roots, aerial parts, and cob of each plant were weighed on a precision balance (Luna LBL 34001e Precision Balance, Milton Keynes, UK). Biomass was expressed as the sum of all plant organs (g plant^{-1} f.w.). Yield was expressed as the fresh weight of ears per plant (g plant^{-1} f.w.).

2.4.2. Nitrate Reductase Enzyme Activity (NR Activity) "In Vivo"

The nitrate reductase enzyme (NR, EC 1.7.1.1) activity was determined by the methodology suggested in [54], where 0.1 g of fresh 7 mm diameter leaf discs was weighed and placed in 10 mL of incubation buffer (100 mM K-phosphate buffer, pH 7.5 and 1% (*v*/*v*) propanol). The samples were infiltrated at a pressure of 0.8 bars (NAPCO 5851 vacuum oven, Winchester, VA, USA) and they were incubated at 30 °C in the dark for 1 h (WIG-50 digital incubator, DAIHAN SCIENTIFIC, Seoul, Republic of Korea), after which they were placed in a FELISA FE-371 boiling water bath (Zapopan, Jalisco, Mexico) for 15 min to stop NR activity. Then, 1 mL of enzyme extract was taken, and 2 mL of 1% (*w*/*v*) sulfanilamide in 1.5 M HCl and 2 mL of 0.02% (*w*/*v*) N-(1-naphthyl)-ethylenediamine dichlorohydrate were added (20 mg of NNEDA dissolved in 100 mL of distilled water). Finally, the resulting nitrite concentration was determined spectrophotometrically (Thermo Fisher Scientific, GENESYS™ 10S, Madison, WI, USA) at 540 nm against a NO_2^- standard curve. The results were expressed as µmol of NO_2^- formed g^{-1} f.w. h^{-1}.

2.4.3. SPAD Values

SPAD (Soil Plant Analysis Development) index values were determined using the SPAD-502 portable chlorophyllometer (Konica Minolta Sensing, Inc., Osaka, Japan), which provides a quantitative assessment of leaves' green intensity [55]. The measurements were conducted around midday (during peak light hours), with three random readings taken per plant. Two readings were made with the SPAD-502 m: one on 15 October 2022, before the grain-filling stage, and the second one on 10 November 2022, before the maturation stage.

2.4.4. Photosynthetic Pigments

The methodology of [56] was used for photosynthetic pigment analysis. Leaves sample stalks of 7 mm in diameter, with approximately 0.125 g of weight from each treatment, were placed in test tubes, and 10 mL of methanol was added to each sample. The samples were left to rest for 24 h in the dark. After this time, the samples were read at wavelengths of 470, 653, and 666 nm with a Genesis 10S UV-VIS spectrophotometer (Thermo Fisher Scientific, GENESYS™ 10S, Madison, WI, USA). The results were expressed in µg cm^2 of fresh weight, and the following formulae were used (Equations (1)–(4)):

$$\text{Chl } a^{*} = [15.65(\text{A666}) - 7.34(\text{A653})] \quad (1)$$

$$\text{Chl } a = (\text{Chl } a^{*} \times \text{Vf} \times \text{W}_1)/(\text{W}_2 \times \pi \times \text{r}^2 \times n)$$

$$\text{Chl } b^{*} = [27.05(\text{A653}) - 11.21(\text{A666})] \quad (2)$$

$$\text{Chl } b = (\text{Chl } b^{*} \times \text{Vf} \times \text{W}_1)/(\text{W}_2 \times \pi \times \text{r}^2 \times n)$$

$$\text{Carotenoids}^{*} = [(1000^{*}\text{A470}) - 2.86(\text{Chl } a) - 129.2(\text{Chl } b)]/(221) \quad (3)$$

$$\text{Carotenoids} = (\text{Carotenoids}^{*} \times \text{V} \times \text{W}_1)/(\text{W}_2 \times \pi \times \text{r}^2 \times n)$$

$$\text{Total Chlorophyll} = \text{Chl} = \text{Chl } a + \text{Chl } b \quad (4)$$

where Vf: final volume; W_1: weight per leaf disc; W_2: total weight of leaf discs; r: radius of the leaf discs; n: number of leaf discs.

2.4.5. Number of Leaves and Plant Height

The number of leaves was determined by manual counting. The plant height refers to the measurement taken from the highest point to the base of naturally growing plants, and it was determined with a measuring tape. The results were expressed as cm. Both variables were measured before flowering.

2.5. Statistical Analysis

The normal distribution of the data was examined using the Shapiro–Wilk test ($p \leq 0.05$), and the equality of variances among the analyzed variables was assessed via Levene's test, considering significance at $p \leq 0.05$. Following this, ANOVA was conducted at a significance level of 0.05. The variables were then subjected to Duncan's test to compare the means of the two factor levels (dose and application time) at a significance level of $p \leq 0.05$. The statistical program SAS version 9.0 was used.

3. Results and Discussion

3.1. Biomass and Yield

Biomass is a crucial indicator for crops, as it reflects the total amount of organic matter accumulated by plants, reflecting crop yield potential [57]. In this experiment, all treatments displayed significant differences in biomass compared to the control. The highest biomass occurred in T3 (*Rhizobium* sp.), followed by T2 (algal extracts), T4 (*T. asperellum*), and T5 (combination of the three biofertilizers) (Figure 2). The application of the bacterium *Rhizobium* sp. induced an increment of 13.4% in the results compared to the control. The other treatments (algal extracts, *T. asperellum*, and the combination of all biofertilizers) had increases of 11.39%, 11.18%, and 11.52%, compared to the control, respectively. Several researchers have reported the application of *Rhizobium* sp. as PGPR (Plant Growth-Promoting Rhizobacteria) in other crops [24,58,59], because *Rhizobium* sp., like other rhizobacteria, effectively colonizes the roots of non-leguminous plants. It can solubilize phosphorus, produces siderophores and plant hormones, and also shows antagonistic effects against plant pathogens [60,61]. In the present study, the application of *Rhizobium* sp. improved the biomass. In physiological terms, this phenomenon may arise in maize because, by receiving an additional supply of N through biological N fixation, it can increase its rate of photosynthesis and its capacity to produce more plant matter. The combined treatment (T5) also showed high biomass, suggesting that synergistic interactions between the biofertilizers were optimizing the uptake of N and other nutrients, thus maximizing biomass production. As reported in [62–64], the use of biofertilizers promotes biological N fixation, making a more efficient use of N and increasing nutrient uptake, generating more biomass.

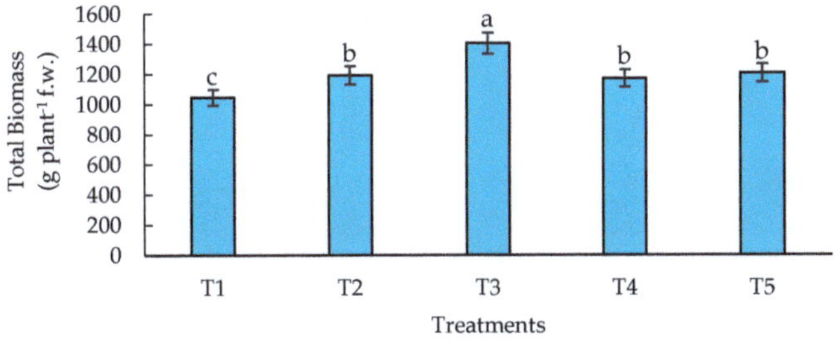

Figure 2. Effects of biofertilizers' application on total biomass: T1 = control; T2 = algal extracts; T3 = *Rhizobium* sp.; T4 = *T. asperellum*; T5 = T2 + T3 + T4. Means with equal letters do not differ according to Duncan's multiple range test ($p \leq 0.05$).

Crop yields are widely recognized as a key indicator for global food security [65]. In this experiment, the highest yields were obtained with *Rhizobium* sp. (T3) and the combination of algal extracts, *Rhizobium* sp., and *T. asperellum* (T5) as a management practice for hybrid maize cultivation. These treatments were significantly different from the rest. Treatments 2 (algal extracts) and 4 (*T. asperellum*) were not different to the control (Figure 3). Additionally, when applying *Rhizobium* sp. (T3) on the hybrid maize crop, an increase in yield of 11.82% was obtained, and when combining the three biofertilizers the increase was 11.60% compared to the control. In the case of maize, which has a high N demand, the presence of *Rhizobium* sp. could be a relevant factor that allows plants to avoid N deficiencies and favor fruit filling [66]. Furthermore, according to [67], the application of *Rhizobium* sp. reduces the internal competition for resources between vegetative and reproductive growth, optimizing both processes and favoring the final yield in terms of grain production. According to [68,69], the use of biofertilizers and their synergy can promote nutrient uptake and, thus, improve yield.

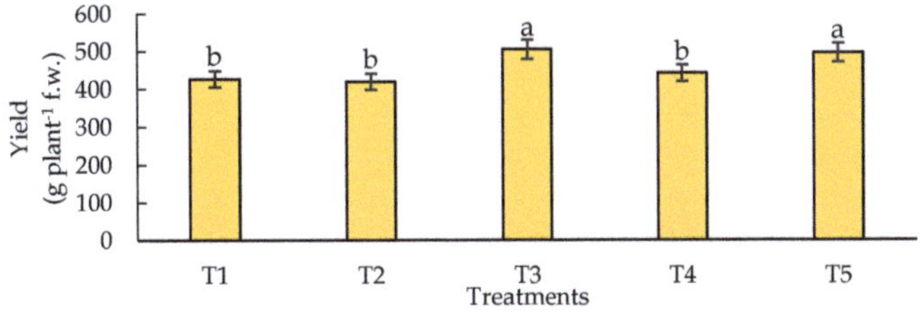

Figure 3. Effects of biofertilizers' application on yield: T1 = control; T2 = algal extracts; T3 = *Rhizobium* sp.; T4 = *T. asperellum*; T5 = T2 + T3 + T4. Means with equal letters do not differ according to Duncan's multiple range test ($p \leq 0.05$).

3.2. Nitrate Reductase Enzyme Activity (NR Activity) "In Vivo"

The NR enzyme is essential for nitrate assimilation in higher plants, so its activity is often a limiting process for N assimilation in plants [70]. In this experiment, the control (T1), algal extracts (T2), and *Rhizobium* sp.(T3) treatments showed no significant differences between them (Figure 4). However, T3 demonstrated the highest NR activity among these treatments. Moreover, there was a 20% increase in activity when compared to T. *asperellum* (T4) and the combination of the three biofertilizers (T5).

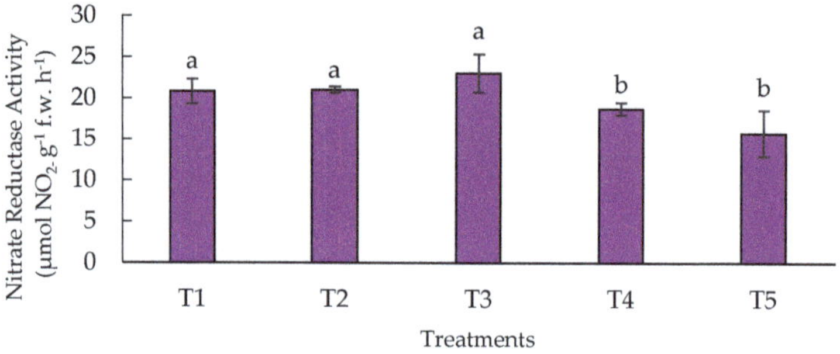

Figure 4. Effects of biofertilizers' application on nitrate reductase activity "In Vivo": T1 = control; T2 = algal extracts; T3 = *Rhizobium* sp.; T4 = *T. asperellum*; T5 = T2+T3+T4. Means with equal letters do not differ according to Duncan's multiple range test ($p \leq 0.05$).

It was observed that as NR activity decreased (particularly in T5), biomass and yield were not compromised; in fact, significant improvements in these parameters were observed (Figures 2 and 3). This suggests that T4 and T5 facilitated inorganic N uptake, making the process more efficient, where NR plays an important role. This is consistent with the findings reported by Shoresh et al. [71], where *Trichoderma* spp. applied to *Sebacinales* spp. and their interaction with other biofertilizers was able to improve N uptake in plants. This may be related to how microorganisms modify the rhizosphere environment and nitrogen assimilation pathways in plants. *T. asperellum* is known for its ability to promote plant growth and improve nutrient uptake efficiency through phosphate solubilization and providing organic N through the degradation of organic matter [72]. The availability of organic forms of nitrogen, such as amino acids or peptides, may reduce the need for inorganic nitrate as the main source of N, which would explain the lower activity of NR in this treatment [73]. In addition, the combination of algal extracts, *Rhizobium* sp., and *T. asperellum* (T5) could increase the availability of organic N and promote direct assimilation of ammonium or reduced forms of N, decreasing the need for the plant to use nitrate and, therefore, reducing NR activity. This synergy between biofertilizers may favor alternative metabolic pathways that are more efficient in nitrogen incorporation, optimizing biomass production without relying so much on nitrate reduction [74]. In the case of algal extracts, Kaushal et al. [75] noted that they provide not only alternative sources of N but also bioactive compounds that improve N assimilation efficiency and promote growth and yield in crops, such as maize, by overexpression of N transporter genes. Regarding T3 (*Rhizobium* sp.), it was observed that N assimilation was favored due to increased NR activity. *Rhizobium* sp. can fix atmospheric N and also produce auxins, which are essential for root growth. According to what was reported in [76], auxins promote cell elongation and the formation of lateral roots, increasing the nutrient uptake area in plants. On the other hand, there are factors that also affect nitrogen assimilation, such as the light/darkness cycle, CO_2 levels, temperature, nitrate and ammonium ions, sugars and sucrose, and genetic variability [55]. These findings reinforce the results obtained in this experiment, where inoculation with *Rhizobium* sp. led to a significant increase in NR activity. Similarly, previous research has shown that different *Rhizobium* species, such as *Mesorhizobium ciceri* and *Rhizobium phaseoli*, promote the growth of legumes and non-leguminous crops such as maize. These studies show that improved nutrient uptake and plant yield are related to the ability of these bacteria to stimulate root growth mechanisms and the assimilation of inorganic N [77–79].

3.3. SPAD Values

The chlorophyll index is a fast and non-destructive way to detect chlorophyll levels in plants and, thus, form an idea of their nutritional situation [80]. The index of chlorophyll in hybrid maize (ASGROU A 75-73) was affected by the application of the biofertilizers used (Figure 5). This variable allows us to know the levels of chlorophyll in the leaves and estimate the efficiency of the photosynthetic process.

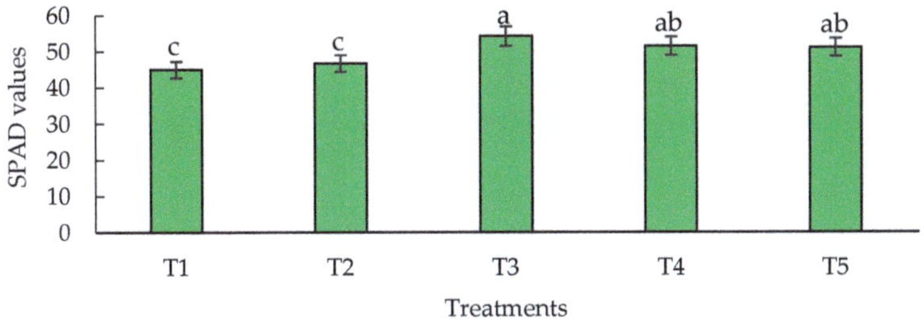

Figure 5. Effects of biofertilizers' application on SPAD values: T1 = control; T2 = algal extracts; T3 = *Rhizobium* sp.; T4 = *T. asperellum*; T5 = T2 + T3 + T4. Means with equal letters do not differ according to Duncan's multiple range test ($p \leq 0.05$).

The highest result was observed with the application of *Rhizobium* sp. compared to the other treatments, with a significant increase of 12% over the control. The treatments with *T. asperellum* and the combination of all biofertilizers did not show significant differences between them, although they were significantly different to the control, showing increases of 11.45% (T4) and 11.35% (T5) over it, respectively. The lowest SPAD values were observed in the control and the algal extracts. Previous research has documented increases in chlorophyll content (SPAD values), photosynthetic pigments, biomass, and yield after the application of biofertilizers to maize crops [81–83]. Referring to the chlorophyll content SPAD values, the highest values in this research were obtained when applying *Rhizobium* sp.; likewise, in terms of total chlorophyll and yield, the values were similar when applying the bacterium and the combination of the three biofertilizers. In this context, [59] inoculated *Rhizobium* sp. in maize crops, resulting in an increase in shoots' relative chlorophyll content, dry weight, photosynthetic rate, leaf area, and grain yield. Hussain et al. [59] reported a significant increase in chlorophyll levels when a rhizobial strain identified as *Rhizobium phaseoli* was used in maize under drought conditions induced by polyethylene glycol. Comparable results were described by Egamberdieva et al. [84] in fodder maize inoculated with rhizobia in field conditions, noting significant increases in chlorophyll content, photosynthesis, and transpiration rate.

3.4. Photosynthetic Pigments

Photosynthetic pigments are molecules that capture light and transfer it to the reaction centers to be converted into chemical energy, enabling photosynthesis [85]. These pigments can be divided into total Chl (Chl a and Chl b) and total carotenoids. Chl is the main pigment for light absorption, while carotenoids play a supporting role. N is essential for the synthesis of photosynthetic pigments; thus, a continuous and adequate supply of N is necessary [86]. In this experiment, the application of *Rhizobium* sp. in maize plants and the combination of the three biofertilizers induced higher Chl contents, with significant differences compared to the other treatments. Likewise, the highest values for carotenes were obtained with *Rhizobium* sp. In both cases, the lowest pigment contents occurred in the control (Figure 6). The increase in Chl content observed in the *Rhizobium* sp. treatment compared to the control was 18.4%, and the treatment (T5) with the combination of biofertilizers had an increase of 15.98% compared to the control. Regarding carotenes, treatment with *Rhizobium* sp. and the combination of biofertilizers (T5) had the highest significant values compared to the control (13.6%). Photosynthesis is one of the most important processes in plants, determining biomass and yield [81,82]. This increase could be attributed to improved NR activity (Figure 4) but also to increased N fixation and uptake. It is also possible that siderophores (iron-chelating molecules) released by *Rhizobium* sp. increased Fe's accessibility for plants [87]. Fe is also a key component for Chl synthesis, redox reactions, and several biochemical reactions, e.g., N fixation. Algal effects on crops include increased N availability from soil and improved photosynthetic pigment content due to the presence of phytohormones that share similar regulatory roles in higher plants [12,88]. *T. asperellum* could upregulate the photosynthetic pigment concentration by the upregulation of genes modulating Chl and carotenoid biosynthesis or the response against reactive oxygen species [89,90]. The secondary metabolites released by *Trichoderma* sp. may influence plants' nutrient uptake and growth by activating phytohormones consisting of indole acetic acid or auxin [91].

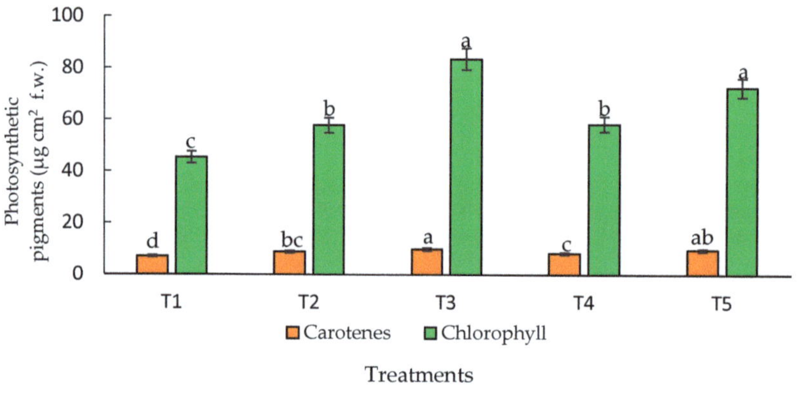

Figure 6. Effects of the application of biofertilizers on photosynthetic pigment activity: T1 = control; T2 = algal extracts; T3 = *Rhizobium* sp.; T4 = *T. asperellum*; T5 = T2 + T3 + T4. Means with equal letters do not differ according to Duncan's multiple range test ($p \leq 0.05$).

3.5. Number of Leaves and Plant Height

Leaves play key roles in photosynthesis, carbon fixation, respiration, and transpiration [92]. The number of leaves in hybrid maize plants (ASGROU A 75-73) was higher with the application of *Rhizobium* sp., with significant differences compared to the other treatments (Figure 7).

The hybrid maize plants increased their leaf numbers by 11.4% compared to the control when *Rhizobium* sp. was used (T3). The combination of the three biofertilizers (T5) increased this variable by 11.1%, while the application of *T. asperellum* (T4) showed a 10% increase. Regarding plant height, the results did not show significant differences when using *Rhizobium* sp. (T3) and the three combined biofertilizers (T5) (Figure 8), obtaining increases of 11.27% and 11.21% when compared to the control, respectively. The lowest value was obtained with the algal extracts (T2). On the other hand, the control (T1) and the plants treated with *T. asperellum* (T4) were statistically similar.

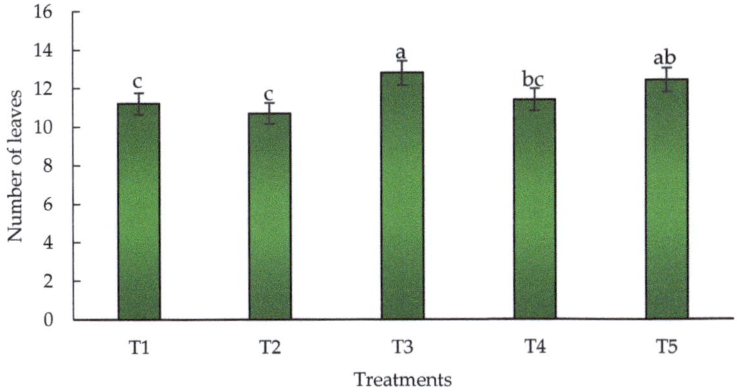

Figure 7. Effects of biofertilizers' application on leaf numbers: T1 = control; T2 = algal extracts; T3 = *Rhizobium* sp.; T4 = *T. asperellum*; T5 = T2 + T3 + T4. Means with equal letters do not differ according to Duncan's multiple range test ($p \leq 0.05$).

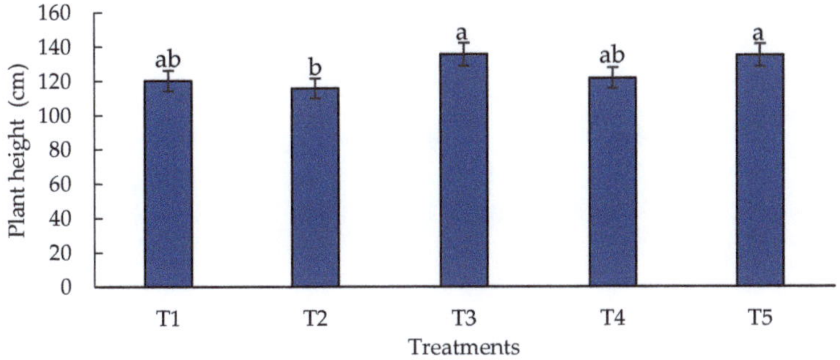

Figure 8. Effects of biofertilizers' application on plant height: T1 = control; T2 = algal extracts; T3 = *Rhizobium* sp.; T4 = *T. asperellum*; T5 = T2 + T3 + T4. Means with equal letters do not differ according to Duncan's multiple range test ($p \leq 0.05$).

In this investigation, the application of *Rhizobium* sp. to the soil increased the number of leaves and the height of the maize plants, most likely because *Rhizobium* has different mechanisms to promote plant growth. These include direct mechanisms such as N fixation, phosphate solubilization, and the production of phytohormones [85]. An increase in N fixation due to *Rhizobium* can improve plant growth, leading to more biomass accumulation, as noted in alfalfa [93]. Another trait of *Rhizobium* is the capacity to transform soluble phosphate from its insoluble form via the release of different organic acids (acetic acid, oxalic acid, and gluconic acid) or by proton (H^+) extrusion. P is not only essential for the formation of organic compounds (DNA, RNA, ATP), it is also involved in cell division and energy transfer processes [94]. The production of phytohormones by various bacteria is also an important trait, and these comprise gibberellins, indole acetic acid, auxins, cytokinins, and abscisic acid, which are crucial for plant and leaf growth and stem elongation in plants [84]. Apart from phytohormones, enzymes including phytase and protease can also be synthesized, enhancing nutrient uptake [95]. Moreover, *Rhizobium* can stimulate root growth, facilitating the absorption of crucial elements such as Fe and P [22]. Similar results were found in [96], where the application of *Rhizobium* (*R. etli* bv. *Phaseoli*, *R. leguminosarum* bv. *Trifolii*) and *Sinorhizobium* sp. improved the plant height, corn growth, and grain yield of maize in several areas. In another study, the inoculation of maize with several *Rhizobium* species (Cp3, Lt2, and Br3) boosted the plant height from 101.0 to 104.0 cm [97]. In addition,

Yalçın et al. [33] reported an increase in plant height, stem diameter, ear dry weight, and ear weight without husks upon the application of *Rhizobium* sp. in maize. In the present experiment, the maximum height of the plants ranged from 135 to 134 cm (T3–T5).

3.6. Correlation Analysis and Heatmap

The Pearson heatmap and correlation (Figure 9) allowed us to observe the dependence or interdependence relationships between the variables analyzed. The color scale indicates that the shades closer to red are strong positive correlations (close to +1), while the bluish shades show negative correlations (close to −1) and white or light color suggests no or very low correlation.

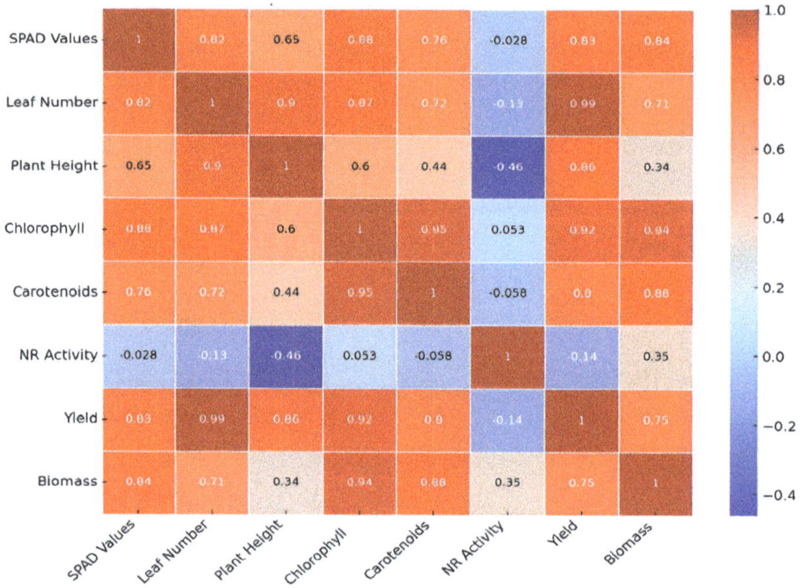

Figure 9. The figure shows the Pearson correlation analysis between the variables of biomass, yield, nitrate reductase activity, SPAD values, photosynthetic pigments, number of leaves, and plant height, evaluating the effects of the biofertilizers. The colors in the heatmap represent the magnitude and direction of the correlations, with red tones indicating strong positive correlations and blue tones indicating negative or weak correlations. The numerical values within each cell indicate the correlation coefficient between pairs of variables, providing a clear view of the interdependent relationships among the measured variables.

The map shows that leaf number and yield have an almost perfect correlation (0.99), indicating that a higher the number of leaves increases the yield significantly. Total Chl and carotenoids have a very high correlation (0.95), suggesting that the presence of one of these pigments is closely related to that of the other. Total Chl and biomass also have a strong correlation (0.94), indicating that biomass accumulation is directly related to the amount of Chl in plants.

The number of leaves and plant height (0.9) are strongly correlated, indicating that the higher the number of leaves, the greater the plant height. Plant height and yield (0.86) and plant height and number of leaves (0.9) have strong positive correlations, suggesting that taller plants tend to generate higher yields and have more leaves. Biomass is also strongly correlated with yield (0.75), indicating that an increase in biomass is related to higher plant yields. NR activity had weak negative correlations with variables such as plant height (−0.46), number of leaves (−0.13), yield (−0.14), and biomass (0.35), suggesting that an increase in the activity of this enzyme may be slightly related to a decrease in these variables, probably due to the effects of the biofertilizers applied.

3.7. Venn and Interaction Diagram

Figure 10 is a clear and concise representation of how different biofertilizer treatments (algal extracts, *Rhizobium* sp., *Trichoderma asperellum*, and their combination) directly impact the growth, physiology, and yield of hybrid maize.

Part (a): Venn diagram

This diagram illustrates the synergy between the different biological treatments and how, by combining them, outstanding effects on critical variables in plant growth can be obtained. The intersections of the diagram show that T2 is particularly associated with an increase in NR activity, which is crucial for nitrogen assimilation and, thus, for plant growth. *Rhizobium* sp. stands out in terms of the increase in SPAD values, which are related to the amount of Chl in the leaves and, therefore, to photosynthetic efficiency. *Trichoderma asperellum* favors the production of photosynthetic pigments, which are essential for light absorption and efficiency in the photosynthetic process. The central zone of the diagram, which combines the three treatments (T5), shows improvements in biomass, number of leaves, plant height, and yield, suggesting that the joint application of these microorganisms and plant extracts could be more effective than the individual treatments. This combination not only optimizes resource use but also tends to maximize plant growth and productivity.

Part (b): Bar chart

This bar chart quantifies the impact of each treatment (T1 to T5) on key variables. Analyzing biomass, treatment T3 generated the greatest increase, implying that the application of *Rhizobium* sp. on hybrid corn significantly enhances overall plant growth. In yield, as with biomass, the T3 treatment offered the best results, representing a crucial contribution to sustainable agricultural productivity. In NR activity, a significant increase was observed with the T2 and T3 treatments, indicating that these biofertilizers improve the efficiency of nitrogen assimilation, an essential process for plant metabolism. In SPAD values, T3 and T5 presented the greatest increases, suggesting an improvement in photosynthetic efficiency, crucial to biomass production and yield. For the variable photosynthetic pigments, the treatments with *Rhizobium* sp. (T3) and the combination of all biofertilizers (T5) promoted the production of essential pigments for photosynthesis, which also contributed to better growth and yield. As for the number of leaves and plant height, all biofertilizers except for T2 increased these variables considerably.

(a)

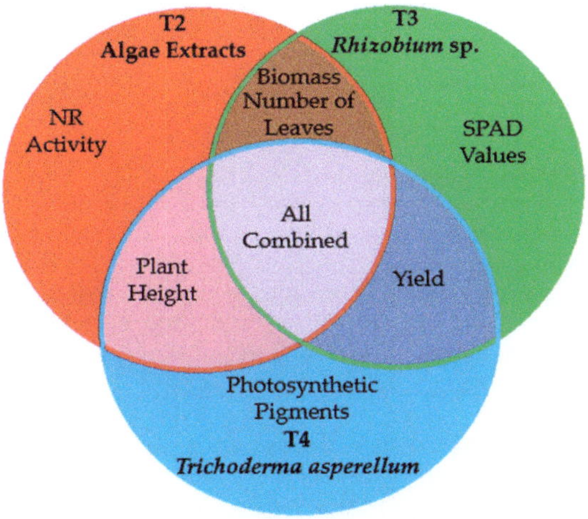

Figure 10. *Cont.*

(**b**)

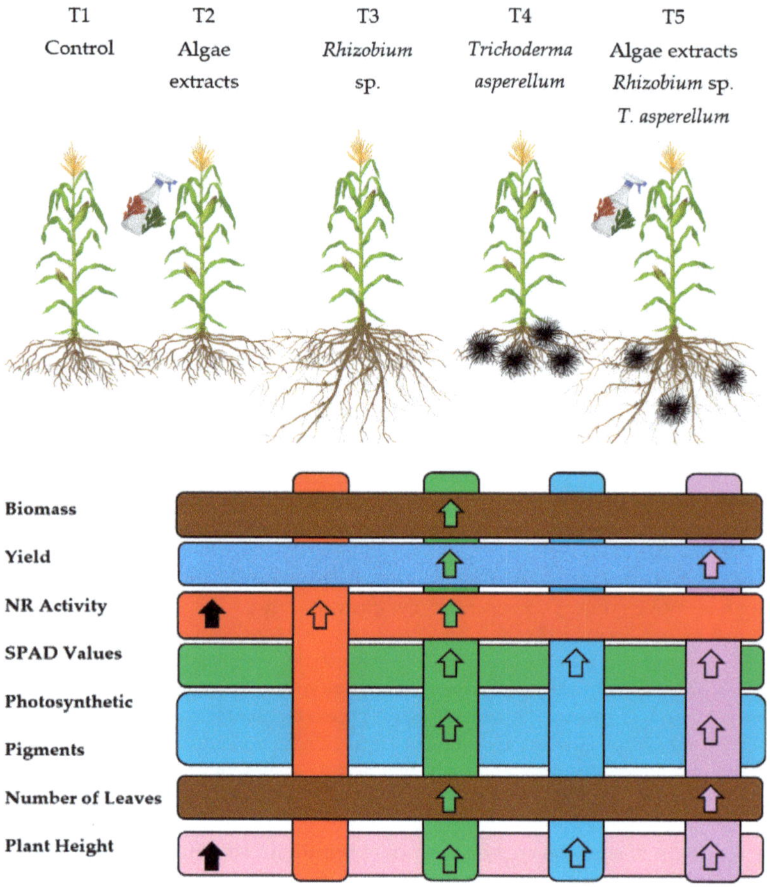

Figure 10. (**a**) Venn diagram representing the effects of different treatments on plant growth and physiological variables. (**b**) Interaction diagram: The bar chart shows how the different variables (biomass, yield, nitrate reductase activity, SPAD values, photosynthetic pigments, number of leaves, and plant height) respond to each treatment (T1 to T5). The arrows indicate an increase in these variables compared to the control (T1).

4. Conclusions

In this experiment, the application of *Rhizobium* sp. to the rhizosphere in hybrid maize plants had a greater effect on nitrogen assimilation, thanks to the symbiotic relationship generated with the maize. It is important to mention that treatment T5, which was the combination of algal extracts, *Rhizobium* sp., and *Trichoderma asperellum*, also had very positive effects on nitrogen assimilation. However, more studies are needed on the synergistic relationship that these biofertilizers form in terms of nitrogen assimilation and nitrogen-use efficiency. Finally, the use of *Rhizobium* sp. and the combination of algal extracts, *Rhizobium* sp., and *Trichoderma asperellum*, in relation to N assimilation and N-use efficiency, has not been sufficiently explored; thus, this research adds a new dimension to the field of study by demonstrating that these biofertilizers produce a very effective synergistic effect on nitrogen assimilation, which promises to be a key mechanism for sustainable agricultural practices in the future.

Author Contributions: Conceptualization, S.P.-Á. and E.S.; methodology, S.P.-Á., E.H.O.-C., and J.C.A.-P.; formal analysis, C.M.E.-B.; investigation, S.P.-Á.; writing—original draft preparation, S.P.-Á. and E.H.O.-C.; writing—review and editing, J.R.-S. and M.A.M.-T.; visualization, R.H.-C. and E.H.O.-C.; supervision, L.P.U.-V. All authors have read and agreed to the published version of the manuscript.

Funding: This research received no external funding.

Data Availability Statement: The original contributions presented in this study are included in the article; further inquiries can be directed to the corresponding author.

Acknowledgments: We would like to thank Micah Royan Isaac of the Olmix Company for the algal extracts supplied for the completion of this work.

Conflicts of Interest: The authors declare no conflicts of interest.

References

1. Rosas-Castor, J.M.; Guzmán-Mar, J.L.; Hernández-Ramírez, A.; Garza-González, M.T.; Hinojosa-Reyes, L. Arsenic accumulation in maize crop (*Zea mays*): A review. *Sci. Total Environ.* **2014**, *488–489*, 176–187. [CrossRef] [PubMed]
2. Statista. Available online: https://es.statista.com/estadisticas/613419/prinicpales-productores-de-maiz-en-el-mundo/ (accessed on 30 May 2024).
3. Hou, P.; Gao, Q.; Xie, R.; Li, S.; Meng, Q.; Kirkby, E.A.; Chen, X. Grain yields in relation to N requirement: Optimizing nitrogen management for spring maize grown in China. *Field Crops Res.* **2012**, *129*, 1–6. [CrossRef]
4. Lai, Z.; Zhang, H.; Ding, X.; Liao, Z.; Zhang, C.; Yu, J.; Fan, J. Ridge-furrow film mulch with nitrogen fertilization improves grain yield of dryland maize by promoting root growth, plant nitrogen uptake and remobilization. *Soil Tillage Res.* **2024**, *241*, 106118. [CrossRef]
5. Mejías, J.H.; Salazar, F.; Pérez Amaro, L.; Hube, S.; Rodriguez, M.; Alfaro, M. Nanofertilizers: A cutting-edge approach to increase nitrogen use efficiency in grasslands. *Front. Environ. Sci.* **2021**, *9*, 635114. [CrossRef]
6. Bai, F.; Qi, X.; Li, P.; Du, Z.; Guo, W. Groundwater depth and nitrogen application amount jointly regulate the water and residual soil nitrate accumulation in agricultural soil profile. *Agronomy* **2023**, *13*, 1163. [CrossRef]
7. Wang, Y. Effect of NPK on NR and NiR activity of sugar beet. *J. Nucl. Agric. Sci.* **2012**, *26*, 0803–0808.
8. Liu, X.; Hu, B.; Chu, C. Nitrogen assimilation in plants: Current status and futurs prospects. *J. Genet. Genomics.* **2022**, *49*, 394–404. [CrossRef]
9. Zhang, W.; Ni, K.; Long, L.; Ruan, J. Nitrogen transport and assimilation in tea plant (*Camellia sinensis*): A review. *Front. Plant Sci.* **2023**, *14*, 1249202. [CrossRef]
10. Yadav, M.R.; Kumar, S.; Lal, M.K.; Kumar, D.; Kumar, R.; Yadav, R.K.; Kumar, S.; Nanda, G.; Singh, J.; Udawat, P.; et al. Mechanistic Understanding of Leakage and Consequences and Recent Technological Advances in Improving Nitrogen Use Efficiency in Cereals. *Agronomy* **2023**, *13*, 527. [CrossRef]
11. Nosheen, S.; Ajmal, I.; Song, Y. Microbes as Biofertilizers, a Potential Approach for Sustainable Crop Production. *Sustainability* **2021**, *13*, 1868. [CrossRef]
12. Ammar, E.E.; Aioub, A.A.A.; Elesawy, A.E.; Karkour, A.M.; Mouhamed, M.S.; Amer, A.A.; EL-Shershaby, N.A. Algae as Bio-fertilizers: Between current situation and future prospective. *Saudi J. Biol. Sci.* **2022**, *29*, 3083–3096. [CrossRef] [PubMed]
13. Ronga, D.; Biazzi, E.; Parati, K.; Carminati, D.; Carminati, E.; Tava, A. Microalgal Biostimulants and Biofertilisers in Crop Productions. *Agronomy* **2019**, *9*, 192. [CrossRef]
14. Tolisano, C.; Del Buono, D. Biobased: Biostimulants and Biogenic nanoparticles enter the scene. *Sci. Total Environ.* **2023**, *885*, 163912. [CrossRef] [PubMed]
15. Olanrewaju, O.S.; Glick, B.R.; Babalola, O.O. Mechanisms of action of plant growth promoting bacteria. *World J. Microbiol. Biotechnol.* **2017**, *33*, 197. [CrossRef] [PubMed]
16. Zhang, C.; Zhang, Y.; Ding, Z.; Bai, Y. Contribution of Microbial Inter-Kingdom Balance to Plant Health. *Mol. Plant* **2019**, *12*, 148–149. [CrossRef]
17. El-Maraghy, S.S.; Tohamy, A.T.; Hussein, K.A. Plant Protection Properties of the Plant Growth Promoting Fungi (PGPF): Mechanisms and Potentiality. *Curr. Res. Environ. Appl. Mycol.* **2021**, *11*, 391–415. [CrossRef]
18. Cao, M.; Narayanan, M.; Shi, X.; Chen, X.; Li, Z.; Ma, Y. Optimistic Contributions of Plant Growth-Promoting Bacteria for Sustainable Agriculture and Climate Stress Alleviation. *Environ. Res.* **2023**, *15*, 217. [CrossRef]
19. Khan, M.S.; Zaidi, A.; Wani, P.A. Role of Phosphate-Solubilizing Microorganisms in Sustainable Agriculture—A Review. *Agron. Sustain. Dev.* **2007**, *27*, 29–43. [CrossRef]
20. Bhattacharyya, P.N.; Jha, D.K. Plant Growth-Promoting Rhizobacteria (PGPR): Emergence in Agriculture. *World J. Microbiol. Biotechnol.* **2012**, *28*, 1327–1350. [CrossRef]
21. Saharan, B.S.; Nehra, V. Plant Growth Promoting Rhizobacteria: A Critical Review. *Life Sci. Med. Res.* **2011**, *2011*, 1–30.
22. Glick, B.R. Plant Growth-Promoting Bacteria: Mechanisms and Applications. *Scientifica* **2012**, *2012*, 963401. [CrossRef] [PubMed]

23. Deka, H.; Deka, S.; Baruah, C. Plant Growth Promoting Rhizobacteria for Value Addition: Mechanism of Action. In *Plant-Growth-Promoting Rhizobacteria (PGPR) and Medicinal Plants*, 1st ed.; Egamberdieva, D., Shrivastava, S., Varma, A., Eds.; Springer: New York, NY, USA, 2015; pp. 305–321.
24. Dodd, I.C.; Zinovkina, N.Y.; Safronova, V.I.; Belimov, A.A. Rhizobacterial Mediation of Plant Hormone Status. *Ann. Appl. Biol.* **2010**, *157*, 361–379. [CrossRef]
25. Abdel-Raouf, N.; Al-Homaidan, A.A.; Ibraheem, I.B. Agricultural Importance of Algae. *Afr. J. Biotechnol.* **2012**, *11*, 11648–11658. [CrossRef]
26. Barone, G.D.; Cernava, T.; Ullmann, J.; Liu, J.; Lio, E.; Germann, A.T.; Nakielski, A.; Russo, D.A.; Chavkin, T.; Knufmann, K.; et al. Recent Developments in the Production and Utilization of Photosynthetic Microorganisms for Food Applications. *Heliyon* **2023**, *9*, e14708. [CrossRef] [PubMed]
27. Al-Ghazal, S.A.Y.; Aziz, M.M.; AL-Juheiehy, W.K.S. Response of Growth and Yield of Corn (*Zea mays* L.) to Biofertilizer and Sea-Algae Extract. *Int. J. Agric. Stat. Sci.* **2023**, *19*, 161–165.
28. Basavaraja, P.K.; Yogendra, N.D.; Zodape, S.T.; Prakash, R.; Ghosh, A. Effect of Seaweed Sap as Foliar Spray on Growth and Yield of Hybrid Maize. *J. Plant Nutr.* **2018**, *41*, 1851–1861. [CrossRef]
29. Layek, J.; Das, A.; Ramkrushna, G.I.; Ghosh, A.; Panwar, A.S.; Krishnappa, R.; Ngachan, S.V. Effect of Seaweed Sap on Germination, Growth and Productivity of Maize (*Zea mays*) in North Eastern Himalayas. *Indian J. Agron.* **2016**, *61*, 354–359. [CrossRef]
30. Pal, A.; Dwivedi, S.K.; Maurya, P.K.; Kanwar, P. Effect of Seaweed Saps on Growth, Yield, Nutrient Uptake and Economic Improvement of Maize (Sweet Corn). *J. Appl. Nat. Sci.* **2015**, *7*, 970–975. [CrossRef]
31. Mansori, M.; Chernane, H.; Latique, S.; Benaliat, A.; Hsissou, D.; El Kaoua, M. Seaweed extract effect on water deficit and antioxidative mechanisms in bean plants (*Phaseolus vulgaris* L.). *J. Appl. Phycol.* **2015**, *27*, 1689–1698. [CrossRef]
32. Spain, O.; Hardouin, K.; Bourgougnon, N.; Michalak, I. Enzyme-assisted extraction of red seaweed *Solieria chordalis* (C. Agardh) J. Agardh 1842—The starting point for the production of biostimulants of plant growth and biosorbents of metal ions. *Biomass Convers. Biorefin.* **2024**, *14*, 1621–1635. [CrossRef]
33. Yalçın, S.; Okudan, E.Ş.; Karakaş, Ö.; Önem, A.N.; Başkan, K.S. Identification and quantification of some phytohormones in seaweeds using UPLC-MS/MS. *J. Liq. Chromatogr. Relat. Technol.* **2019**, *42*, 475–484. [CrossRef]
34. Benítez, I.; Dueñas, A.K.; Martínez, E.; Salazar, J.A.; Carrera, E.; Osuna, I. Identification and Quantification of Plant Growth Regulators and Antioxidant Compounds in Aqueous Extracts of *Padina durvillaei* and *Ulva lactuca*. *Agronomy* **2020**, *10*, 866. [CrossRef]
35. Santoyo, G.; Moreno-Hagelsieb, G.; Orozco-Mosqueda, M.C.; Glick, B.R. Plant Growth-Promoting Bacterial Endophytes. *Microbiol. Res.* **2016**, *183*, 92–99. [CrossRef] [PubMed]
36. Zhou, S.; Xia, P.; Chen, J.; Xiong, Q.; Li, G.; Tian, J.; Wu, B.; Zhou, F. Optimizing Nitrogen Application Position to Change Root Distribution in Soil and Regulate Maize Growth and Yield Formation in a Wide–Narrow Row Cropping System: Pot and Field Experiments. *Front. Plant Sci.* **2024**, *15*, 1298249. [CrossRef] [PubMed]
37. Widodo, T.W.; Muhklisin, I.; Nugroho, S.A.; Wardana, R.; Ummah, U.S.A. Growth and Yield of Maize Applicated by *Rhizobium* spp. from Legume and Non-legume Rhizosphere. *J. Agric. Appl. Biol.* **2023**, *4*, 151–160. [CrossRef]
38. Marks, B.B.; Megías, M.; Ollero, F.J.; Nogueira, M.A.; Araujo, R.S.; Hungria, M. Maize Growth Promotion by Inoculation with *Azospirillum brasilense* and Metabolites of *Rhizobium tropici* Enriched on Lipo-chitooligosaccharides (LCOs). *AMB Express* **2015**, *5*, 71. [CrossRef]
39. Pessoa Cavalcanti, M.I.; de Carvalho Nascimento, R.; Ribeiro, D.R.; Costa Escobar, I.E.; Resende, A.C.F.; Pereira, A.S.; Santiago, A.D.F.; Abrahão, R.S.N.; Fernandes-Júnior, P.V. Maize Growth and Yield Promoting Endophytes Isolated into a Legume Root Nodule by a Cross-Over Approach. *Rhizosphere* **2020**, *15*, 100211. [CrossRef]
40. Kubiak, A.; Wolna-Maruwka, A.; Pilarska, A.A.; Niewiadomska, A.; Piotrowska-Cyplik, A. Fungi of the *Trichoderma Genus*: Future Perspectives of Benefits in Sustainable Agriculture. *Appl. Sci.* **2023**, *13*, 6434. [CrossRef]
41. Alfiky, A.; Weisskopf, L. Deciphering *Trichoderma*–Plant–Pathogen Interactions for Better Development of Biocontrol Applications. *J. Fungi* **2021**, *7*, 61. [CrossRef]
42. Chagas, L.F.B.; Castro, H.G.; Colonia, B.S.O.; Carvalho-Filho, M.R.; Miller, L.O.; Chagas-Junior, A.F. Efficiency of the Inoculation of *Trichoderma asperellum* UFT-201 in *Cowpea* Production Components under Growth Conditions in Field. *Rev. Ciênc. Agrár.* **2016**, *39*, 413–421. [CrossRef]
43. Wei, Y.; Yang, H.; Hu, J.; Li, H.; Zhao, Z.; Wu, Y.; Li, J.; Zhou, Y.; Yang, K.; Yang, H. *Trichoderma harzianum* Inoculation Promotes Sweet Sorghum Growth in the Saline Soil by Modulating Rhizosphere Available Nutrients and Bacterial Community. *Front. Plant Sci.* **2023**, *14*, 1258131. [CrossRef] [PubMed]
44. Cai, W.J.; Ye, T.T.; Wang, Q.; Cai, B.D.; Feng, Y.Q. A Rapid Approach to Investigate Spatiotemporal Distribution of Phytohormones in Rice. *Plant Methods* **2016**, *12*, 47. [CrossRef] [PubMed]
45. Jaroszuk-Ściseł, J.; Tyśkiewicz, R.; Nowak, A.; Ozimek, E.; Majewska, M.; Hanaka, A.; Tyśkiewicz, K.; Pawlik, A.; Janusz, G. Phytohormones (Auxin, Gibberellin) and ACC Deaminase In Vitro Synthesized by the Mycoparasitic *Trichoderma* DEMTkZ3A0 Strain and Changes in the Level of Auxin and Plant Resistance Markers in Wheat Seedlings Inoculated with This Strain Conidia. *Int. J. Mol. Sci.* **2019**, *20*, 4923. [CrossRef] [PubMed]

46. Díaz, G.; Rodríguez, G.; Montana, L.; Miranda, T.; Basso, C.; Arcia, M. Efecto de la aplicación de bioestimulantes y *Trichoderma* sobre el crecimiento en plántulas de maracuyá (*Passiflora edulis Sims*) en vivero. *Bioagro* **2020**, *32*, 195–204.
47. Sabre, W.I.; Ghoneem, K.M.; Rashad, Y.M.; Al-Askar, A.A. *Trichoderma harzianum* WKY1: An Indole Acetic Acid Producer for Growth Improvement and Anthracnose Disease Control in Sorghum. *Biocontrol Sci. Technol.* **2017**, *27*, 654–676. [CrossRef]
48. Halifu, S.; Deng, X.; Song, X.; Song, R. Effects of Two *Trichoderma* Strains on Plant Growth, Rhizosphere Soil Nutrients and Fungal Community of *Pinus sylvestris* var. mongolica Annual Seedlings. *Forests* **2019**, *10*, 758. [CrossRef]
49. Karuppiah, V.; Vallikkannu, M.; Li, T.; Chen, J. Simultaneous and Sequential-Based Co-Fermentations of *Trichoderma asperellum* GDFS1009 and *Bacillus amyloliquefaciens* 1841: A Strategy to Enhance the Gene Expression and Metabolites to Improve the Biocontrol and Plant Growth Promoting Activity. *Microb. Cell Factories* **2019**, *18*, 185. [CrossRef]
50. Petcu, V.; Bubueanu, C.; Casarica, A.; Săvoiu, G.; Stoica, R.; Bazdoaca, C.; Lazăr, D.A.; Iordan, H.L.; Horhocea, D. Efficacy of *Trichoderma harzianum* and *Bacillus subtilis* as Seed and Vegetation Application Combined with Integrated Agroecology Measures on Maize. *Romanian Agric. Res.* **2023**, *40*, 1–10. [CrossRef]
51. Martyniuk, S.; Oron, J. Use of potato extract broth for culturing root-nodule bacteria. *Pol. J. Microbiol.* **2011**, *60*, 323. [CrossRef]
52. Díaz, F.G.; Gutiérrez, R.T.; Mora, K.G.; García, M.C.N. Isolation and Characterization of Rhizobia from *Crotalaria* sp. in Southern Ecuador. *Cultivos Tropicales* **2016**, *37*, 40–47.
53. Andrzejak, R.; Janowska, B.; Renska, B.; Kosiada, T. Effect of *Trichoderma* spp. and Fertilization on the Flowering of *Begonia* × *tuberhybrida Voss. 'Picotee Sunburst'*. *Agronomy* **2022**, *11*, 1278. [CrossRef]
54. Sánchez, E.; Rivero, R.M.; Ruiz, J.M.; Romero, L. Changes in Biomass, Enzymatic Activity and Protein Concentration in Roots and Leaves of Green Bean Plants (*Phaseolus vulgaris* L. cv. Strike) under High NH_4NO_3 Application Rates. *Sci. Hortic.* **2004**, *99*, 237–248. [CrossRef]
55. Cunha, A.R.D.; Katz, I.; Sousa, A.D.P.; Martinez-Uribe, R.A. Índice SPAD en el Crecimiento y Desarrollo de Plantas de Lisianthus en Función de Diferentes Dosis de Nitrógeno en Ambiente Protegido. *Idesia* **2015**, *33*, 97–105. [CrossRef]
56. Wellburn, A.R. The Spectral Determination of Chlorophylls *a* and *b*, as well as Total Carotenoids, Using Various Solvents with Spectrophotometers of Different Resolution. *J. Plant Physiol.* **1994**, *144*, 307–313. [CrossRef]
57. Bakuei, N.; Amini, G.; Njafpour, G.D.; Jahanshahi, M. Total and Sustainable Utilization of Biomass Resources: A Perspective. *Front. Sustain. Energy* **2015**, *8*, 546.
58. Razafintsalama, H.; Trap, J.; Rabary, B.; Razakatiana, A.T.E.; Ramanankierana, H.; Rabeharisoa, L.; Becquer, T. Effect of *Rhizobium* Inoculation on Growth of Common Bean in Low-Fertility Tropical Soil Amended with Phosphorus and Lime. *Sustainability* **2022**, *14*, 4907. [CrossRef]
59. Kandil, A.E.; Özdamar Ünlü, H. Effect of *Rhizobium* Inoculation on Yield and Some Quality Properties of Fresh *Cowpea*. *Cogent Food Agric.* **2023**, *9*, 2275410. [CrossRef]
60. Díez-Méndez, A.; Menéndez, E. Rhizobium Presence and Functions in Microbiomes of Non-Leguminous Plants. In *Symbiotic Soil Microorganisms*, 1st ed.; Shrivastava, N., Mahajan, S., Varma, A., Eds.; Springer: Cham, Switzerland, 2021; pp. 241–266.
61. Beltran-Medina, J.I.; Romero-Perdomo, F.; Molano-Chavez, L.; Silva, A.M.M.; Estrada-Bonilla, G.A. Differential Plant Growth Promotion Under Reduced Phosphate Rates in Two Genotypes of Maize by a Rhizobial Phosphate-Solubilizing Strain. *Front. Sustain. Food Syst.* **2022**, *6*, 955473. [CrossRef]
62. Fageria, N.K.; Baligar, V.C. Enhancing Nitrogen Use Efficiency in Crop Plants. *Adv. Agron.* **2005**, *88*, 97–185.
63. Canfield, D.E.; Glazer, A.N.; Falkowski, P.G. The Evolution and Future of Earth's Nitrogen Cycle. *Science* **2010**, *330*, 192–196. [CrossRef]
64. Chalk, P.M.; Souza, R.F. The Role of Biological Nitrogen Fixation in the Nitrogen Cycle of Sugarcane and Maize Cropping Systems. *Symbiosis* **2019**, *78*, 105–118.
65. Albahri, G.; Alyamani, A.A.; Badran, A.; Hijazi, A.; Nasser, M.; Maresca, M.; Baydoun, E. Enhancing Essential Grains Yield for Sustainable Food Security and Bio-Safe Agriculture Through Latest Innovative Approaches. *Agronomy* **2023**, *13*, 1709. [CrossRef]
66. Meng, L.; Zhang, A.; Wang, F.; Han, X.; Wang, D.; Li, S. Arbuscular Mycorrhizal Fungi and *Rhizobium* Facilitate Nitrogen Uptake and Transfer in Soybean/Maize Intercropping System. *Front. Plant Sci.* **2015**, *6*, 339. [CrossRef] [PubMed]
67. Zheng, F.; Tan, D. Enhancing Maize Yield and Nutrient Utilization Through Improved Soil Quality Under Reduced Fertilizer Use: The Efficacy of Organic–Inorganic Compound Fertilizer. *Agriculture* **2024**, *14*, 1482. [CrossRef]
68. Kumar, V.; Prasad, R. Role of Biofertilizers in Improving Soil Fertility and Crop Productivity: A Review. *J. Pharmacogn. Phytochem.* **2021**, *10*, 634–641.
69. Xie, J.; Hu, L.; Wang, Z.; Yan, C.; Bao, F. Synergistic Effects of Biofertilizers on Crop Yield and Nutrient Uptake in Maize. *Agron. J.* **2017**, *109*, 2732–2740.
70. Campbell, W.H. Nitrate Reductase Structure, Function and Regulation: Bridging the Gap Between Biochemistry and Physiology. *Annu. Rev. Plant Biol.* **1999**, *50*, 277–303. [CrossRef]
71. Shoresh, M.; Harman, G.E.; Mastouri, F. Induced Systemic Resistance and Plant Responses to Fungal Biocontrol Agents. *Annu. Rev. Phytopathol.* **2010**, *48*, 21–43. [CrossRef]
72. Altomare, A.; Burla, M.C.; Camalli, M.; Carrozzini, B.; Cascarano, G.L.; Giacovazzo, C.; Guagliardi, A.; Moliterni, A.G.G.; Polidori, G.; Rizzi, R. EXPO: A Program for Full Powder Pattern Decomposition and Crystal Structure Solution. *J. Appl. Crystallogr.* **1999**, *32*, 339–340. [CrossRef]

73. Yao, R.; Bai, R.; Yu, Q.; Bao, Y.; Yang, W. The Effect of Nitrogen Reduction and Applying Bio-Organic Fertilisers on Soil Nutrients and Apple Fruit Quality and Yield. *Agronomy* **2024**, *14*, 345. [CrossRef]
74. Khan, A.R.; Ullah, I.; Khan, S.; Fahad, S.; Khan, A.H.; Ali, S.; Shah, A.; Zhou, W. Plant Growth-Promoting Rhizobacteria as an Alternative Tool for Boosting Crop Productivity in Stressful Environments: A Review. *Front. Plant Sci.* **2018**, *9*, 1801.
75. Kaushal, M.; Wani, S.P. Rhizobacterial-Plant Interactions: Strategies Ensuring Plant Growth Promotion Under Drought and Salinity Stress. *Agric. Ecosyst. Environ.* **2016**, *231*, 68–78. [CrossRef]
76. Velasquez, S.M.; Barbez, E.; Kleine-Vehn, J.; Estevez, J.M. Auxin and Cellular Elongation. *Plant Physiol.* **2016**, *170*, 1206–1215. [CrossRef] [PubMed]
77. Li, X.; Li, Z. What Determines Symbiotic Nitrogen Fixation Efficiency in *Rhizobium*: Recent Insights into *Rhizobium leguminosarum*. *Arch. Microbiol.* **2023**, *205*, 300. [CrossRef] [PubMed]
78. Maitra, S.; Praharaj, S.; Brestic, M.; Sahoo, R.K.; Sagar, L.; Shankar, T.; Palai, J.B.; Sahoo, U.; Pramanick, B.; Nath, S.; et al. *Rhizobium* as Biotechnological Tools for Green Solutions: An Environment-Friendly Approach for Sustainable Crop Production in the Modern Era of Climate Change. *Curr. Microbiol.* **2023**, *80*, 219. [CrossRef]
79. Mehboob, I.; Zahir, Z.A.; Arshad, M.; Tanveer, A.; Khalid, M. Comparative Effectiveness of Different *Rhizobium* sp. for Improving Growth and Yield of Maize (*Zea mays* L.). *Soil Environ.* **2012**, *31*, 37–46.
80. Shrestha, S.; Brueck, H.; Asch, F. Chlorophyll Index, Photochemical Reflectance Index and Chlorophyll Fluorescence Measurements of Rice Leaves Supplied with Different N Levels. *J. Photochem. Photobiol. B Biol.* **2012**, *113*, 7–13. [CrossRef]
81. Ma, L.S.; Li, Y.J.; Wu, P.T.; Zhao, X.N.; Gao, X.D.; Chen, X.L. Recovery Growth and Water Use of Intercropped Maize Following Wheat Harvest in Wheat/Maize Relay Strip Intercropping. *Field Crops Res.* **2020**, *256*, 107924. [CrossRef]
82. Hu, W.; Lu, Z.; Meng, F.; Li, X.; Cong, R.; Ren, T.; Sharkey, T.D.; Lu, J. The Reduction in Leaf Area Precedes That in Photosynthesis Under Potassium Deficiency: The Importance of Leaf Anatomy. *New Phytol.* **2020**, *227*, 1749–1763. [CrossRef]
83. Hussein, M.H.; Eltanahy, E.; Al Bakry, A.F.; Elsafty, N.; Elshamy, M.N. Seaweed Extracts as Prospective Plant Growth Biostimulant and Salinity Stress Alleviator for Vigna sinensis and *Zea mays*. *J. Appl. Phycol.* **2021**, *33*, 1273–1291. [CrossRef]
84. Egamberdieva, D.; Wirth, S.J.; Alqarawi, A.A.; Abd_Allah, E.F.; Hashem, A. Phytohormones and Beneficial Microbes: Essential Components for Plants to Balance Stress and Fitness. *Front. Microbiol.* **2017**, *8*, 2104. [CrossRef] [PubMed]
85. Lichtenthaler, H.K. Chlorophylls and Carotenoids: Pigments of Photosynthetic Biomembranes. *Methods Enzymol.* **1987**, *148*, 350–382.
86. Zhao, J.; Chen, N.; Zhu, T.; Zhao, X.; Yuan, M.; Wang, Z.; Du, H. Simultaneous Quantification and Visualization of Photosynthetic Pigments in *Lycopersicon esculentum* Mill. Under Different Levels of Nitrogen Application with Visible-Near Infrared Hyperspectral Imaging Technology. *Plants* **2023**, *12*, 2956. [CrossRef] [PubMed]
87. Gen-Jiménez, A.; Flores-Félix, J.D.; Rincón-Molina, C.I.; Manzano-Gomez, L.A.; Rogel, M.A.; Ruíz-Valdiviezo, V.M.; Rincón-Rosales, R. Enhance of Tomato Production and Induction of Changes on the Organic Profile Mediated by *Rhizobium* Biofortification. *Front. Microbiol.* **2023**, *14*, 1235930. [CrossRef] [PubMed]
88. Lu, Y.; Xu, J. Phytohormones in Microalgae: A New Opportunity for Microalgal Biotechnology? *Trends Plant Sci.* **2015**, *20*, 273–282. [CrossRef] [PubMed]
89. Doni, F.; Fathurrahman, F.; Mispan, M.S.; Suhaimi, N.S.M.; Yusoff, W.M.W.; Uphoff, N. Transcriptomic Profiling of Rice Seedlings Inoculated with the Symbiotic Fungus *Trichoderma asperellum* SL2. *J. Plant Growth Regul.* **2019**, *38*, 1507–1515. [CrossRef]
90. Harman, G.; Khadka, R.; Doni, F.; Uphoff, N. Benefits to Plant Health and Productivity from Enhancing Plant Microbial Symbionts. *Front. Plant Sci.* **2021**, *11*, 610065. [CrossRef]
91. Vinale, F.; Sivasithamparam, K. Beneficial effects of *Trichoderma* secondary metabolites on crops. *Phytother. Res.* **2020**, *34*, 2835–2842. [CrossRef]
92. Liu, W.; Zheng, L.; Qi, D. Variation in Leaf Traits at Different Altitudes Reflects the Adaptive Strategy of Plants to Environmental Changes. *Ecol. Evol.* **2020**, *10*, 8166–8175. [CrossRef]
93. Fang, L.; Ju, W.; Yang, C.; Jin, X.; Liu, D.; Li, M.; Zhang, C. Exogenous Application of Signaling Molecules to Enhance the Resistance of Legume-*Rhizobium* Symbiosis in Pb/Cd-Contaminated Soils. *Environ. Pollut.* **2020**, *265*, 114744. [CrossRef]
94. Sharma, D.; Gahtyari, N.C.; Chhabra, R.; Kumar, D. Role of Microbes in Improving Plant Growth and Soil Health for Sustainable Agriculture. In *Advances in Plant Microbiome and Sustainable Agriculture: Diversity and Biotechnological Applications*; Springer: Cham, Switzerland, 2020; pp. 207–256.
95. Fahde, S.; Boughribil, S.; Sijilmassi, B.; Amri, A. Rhizobia: A Promising Source of Plant Growth-Promoting Molecules and Their Non-Legume Interactions: Examining Applications and Mechanisms. *Agriculture* **2023**, *13*, 1279. [CrossRef]
96. Hayat, R.; Ahmed, I.; Sheirdil, R.A. An Overview of Plant Growth Promoting Rhizobacteria (PGPR) for Sustainable Agriculture. In *Crop Production*; Springer: Cham, Switzerland, 2016.
97. Qureshi, M.A.; Shahzad, H.; Imran, Z.; Mushtaq, M.; Akhtar, N.; Ali, M.A.; Mujeeb, F. Potential of *Rhizobium* Species to Enhance Growth and Fodder Yield of Maize in the Presence and Absence of L-Tryptophan. *J. Anim. Plant Sci.* **2013**, *23*, 1448–1454.

Nitrogen

MDPI

Article

Iodine-Enriched Urea Reduces Volatilization and Improves Nitrogen Uptake in Maize Plants

João Victor da Costa Cezar [1], Everton Geraldo de Morais [1], Jucelino de Sousa Lima [2], Pedro Antônio Namorato Benevenute [1] and Luiz Roberto Guimarães Guilherme [1,*]

1 Department of Soil Science, Federal University of Lavras, University Campus, P.O. Box 3037, Lavras 37203-202, Minas Gerais, Brazil
2 Department of Biology, Institute of Natural Sciences, Federal University of Lavras, University Campus, P.O. Box 3037, Lavras 37203-202, Minas Gerais, Brazil
* Correspondence: guilherm@ufla.br

Abstract: Urea is the primary source of nitrogen (N) used in agriculture. However, it has a high N loss potential through volatilization. Various mechanisms can be employed to reduce N volatilization losses by inhibiting urease. When added to urea, iodine (I) has high potential for this purpose. Thus, this study aimed to determine whether adding I to urea reduces volatilization losses and increases N uptake in maize plants. Maize plants were cultivated in greenhouse conditions for 36 days. Urea treatments were applied at 15 days of testing, including iodine-enriched urea, conventional urea, and no urea application. Additionally, a study concerning N volatilization from urea was conducted using the same treatments under the same environmental conditions. Iodine was incorporated and adhered to urea, at an I concentration of 0.2%, using potassium iodate (KIO_3). Under controlled conditions and over a short period of time, it was observed that the application of iodine-enriched urea increased the chlorophyll *b* content, root N accumulation, and total N accumulation in maize plants compared with conventional urea. Moreover, iodine-enriched urea reduced N losses from volatilization by 11% compared with conventional urea. The reduction in N volatilization correlated positively with the increased chlorophyll *b*, total chlorophyll, root N accumulation, and total N accumulation favored by the iodine-enriched urea treatment. Our findings demonstrated that adding I to urea is an efficient and promising strategy to reduce N losses and increase N uptake in plants.

Keywords: ammonia; chlorophyll; nitrogen losses; plant metabolism; urease

1. Introduction

Fertilizers are essential for the adequate supply of nutrients to plants, ensuring the productivity of agricultural systems [1,2]. Among the nutrients applied through fertilizers, nitrogen (N), phosphorus, and potassium are the most widely used, with global production estimated at approximately 116.68, 48.27, and 46.64 million tons, respectively [3]. Regarding N fertilizers, less than 50% of the N applied via fertilizers in agroecosystems is absorbed by plants [1,4]. Nitrogen uptake by plants from fertilizers depends highly on environmental conditions during application, soil types, and the sources' properties [1,2,4]. Different sources are used to provide N to plants, such as organic residues, ammonium nitrate (NH_4NO_3), sulfates ($(NH_4)_2SO_4$), nitrates ($NaNO_3$, $Ca(NO_3)_2$, KNO_3), phosphates ($NH_4H_2PO_4$, $(NH_4)_2HPO_4$), aqua ammonia (NH_4OH), anhydrous ammonia ($2NH_3$), and urea ($CO (NH_2)_2$), with urea being the primary source used in agriculture, accounting for 56% of the current N consumption [1].

The widespread use of urea is mainly due to its lower cost than other sources and high N content (approximately 46% N) [1,2,5]. When applied to the soil, urea undergoes hydrolysis promoted by the enzyme urease, and consequently, there can be significant losses of N, as the gas ammonia (NH_3), through volatilization [1,4]. Although ammonia is not a greenhouse gas (GHG), it contributes indirectly to emissions [6,7]. Nitrogen losses from urea through volatilization can exceed 60% of the N applied [8].

Several strategies can be used to minimize N losses through volatilization, such as management practices in the field or materials that can inhibit the urease enzyme in the soil, thereby reducing its activity and consequently reducing ammonia production [9–13]. Among the compounds used as urease inhibitors, currently, only N-(n-butyl) thiophosphoric triamide (NBPT) has been used worldwide, with its market growing 16% annually over the past ten years [5,10–13]. However, several other materials and elements, especially metals, can also reduce urease activity and N volatilization from urea [1,13–15]. Among these elements, iodine (I) has shown a high capacity to reduce N losses through volatilization in urea [16].

In rare studies, it has been shown that I can reduce urease activity, primarily because it acts on specific functional groups, mainly thiol groups (-SH), oxidizing and destroying them, resulting in a drastic reduction in or complete inhibition of urease activity [17–19]. Once urease activity is reduced, N losses from urea through volatilization decrease, increasing plants' N uptake [1,5,20,21]. Additionally, in some studies, whether through the application of I to the soil along with fertilizers or isolated iodine applications to the soil, it has been shown that iodine can influence the metabolism of N, increasing its uptake by plants and also affecting the photosynthetic process in plants through different mechanisms [22–30].

Despite the potential of I to reduce urease activity and influence N metabolism, no current literature has shown that when it is incorporated into urea, it can reduce N losses through volatilization and increase N uptake by plants. This research gathers the primary information on using I to increase the efficiency of urea as a source of N in agriculture. Innovatively, the paper explores the potential of iodine-enriched urea to reduce losses due to N volatilization, a significant problem with conventional urea fertilizers. It also finds an improved N uptake and correlation with plant growth parameters. Thus, our study hypothesized that synthesizing iodine-enriched urea could produce N-enhanced efficiency fertilizers. The aims of the study were as follows: (i) to determine whether the use of iodine-enriched urea improves N accumulation in maize plants compared with conventional urea; (ii) to verify whether iodine-enriched urea reduces N volatilized as NH_3; (iii) to evaluate whether the application of iodine via urea increases chlorophyll contents in maize plants; and (iv) to examine whether the reduction in volatilization promoted by iodine-enriched urea has a positive correlation with increased chlorophyll contents and N accumulation in maize plants.

2. Materials and Methods

2.1. *Synthesis of Iodine-Enriched Urea*

In synthesizing iodine-enriched urea, urea (carbamide = CH_4N_2O) (reagent-grade, Synth, Diadema, São Paulo, Brazil), triethanolamine, organic dye, and an I source were used. The I source used was potassium iodate (KIO_3) (reagent-grade, Synth, Diadema, São Paulo, Brazil). Triethanolamine (surface-active agent) was incorporated and adhered I to the urea granule surface. The dye was used to ensure that all urea granules were impregnated with iodine.

The synthesis process started with KIO_3, triethanolamine, and the organic dye being initially added and homogenized in a beaker. Subsequently, urea was added to the initial mixture, and the material was homogenized again until all the urea was uniformly impregnated with the mixture and no material was retained in the beaker. The final concentration of I in the enriched urea was 0.2%, calculated from the amount of I added concerning urea's final weight. Thus, based on the molarity of I and KIO_3, the final amount of KIO_3 in urea was 0.34%. The concentration of I in urea was determined according to previous studies. After this, the samples were dried at room temperature for 24 h and stored for later use (Figure 1).

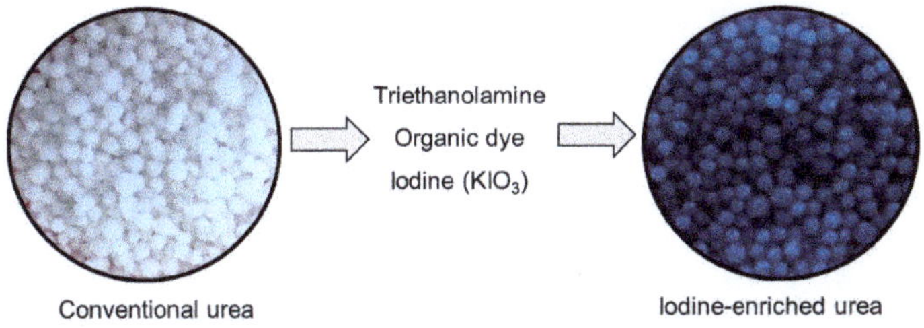

Figure 1. Conventional urea and iodine-enriched urea photos.

2.2. Maize Growing and Treatments

The study was carried out using samples of an Oxisol collected at the 0–0.20 m layer. According to the Brazilian soil classification system, the soil was classified as Latossolo Vermelho [31], which corresponds to Ferralsols [32] and Oxisols in the soil taxonomy [33]. The collected soil was dried and characterized (Table 1) and then incubated with $CaCO_3$ and $MgCO_3$ in a 3:1 molar ratio to increase the contents of Ca and Mg and to increase soil pH to ~5.5–6.0. The amount of the $CaCO_3$:$MgCO_3$ mixture used was based on the method of incubating soils with increasing doses of $CaCO_3$. The samples were then incubated for 30 days, maintaining the soil moisture at 70% of the maximum water retention capacity (MWRC). Next, the samples were dried and sieved (<4 mm). The pH values in water at a 1:2.5 (w/v) after incubation were 5.6 $\pm$ 0.03. Tropical soils under natural conditions have low availabilities of Ca and Mg and most of them have an acidic pH [34]. These conditions promote low N uptake by plants and other nutrients supplied through fertilizers [34,35]. Therefore, the correction of soil pH, as performed in our study, is essential to optimize the use of soil-added nutrients and plant yield.

Table 1. Chemical, physicochemical, and soil particle size distributions of the soil used.

Parameters	Values	Protocol
pH (1 g of soil: 2.5 mL of water)	4.5	pH determined in water
Soil organic matter (g kg^{-1})	24.9	Walkley–Black method
Total nitrogen (g kg^{-1})	2.3	Kjeldahl method
Clay (g kg^{-1})	670	Boyoucos method
Silt (g kg^{-1})	130	Boyoucos method
Sand (g kg^{-1})	200	Boyoucos method
Exchangeable calcium $^{2+}$ ($cmol_c$ kg^{-1})	0.4	1 mol L^{-1} KCl solution–soil test
Exchangeable magnesium^{2+} ($cmol_c$ kg^{-1})	0.2	1 mol L^{-1} KCl solution–soil test
Available phosphorus (mg kg^{-1})	0.4	Mehlich-1 soil test
Available potassium (mg kg^{-1})	24.8	Mehlich-1 soil test
Available zinc (mg kg^{-1})	0.2	Mehlich-1 soil test
Available iron (mg kg^{-1})	38.0	Mehlich-1 soil test
Available manganese (mg kg^{-1})	3.4	Mehlich-1 soil test
Available copper (mg kg^{-1})	1.2	Mehlich-1 soil test
Available boron (mg kg^{-1})	0.01	Hot-water extraction method
Available sulfur (mg kg^{-1})	2.9	Monocalcium phosphate diluted in acetic acid method

All methodologies are described in Teixeira et al. [36].

Pots were filled with 0.5 kg of soil, and 200 mg kg^{-1} of P and 90.4 mg kg^{-1} of N were homogeneously applied in the whole soil using $NH_4H_2PO_4$ (reagent-grade, Synth, Diadema, São Paulo, Brazil). Then, ten maize seeds (*Zea mays* L.) were sown in each pot, and the remaining planting fertilization was performed using 100, 41, 0.8, 1.33, 3.66, 0.15, 4, and 1.55 mg kg^{-1} of K, S, B, Cu, Mn, Mo, Zn, and Fe, respectively, using the following reagents: K_2SO_4, H_3BO_3, $CuSO_4.5H_2O$, $MnSO_4 \cdot H_2O$, $Na_2MoO_4 \cdot 2H_2O$, $ZnSO_4 \cdot 7H_2O$, and $FeCl_3 \cdot 6H_2O$ (reagent-grade, Synth, Diadema, São Paulo, Brazil). Seven days after planting, thinning was performed, leaving two plants per pot until harvest.

Top-dress fertilization related to the treatments tested was performed 15 days after planting. Urea fertilization was the factor studied and three urea application conditions were tested: iodine-enriched urea, conventional urea, and no urea application. The experiment was conducted using a completely randomized design with five replicates, totaling 15 experimental units. Urea was applied superficially at a rate of 300 mg of N kg^{-1} of soil. Immediately after urea application, the pots were irrigated to incorporate the urea. Irrigation was carried out since urea is generally applied before a rain event; thus, irrigation aims to simulate the effect of incorporating urea into the soil [37].

The maize plants were grown until 36 days, reaching the V6 stage. During the maize cultivation in the greenhouse, the average daytime temperature (12 h) was maintained at 28 °C and the average nighttime temperature (12 h) at 18 °C. At 36 days, the chlorophyll *a* and *b* contents of the newly expanded leaf from each plant in each pot were determined using a portable chlorophyll meter (ClorofiLog®). The total chlorophyll content was obtained by summing the chlorophyll *a* and *b* contents. After this, the plants were harvested and divided into roots and shoots. The roots were washed with distilled water to remove any adhering soil particles. The root and shoot samples were then dried in an oven at 45 °C until constant weight was achieved, and the dry matter contents of the shoot (SDM) and root (RDM) were determined. The total dry mass was obtained by summing SDM and RDM. The Kjeldahl method determined the total N content in dried root and shoot samples [38,39].

2.3. Nitrogen Volatilization

The study used the same experimental design, type of soil (Table 1), and environmental conditions adopted for maize cultivation, evaluating three urea application conditions: iodine-enriched urea, conventional urea, and without urea application. However, although this study was conducted simultaneously, it involved different experimental units. The experiment also used a completely randomized design with five replicates, totaling 15 experimental units. The N volatilization study was conducted using mini-lysimeters. A 0.45 μm pore filter was inserted from the bottom to the upper chamber of the mini-lysimeters, which were then filled with 0.5 cm of glass wool and 0.5 kg of soil (pH-corrected soil).

On top of the soil, urea was applied at a rate of 300 mg of N kg^{-1} of soil as in the maize experiment. Immediately after urea application, the pots were irrigated to incorporate the urea, maintaining soil moisture at 70% of the maximum water retention capacity (MWRC) throughout the experimental period. A sponge (0.02 g cm^{-3} density) soaked with a phosphoric acid (75 mL L^{-1}) solution and glycerin (50 mL L^{-1}) was then inserted to capture volatilized N as ammonia (N-NH_3) [20,21]. The sponge with phosphoric acid and glycerin was placed about 1.5 cm above the soil. Another sponge was placed above it to protect the system and prevent contamination of the ammonia-capturing sponge.

The sponge soaked with glycerin and phosphoric acid was maintained throughout the experimental period, with sponges being replaced on the collection days. Ammonia volatilized in the sponge was collected on days 1, 2, 3, 5, 7, 10, 14, and 21 after urea application. These times were chosen based on the interval between urea top-dressing in the maize experiment and the harvest. The size of each pot limited the maximum cultivation period.

After collection, the sponges were washed with distilled water in a volumetric flask up to a final volume of 500 mL. Aliquots of 20 mL were then removed from this solution to determine the total N content through distillation, followed by titration using the Kjeldahl method [39]. In the distillation and titration procedure for N in the aliquots, 10 mL of 13 mol L^{-1} NaOH was used in each tube to promote the distillation of N as N-NH_3, which was captured in a boric acid solution (20 g boric acid L^{-1}) until 50 mL of distillate had accumulated in the Erlenmeyer flask. Then, the titration phase was performed using 0.07143 mol L^{-1} HCl. The cumulative volatilized N was calculated based on the total N determined in the aliquots, the total volume of the solutions, and the number of days between collections. The mini-lysimeter setup and the steps for determining the collected N are shown in Figure 2.

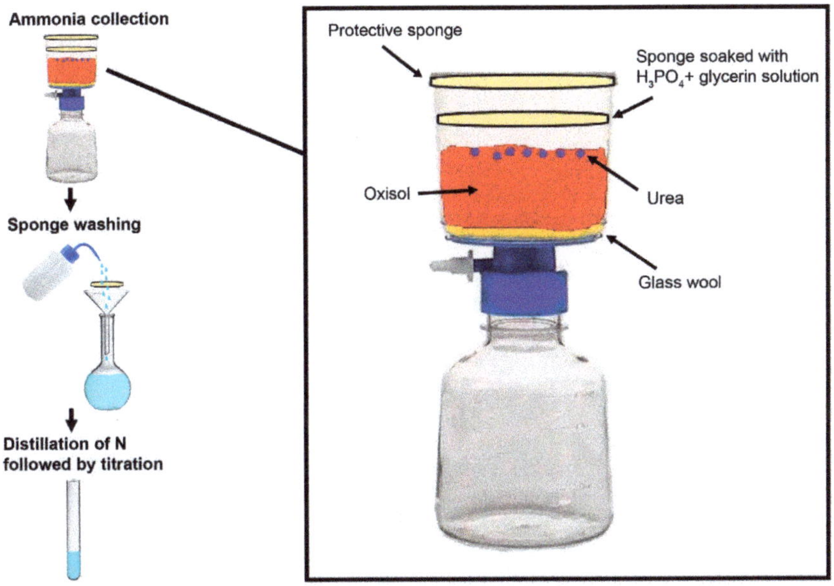

Figure 2. A scheme for the nitrogen volatilization study that was carried out.

2.4. Statistical Analysis

All statistical analyses were performed using the R software version 4.3.1 with the following packages: stats, base, tidyverse, agricolae, factoextra, FactoMineR, and corrplot [40–44]. The cumulative N-volatilization-amount dataset was adjusted to different nonlinear mathematical models to show the relationship between this variable and increasing evaluation days [45–47]. The following mathematical models were adjusted: linear model (Equation (1)), Elovich model (Equation (2)), exponential simple (Equation (3)) model, power function (Equation (4)) model, and hyperbolic model (Equation (5)).

$$N = a + b\ x\ \text{days} \tag{1}$$

$$N = a + b\ \ln \text{days} \tag{2}$$

$$N = a\ x\ \left(1 - e^{b \times \text{days}}\right) \tag{3}$$

$$N = a\ x\ \text{days}^{b} \tag{4}$$

$$N = \frac{V0 \times \text{days}}{V0\ x\ b\ x\ \text{days}} \tag{5}$$

In these equations, N is the cumulative N volatilization amount from each urea condition in the days evaluated, a is the initial value of variable used from each urea condition on one day, b is the variable rate constant for each urea condition, days are days evaluated, and V0 is the maximum value of the variable used from urea condition across the day range investigated. The best model was selected based on the highest value of the coefficient of determination (R^2), the lowest value of root-mean-squared error (RMSE) (Equation (6)), and the lowest value of the Akaike information criterion (AIC) (Equation (7)) [48]. The comparison of mathematical models fitted to each cumulative N volatilization amount over days evaluated was conducted using a 95% confidence interval created through 1000 bootstrap interactions.

$$\text{RMSE} = \sqrt{\sum_{i=1}^{n} \frac{(\text{Predicted values} - \text{Observed values})^2}{\text{Number of observations}}} \quad (6)$$

$$\text{AIC} = \text{Number of observations } x \ln\left(\frac{\text{Sum square of errors}}{\text{Number of observations}}\right) + (2\ x \text{number of parameters}) \quad (7)$$

Additionally, a principal component analysis (PCA) and Pearson correlation analysis were performed to determine the relation between variables analyzed in maize and volatilization studies. In the PCA and correlation analysis, data related to the lack of urea application were removed from the dataset as there was no N volatilization due to the lack of urea application. Furthermore, these analyses aimed to differentiate the effects of iodine-enriched urea versus conventional urea. The data were initially scaled in the PCA analysis, and the PCA analysis was performed.

3. Results

3.1. Maize Cultivation

When compared with plants that had received top-dressing fertilization with urea, regardless of the addition of I (iodine-enriched urea and conventional urea), the no-urea-application group showed reduced total chlorophyll (−19%) and chlorophyll *a* (−15%) contents (Figure 3a). There was no difference ($p > 0.05$) in the total chlorophyll and chlorophyll *a* contents between the application of conventional urea and iodine-enriched urea. However, for chlorophyll *b* content, fertilization with iodine-enriched urea increased the values by 10% compared with conventional urea. Both fertilization with urea had increased chlorophyll *b* contents compared with the lack of urea application, with 67% and 53% increments for iodine-enriched urea and conventional urea, respectively.

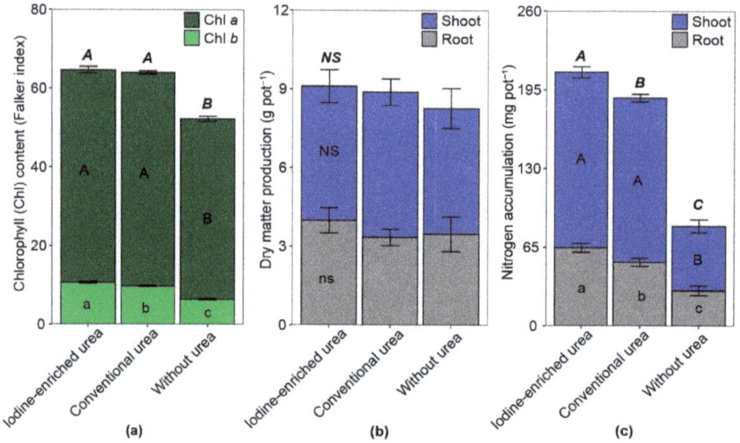

Figure 3. Chlorophyll content, dry matter production, and nitrogen accumulation in maize according to the urea fertilization. Full bar: total chlorophyll (**a**) or total biomass production (**b**) or total nitrogen accumulation (c). ***NS***, NS, and ns: without a statistical difference in total, shoot, and root dry matter production ($p > 0.05$). The means in Figure 3a follow the same scheme with bold–italic capitals, capital letters, and minuscule letters, not differentiating the treatments for total chlorophyll, chlorophyll *a* (Chl *a*), and chlorophyll *b* (Chl *b*) contents, respectively ($p > 0.05$). The means in Figure 3c follow the same scheme with bold–italic capitals, capital letters, and minuscule letters, not differentiating the treatments for total, shoot, and root nitrogen accumulations, respectively ($p > 0.05$).

There was no difference in maize dry matter production (total, shoot, and root) when comparing the urea application treatments with the no-urea-application treatment (Figure 3b). In the maize shoots, N accumulation did not differ for the types of urea tested ($p > 0.05$). Still, both urea treatments tested had increased N accumulation in the shoots by 162% compared with the no-urea-application treatment (Figure 3c). Nitrogen accumulation in the root and the total N accumulation in the plants had increased with urea compared with the lack of urea application. Among the types of urea evaluated, iodine-enriched urea had increased N accumulation in the roots and total N accumulation by 23% and 11%, respectively, compared with conventional urea.

3.2. Nitrogen Volatilization

The best model for N volatilization by both types of urea was the exponential model, as described in Table 2. The exponential models showed the lowest RMSE and AIC values and the highest R^2 values compared with the other models. For the treatment where urea was not applied, the models did not fit because no significant amounts of N had been volatilized.

Table 2. Coefficients of the mathematical regression models adjusted for cumulative nitrogen volatilization according to the urea fertilization treatments.

Models	Iodine-Enriched Urea			Conventional Urea			Without Urea		
	R^2	RMSE	AIC	R^2	RMSE	AIC	R^2	RMSE	AIC
Linear	0.48	5.07	249.32	0.55	5.16	250.76	0.19	0.23	1.75
Elovich	0.75	3.48	219.18	0.84	3.11	210.25	0.16	0.23	3.00
Exponential	0.80	3.22	213.14	0.87	2.80	201.96	0.19	0.23	1.91
Power function	0.66	4.11	232.60	0.74	3.96	229.56	0.18	19.65	229.56
Hyperbolic	0.78	3.35	216.31	0.86	2.99	207.17	0.19	0.23	1.91

R^2: regression coefficient; RMSE: root-mean-squared error; AIC: Akaike Information Criterion.

Generally, for both ureas tested, it was observed that starting from day 10, the amount of N loss through the volatilization process tended to stabilize, with the most significant N loss observed two days after urea application (Figure 4). Compared with conventional urea, it was observed that iodine-enriched urea, from day 6.5 onwards, had had a lower amount of accumulated N volatilized up to 21 days of evaluation. The final cumulative N volatilization at 21 days for conventional urea was 26.3 mg of N (17.5% of the N added), while for iodine-enriched urea, the N volatilized was 23.5 mg of N (15.7% of the N added), a reduction of 11% compared with conventional urea.

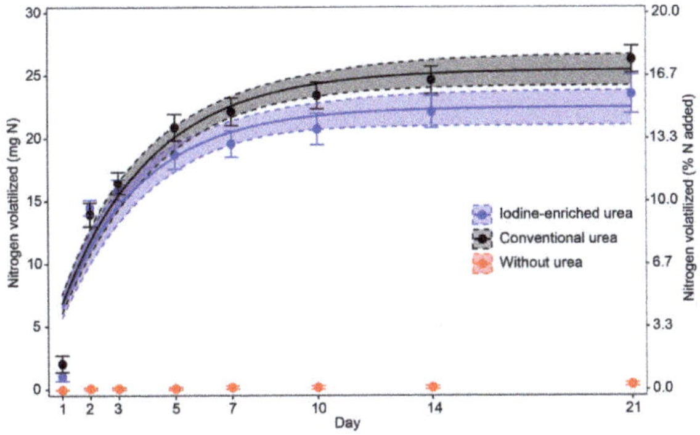

Figure 4. Cumulative nitrogen (N) volatilization according to the urea fertilization treatments. Treatments whose bootstrap-generated confidence intervals do not overlap in the figure were statistically different ($p < 0.05$). The amount of N added was 150 mg per mini-lysimeter.

3.3. Multivariate and Correlation Analysis

The chlorophyll *b* and total chlorophyll contents, root N accumulation, and total N accumulation had positive correlations (Figure 5b), favored by iodine-enriched urea application compared with conventional urea application (Figure 5a). Furthermore, the application of iodine-enriched urea showed a negative relationship with the amount of volatilized N. The amount of volatilized N exhibited a negative correlation with the chlorophyll *b* content (r: −0.89), total chlorophyll content (r: −0.73), root N accumulation (r: −0.91), and total N accumulation (r: −0.77). There was no correlation between volatilized N and chlorophyll *a* content; total, shoot, and root dry matter; or the influence of the types of urea tested.

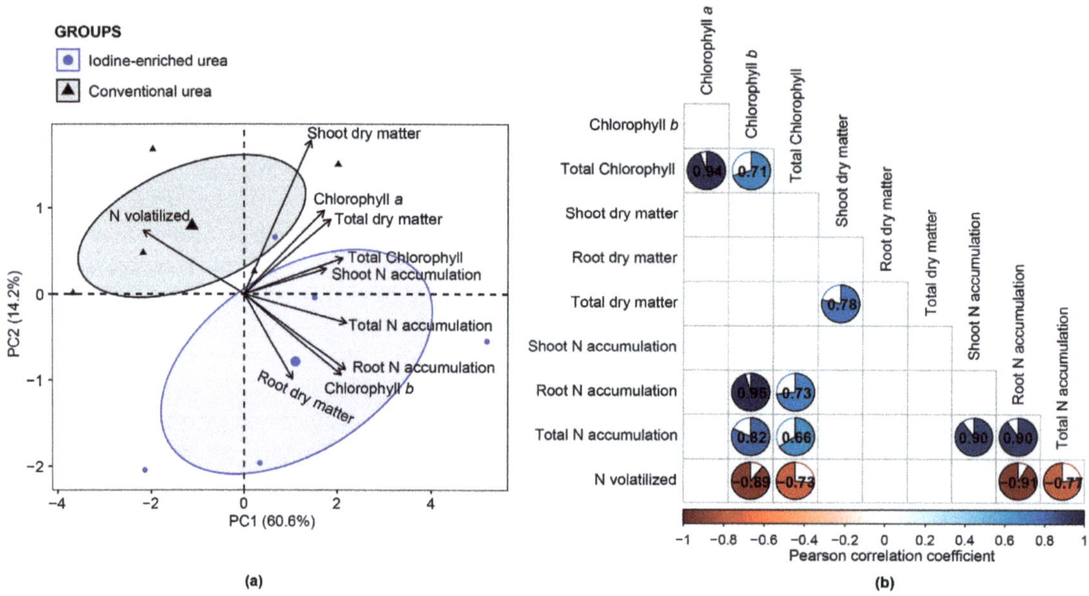

Figure 5. Principal component analysis (**a**) and Pearson correlation (**b**) for variables analyzed in the maize cultivation and N volatilization studies. Significant correlation coefficients ($p < 0.05$) are indicated by bold numbers, with positive and negative correlations distinguished by red and blue, respectively. White boxes indicate non-significance without numbers.

4. Discussion

The main source of N used in agriculture is urea (56% of all), which has significant losses due to volatilization [1]. However, we have shown that adding I to urea can reduce these losses. When hydrolyzed by the enzyme urease, urea ($CO(NH_2)_2$) forms NH_4^+ ions in the soil as follows: $CO(NH_2)_2 + 2H^+ + 2H_2O \longrightarrow 2NH_4^+ + CO_2 + OH^-$ [1,14]. Moreover, urea application raises the pH around the urea granule, and this condition promotes the conversion of ammonium (NH_4^+) to ammonia (NH_3), which is rapidly lost as a gas to the atmosphere through the volatilization process [1,13]. Urea N losses due to volatilization are influenced by several factors, such as pH, moisture, temperature, soil buffering capacity, and soil organic matter [1,12,13]. Surface-applied urea can result in volatilization losses exceeding 60% of the applied N [8]. It has been estimated that global urea N losses due to volatilization amount to USD 74.2 billion, assuming an average global urea N volatilization loss of 14% [1]. Various strategies have been adopted to reduce N losses, such as incorporating urea into the soil [9] or adding compounds inhibiting urease activity, slowing urea hydrolysis, and reducing volatilization losses [5,10,11,13].

The most common compounds applied with urea to reduce volatilization losses by decreasing urease activity include triazoles, quinones, fluorides, flavonoids, phosphorus compounds, imidazoles, Schiff base ligan metal complexes, polyphenols, hydroxamic acids, and Schiff bases [12,14]. However, an element with high potential to reduce volatilization losses is iodine, as demonstrated in our study, under controlled conditions and over a short period of time, where the application of iodine-enriched urea reduced N losses from volatilization compared with conventional urea (Figures 4 and 5).

Iodine can reduce urea volatilization since it can decrease or inactivate the urease activity [17,18], consequently lowering the conversion of urea to ammonia [13,14,18]. This condition occurs because I can act on characteristic groups in the urease molecule, such as thiol groups (-SH), oxidizing and destroying them, and it may also substitute amino ($-NH_2$) and amide (-NH-) functional groups in the urease molecule [17,18]. These structural modifications in urease directly interfere with the interactions between urease and its substrate (urea), altering and inactivating the urease [11,13,14,17,18,49]. The effect of I on urease was observed in a study conducted with *Vicia sativa* seeds, estimating the urease activity by using a medium where 1 mL of 0.1% (w/v) urea was added along with an enzyme solution (50 μL enzyme + 100 μL metal + 1.85 mL 0.2 M buffer). It was observed that the addition of I was able to reduce urease activity by 50%, an effect directly related to the thiol (-SH) group in the enzyme's active site [19].

Treatments that reduce urease activity and consequently decrease N volatilization maximize N uptake by plants, increasing the uptake of ammonium-carrying N fertilizers [1,10,12,13]. This effect was observed in our study under controlled conditions and over a short period of time, where there was a negative correlation between higher rates of N volatilization and the root N accumulation and the total N accumulation in maize plants favored by a greater efficiency of N use due to the application of iodine-enriched urea (Figures 3–5) as the apparent recovery of applied N for the iodine-enriched urea was 70%, compared with 63% for conventional urea. Additionally, applying I to plants has shown a synergistic effect on N uptake by plants, as demonstrated by several studies [26–30]. This synergistic effect of iodine application on N metabolism is poorly understood. However, in research conducted with the application of KIO_3, an increase in growth was observed in lettuce plants due to the stimulation by iodine of nitrate reductase activity and the Glutamine Synthetase/Glutamate Synthase (GS/GOGAT) cycle [25].

In a study conducted with fenugreek, the application of I increased N accumulation and the protein content [29]. In tomatoes, applying KIO_3 via soil [29] or foliar spray [46] also increased total N uptake in plants and tomato fruits. In spinach, using doses similar to our study (2 kg of I ha^{-1}, considering a depth of 0–0.20 m), the soil application of 1 and 2 mg of I dm^{-3}, equivalent to 2 and 4 kg ha^{-1} (0–0.20 m depth), increased both the N content as well as the amino acid content in spinach leaves [26]. Our study also showed an increase in N uptake with the application of iodine-enriched urea compared with no iodine application (conventional urea) (Figure 3). Similar to our study using the same amount and source of I (2 kg ha^{-1} of I supplied as KIO_3), the combined application with an N fertilizer significantly improved N utilization from mineral fertilizers for carrots [27]. In our study, this increase in N accumulation due to I application was more pronounced in the roots compared with the total accumulation, with no effect being observed in the shoot for the different urea treatments (Figure 3).

The improvement in N nutritional status due to I application can affect both N-rich cellular components, such as chlorophylls, and the photosynthetic process in plants [22–25,28,50,51]. Iodine participates in photosynthesis as a chlorophyll component and promotes the synthesis of carbohydrates and proteins [50]. This effect of iodine on chlorophyll was observed in a study with rye plants cultivated under field conditions, where the application of iodine to the soil at 2.5 and 5 kg ha^{-1} increased chlorophyll *a* and *b* contents with an increasing I concentration [28]. Evaluating coffee plants under water deficit conditions, the application of I via soil (2.5 kg ha^{-1} of KIO_3) compared with a lack of I fertilization mitigated the water deficit and also increased chlorophyll *b* (+45%) and total chlorophyll (+23%) contents, without affecting the chlorophyll *a* content. Similar to what was noted in the rye and coffee studies, applying I via urea in our study increased the chlorophyll *b* content in maize plants at doses comparable to those mentioned (Figure 3). Furthermore, in our study, this increase in chlorophyll *b* content was directly related to reduced N losses through volatilization (Figure 5). Thus, it was observed that adding iodine to urea was an effective strategy to increase the uptake of N from fertilizers. However, experiments under field conditions evaluating agronomic efficiency, feasibility, and financial return should be conducted.

5. Conclusions

This research introduces a promising and efficient strategy to enhance nitrogen (N) utilization in agriculture, addressing a key challenge of N loss through volatilization and increasing the sustainability of urea-based fertilizers. Our study, conducted under controlled conditions and over a short period of time, found a direct positive correlation between the highest values of the total accumulation of N, root accumulation of N, total chlorophyll content, and chlorophyll *b* and a negative correlation with the accumulated iodine-enriched urea favoring lower N losses. Compared with conventional urea, iodine-enriched urea reduces N losses through volatilization, increasing N accumulation in maize plants. Additionally, the iodine contained in urea could modify plant metabolism, leading to an increased chlorophyll content. Therefore, coating urea with iodine is an effective strategy for creating fertilizers that improve N uptake. However, the full potential of this technology can only be realized through further field studies. These studies should evaluate N uptake by plants under these conditions, as well as the financial returns of such technology, and are crucial for advancing this research.

Author Contributions: Conceptualization, J.V.d.C.C., E.G.d.M., J.d.S.L., P.A.N.B., and L.R.G.G.; methodology, J.V.d.C.C., and E.G.d.M.; software, E.G.d.M.; validation, J.V.d.C.C., E.G.d.M., and L.R.G.G.; formal analysis, J.V.d.C.C., E.G.d.M., and J.d.S.L.; investigation, J.V.d.C.C., E.G.d.M., J.d.S.L., P.A.N.B., and L.R.G.G.; resources, L.R.G.G.; data curation, J.V.d.C.C., and E.G.d.M.; writing—original draft preparation, J.V.d.C.C., E.G.d.M., J.d.S.L., P.A.N.B., and L.R.G.G.; writing—review and editing, J.V.d.C.C., E.G.d.M., J.d.S.L., P.A.N.B., and L.R.G.G.; visualization, E.G.d.M.; supervision, E.G.d.M., and L.R.G.G.; project administration, L.R.G.G.; funding acquisition, L.R.G.G. All authors have read and agreed to the published version of the manuscript.

Funding: This research was funded by the Minas Gerais State Research Foundation (FAPEMIG) (PRP n° 5/2023), the Coordination for the Improvement of Higher Education Personnel (CAPES) (Grant Code-001), and the National Council for Scientific and Technological Development (CNPq), grant number #153474/2024-6, as well as the National Institute of Science and Technology (INCT) on Soil and Food Security, CNPq grant #406577/2022-6.

Data Availability Statement: The data supporting this study's findings are available on request from the corresponding author.

Acknowledgments: The authors acknowledge the Coordination for the Improvement of Higher Education Personnel (CAPES), the National Council for Scientific and Technological Development (CNPq), and the Minas Gerais State Research Foundation (FAPEMIG) for their financial support and scholarships.

Conflicts of Interest: The authors declare no conflicts of interest. The funders had no role in the study's design, data collection, analysis or interpretation, manuscript writing, or decision to publish the results.

References

1. Cassim, B.M.A.R.; Lisboa, I.P.; Besen, M.R.; Otto, R.; Cantarella, H.; Inoue, T.T.; Batista, M.A. Nitrogen: From discovery, plant assimilation, sustainable usage to current enhanced efficiency fertilizers technologies—A review. *Rev. Bras. Ciênc. Solo* **2024**, *48*, e0230037. [CrossRef]
2. Dimkpa, C.O.; Fugice, J.; Singh, U.; Lewis, T.D. Development of fertilizers for enhanced nitrogen use efficiency—Trends and perspectives. *Sci. Total Environ.* **2020**, *731*, 139113. [CrossRef] [PubMed]
3. Food and Agriculture Organization of the United Nations Fertilizer Production by Nutrient Type, World, 1961 to 2021. Available online: https://ourworldindata.org/grapher/fertilizer-production-by-nutrient-type-npk (accessed on 26 June 2024).
4. Yahaya, S.M.; Mahmud, A.A.; Abdullahi, M.; Haruna, A. Recent advances in the chemistry of nitrogen, phosphorus and potassium as fertilizers in soil: A review. *Pedosphere* **2023**, *33*, 385–406. [CrossRef]

5. Harty, M.A.; McDonnell, K.P.; Whetton, R.; Gillespie, G.; Burke, J.I. Comparison of ammonia-N volatilization losses from untreated granular urea and granular urea treated with NutriSphere-N®. *Soil Use Manag.* **2024**, *40*, e12891. [CrossRef]
6. Awale, R.; Chatterjee, A. Enhanced Efficiency Nitrogen Products Influence Ammonia Volatilization and Nitrous Oxide Emission from Two Contrasting Soils. *Agron. J.* **2017**, *109*, 47–57. [CrossRef]
7. Gorh, D.; Baruah, K.K. Estimation of methane and nitrous oxide emission from wetland rice paddies with reference to global warming potential. *Environ. Sci. Pollut. Res.* **2019**, *26*, 16331–16344. [CrossRef]
8. Pan, B.; Lam, S.K.; Mosier, A.; Luo, Y.; Chen, D. Ammonia volatilization from synthetic fertilizers and its mitigation strategies: A global synthesis. *Agric. Ecosyst. Environ.* **2016**, *232*, 283–289. [CrossRef]
9. Götze, H.; Saul, M.; Jiang, Y.; Pacholski, A. Effect of Incorporation Techniques and Soil Properties on NH_3 and N_2O Emissions after Urea Application. *Agronomy* **2023**, *13*, 2632. [CrossRef]
10. Hube, S.; Alfaro, M.A.; Scheer, C.; Brunk, C.; Ramírez, L.; Rowlings, D.; Grace, P. Effect of nitrification and urease inhibitors on nitrous oxide and methane emissions from an oat crop in a volcanic ash soil. *Agric. Ecosyst. Environ.* **2017**, *238*, 46–54. [CrossRef]
11. Yang, W.; Peng, Z.; Wang, G. An overview: Metal-based inhibitors of urease. *J. Enzym. Inhib. Med. Chem.* **2023**, *38*, 361–375. [CrossRef]
12. Cantarella, H.; Otto, R.; Soares, J.R.; de Brito Silva, A.G. Agronomic efficiency of NBPT as a urease inhibitor: A review. *J. Adv. Res.* **2018**, *13*, 19–27. [CrossRef] [PubMed]
13. Kafarski, P.; Talma, M. Recent advances in design of new urease inhibitors: A review. *J. Adv. Res.* **2018**, *13*, 101–112. [CrossRef] [PubMed]
14. Yadav, N.; Chauhan, S.; Sangwan, S.; Kumar, R. Deepak Recent Advances in Anti-Urease Activity of Schiff Bases and their Metal Complexes. *Asian J. Chem.* **2024**, *36*, 779–794. [CrossRef]
15. Pereira, L.E.T.; Herling, V.R.; Tech, A.R.B. Current Scenario and Perspectives for Nitrogen Fertilization Strategies on Tropical Perennial Grass Pastures: A Review. *Agronomy* **2022**, *12*, 2079. [CrossRef]
16. Pursell, T.; Shirley, A.R., Jr.; Cochran, K.D.; Miller, J.M.; Holt, T.G.; Peeden, G.S. Urea Fertilizer Containing Central Volatilization Inhibitor Particles to Reduce Release of Ammonia and Processes for Making Same. U.S. Patent 8,758,474, 24 June 2014.
17. Hellerman, L.; Perkins, M.E.; Clark, W.M. Urease Activity as Influenced by Oxidation and Reduction. *Proc. Natl. Acad. Sci. USA* **1933**, *19*, 855–860. [CrossRef]
18. Smythe, C.V. The reaction of iodoacetate and of iodoacetamide with various sulfhydryl groups, with urease, and with yeast preparations. *J. Biol. Chem.* **1936**, *114*, 601–612. [CrossRef]
19. Sindi, A.M.; Zaman, U.; Saleh, E.A.M.; Kassem, A.F.; Rahman, K.U.; Khan, S.U.; Alharbi, M.; Rizg, W.Y.; Omar, K.M.; Majrashi, M.A.A.; et al. Biochemical and thermodynamic properties of de novo synthesized urease from Vicia sativa seeds with enhanced industrial applications. *Int. J. Biol. Macromol.* **2024**, *259*, 129190. [CrossRef]
20. Freitas, T.; Bartelega, L.; Santos, C.; Dutra, M.P.; Sarkis, L.F.; Guimarães, R.J.; Dominghetti, A.W.; Zito, P.C.; Fernandes, T.J.; Guelfi, D. Technologies for Fertilizers and Management Strategies of N-Fertilization in Coffee Cropping Systems to Reduce Ammonia Losses by Volatilization. *Plants* **2022**, *11*, 3323. [CrossRef] [PubMed]
21. Chagas, W.F.T.; Guelfi, D.R.; Caputo, A.L.C.; de Souza, T.L.; Andrade, A.B.; Faquin, V. Ammonia volatilization from blends with stabilized and controlled-released urea in the coffee system. *Ciênc. Agrotecnologia* **2016**, *40*, 497–509. [CrossRef]
22. de Sousa Lima, J.; Andrade, O.V.S.; dos Santos, L.C.; de Morais, E.G.; Martins, G.S.; Mutz, Y.S.; Nascimento, V.L.; Marchiori, P.E.R.; Lopes, G.; Guilherme, L.R.G. Soybean Plants Exposed to Low Concentrations of Potassium Iodide Have Better Tolerance to Water Deficit through the Antioxidant Enzymatic System and Photosynthesis Modulation. *Plants* **2023**, *12*, 2555. [CrossRef]
23. de Sousa Lima, J.; Andrade, O.V.S.; de Morais, E.G.; Machado, G.G.L.; dos Santos, L.C.; de Andrade, E.S.; Benevenute, P.A.N.; Martins, G.S.; Nascimento, V.L.; Marchiori, P.E.R.; et al. KI Increases Tomato Fruit Quality and Water Deficit Tolerance by Improving Antioxidant Enzyme Activity and Amino Acid Accumulation: A Priming Effect or Relief during Stress? *Plants* **2023**, *12*, 4023. [CrossRef] [PubMed]
24. Zhang, Y.; Cao, H.; Wang, M.; Zou, Z.; Zhou, P.; Wang, X.; Jin, J. A review of iodine in plants with biofortification: Uptake, accumulation, transportation, function, and toxicity. *Sci. Total Environ.* **2023**, *878*, 163203. [CrossRef] [PubMed]
25. Blasco, B.; Rios, J.J.; Cervilla, L.M.; Sánchez-Rodríguez, E.; Rubio-Wilhelmi, M.M.; Rosales, M.A.; Ruiz, J.M.; Romero, L. Photorespiration Process and Nitrogen Metabolism in Lettuce Plants (*Lactuca sativa* L.): Induced Changes in Response to Iodine Biofortification. *J. Plant Growth Regul.* **2010**, *29*, 477–486. [CrossRef]
26. Smolen, S.; Sady, W. Influence of iodine fertilization and soil application of sucrose on the effectiveness of iodine biofortification, yield, nitrogen metabolism and biological quality of of spinach. *Acta Sci. Pol. Hortorum Cultus* **2011**, *10*, 51–63.
27. Smoleń, S.; Sady, W.; Rożek, S.; Strzetelski, P.; Ledwożyw-Smoleń, I. Preliminary evaluation of the influence of iodine and nitrogen fertilization on the effectiveness of iodine biofortification and mineral composition of carrot storage roots. *J. Elemntology* **2011**, *16*, 275–285. [CrossRef]
28. Krzepiłko, A.; Kościk, B.; Skowrońska, M.; Kuśmierz, S.; Walczak, J.; Prażak, R. Quality of Rye Plants (*Secale cereale*) as Affected by Agronomic Biofortification with Iodine. *Plants* **2022**, *12*, 100. [CrossRef]
29. Ramezani, S.; Yousefshahi, B.; Ramezan, D.; Zargar, M.; Pakina, E.; Bayat, M. Selenium, Iodine, and Supplementary Blue Light Enriched Fenugreek (*Trigonella foenum-gracum* L.) in Terms of Biochemical Quality, Mineral Uptake, and Trace Elements Accumulation in a Hydroponic System. *Agriculture* **2023**, *13*, 2009. [CrossRef]

30. Mageshen, V.; Santhy, P. Effect of chitosan iodate complex biofortification on nutrient uptake in 'shivam' hybrid of tomato (*Solanum lycopersicum* L.). *J. Appl. Nat. Sci.* **2023**, *15*, 549–554. [CrossRef]
31. Santos, H.G.; Jacomine, P.K.T.; Dos Anjos, L.H.C.; De Oliveira, V.A.; Lumbreras, J.F.; Coelho, M.R.; De Almeida, J.A.; de Araujo Filho, J.C.; de Oliveira, J.B.; Cunha, T.J.F. *Sistema Brasileiro de Classificação de Solos*; Embrapa: Brasília, Brasil, 2018.
32. IIUSS Working Group WRB (Ed.) *World Reference Base for Soil Resources 2014, Update 2015 International Soil Classification System for Naming Soils and Creating Legends for Soil Maps*; Food and Agriculture Organization (FAO): Rome, Italy, 2015.
33. Soil Survey Staff (Ed.) *Keys to Soil Taxonomy*, 13th ed.; USDA Natural Resources Conservation Service: Washington, DC, USA, 2022.
34. Lopes, A.S.; Guimarães Guilherme, L.R. A Career Perspective on Soil Management in the Cerrado Region of Brazil. *Adv. Agron.* **2016**, *137*, 1–72. [CrossRef]
35. Agegnehu, G.; Amede, T.; Erkossa, T.; Yirga, C.; Henry, C.; Tyler, R.; Nosworthy, M.G.; Beyene, S.; Sileshi, G.W. Extent and management of acid soils for sustainable crop production system in the tropical agroecosystems: A review. *Acta Agric. Scand. Sect. B Soil Plant Sci.* **2021**, *71*, 852–869. [CrossRef]
36. Teixeira, P.C.; Donagemma, G.K.; Fontana, A.; Teixeira, W.G. *Manual de Métodos de Análise de Solo*, 3rd ed.; Embrapa: Brasília, Brasil, 2017; ISBN 978-85-7035-771-7.
37. Motasim, A.M.; Samsuri, A.W.; Nabayi, A.; Akter, A.; Haque, M.A.; Sukor, A.S.A.; Adibah, A.M. Urea application in soil: Processes, losses, and alternatives—A review. *Discov. Agric.* **2024**, *2*, 42. [CrossRef]
38. Kalra, Y. *Handbook of Reference Methods for Plant Analysis*; Kalra, Y., Ed.; CRC Press: Boca Raton, FL, USA, 1997.
39. Kjeldahl, J. Neue Methode zur Bestimmung des Stickstoffs in organischen Körpern. *Z. Für Anal. Chem.* **1883**, *22*, 366–382. [CrossRef]
40. Arnhold, E. Easyanova: Analysis of Variance and Other Important Complementary Analyses 2019. Available online: https://cran.r-project.org/web/packages/easynova/ (accessed on 2 July 2024).
41. Lê, S.; Josse, J.; Husson, F. FactoMineR: An R Package for Multivariate Analysis. *J. Stat. Softw.* **2008**, *25*, 1–18. [CrossRef]
42. R Core Team R: A Language and Environment for Statistical Computing 2022. Available online: https://www.r-project.org/ (accessed on 2 July 2024).
43. Wei, T.; Simko, V.R. Package "corrplot": Visualization of a Correlation Matrix 2017. Available online: https://cran.r-project.org/web/packages/corrplot/ (accessed on 2 July 2024).
44. Wickham, H.; Averick, M.; Bryan, J.; Chang, W.; McGowan, L.; François, R.; Grolemund, G.; Hayes, A.; Henry, L.; Hester, J.; et al. Welcome to the Tidyverse. *J. Open Source Softw.* **2019**, *4*, 1686. [CrossRef]
45. Zhang, S.; Li, Y.; Singh, B.P.; Wang, H.; Cai, X.; Chen, J.; Qin, H.; Li, Y.; Chang, S.X. Contrasting short-term responses of soil heterotrophic and autotrophic respiration to biochar-based and chemical fertilizers in a subtropical Moso bamboo plantation. *Appl. Soil Ecol.* **2021**, *157*, 103758. [CrossRef]
46. Zhao, W.; Liu, S.; Yin, M.; He, Z.; Bi, D. Co-pyrolysis of cellulose with urea and chitosan to produce nitrogen-containing compounds and nitrogen-doped biochar: Product distribution characteristics and reaction path analysis. *J. Anal. Appl. Pyrolysis* **2023**, *169*, 105795. [CrossRef]
47. Raclavská, H.; Růžičková, J.; Raclavský, K.; Juchelková, D.; Kucbel, M.; Švédová, B.; Slamová, K.; Kacprzak, M. Effect of biochar addition on the improvement of the quality parameters of compost used for land reclamation. *Environ. Sci. Pollut. Res.* **2021**, *30*, 8563–8581. [CrossRef]
48. Akaike, H. A Bayesian Extension of the Minimum AIC Procedure of Autoregressive Model Fitting. *Biometrika* **1979**, *66*, 237. [CrossRef]
49. Suenaga, S.; Takano, Y.; Saito, T. Unraveling Binding Mechanism and Stability of Urease Inhibitors: A QM/MM MD Study. *Molecules* **2023**, *28*, 2697. [CrossRef] [PubMed]
50. Kiferle, C.; Martinelli, M.; Salzano, A.M.; Gonzali, S.; Beltrami, S.; Salvadori, P.A.; Hora, K.; Holwerda, H.T.; Scaloni, A.; Perata, P. Evidences for a Nutritional Role of Iodine in Plants. *Front. Plant Sci.* **2021**, *12*, 616868. [CrossRef] [PubMed]
51. Adrade, O.V.S.; de Sousa Lima, J.; das Neves, T.T.; de Morais, E.G.; Benevenute, P.A.N.; dos Santos, L.C.; Nascimento, V.L.; Guilherme, L.R.G.; Marchiori, P.E.R. The Role of Potassium Iodate inMitigating the Damages of Water Deficit in Coffee Plants. *J. Soil Sci. Plant Nutr.* **2024**, 1–7. [CrossRef]

Nitrogen

MDPI

Article

Influence of Soil Type and Temperature on Nitrogen Mineralization from Organic Fertilizers

Hanna Ibiapina de Jesus [1], Kate Cassity-Duffey [2], Bhabesh Dutta [3], Andre Luiz Biscaia Ribeiro da Silva [4] and Timothy Coolong [2,*]

1 Department of Horticultural Sciences, University of Florida, Fifield Hall, 2550 Hull Road, Gainesville, FL 32611, USA; hannaibiapinadej@ufl.edu

2 Department of Horticulture, University of Georgia, 1111 Miller Plant Sciences, Athens, GA 30602, USA; katecass@uga.edu

3 Department of Plant Pathology, University of Georgia, 2360 Rainwater Rd., Tifton, GA 31793, USA; bhabesh@uga.edu

4 Department of Horticulture, Auburn University, 124 Funchess Hall, Auburn, AL 36849, USA; azb0207@auburn.edu

* Correspondence: tcoolong@uga.edu

Abstract: Organic vegetable producers in Georgia, USA, utilize a range of amendments to supply nitrogen (N) for crop production. However, differences in soil type, fertilizers and environmental conditions can result in variability in N mineralization rates among commonly utilized organic fertilizers in the region. In this study, the effects of temperature on N mineralization from three commercial organic fertilizers [feather meal (FM), pelleted poultry litter (PPL) and a mixed organic fertilizer (MIX)] in two soil types from Georgia, USA (Cecil sandy clay loam and Tifton loamy sand) were evaluated for 120 d. Net N mineralization (Net N_{min}) varied with soil type, fertilizer and temperature. After 120 d, Net N_{min} from the FM fertilizer ranged between 41% and 77% of total organic N applied, the MIX fertilizer ranged between 26% and 59% and the PPL fertilizer ranged between 0% and 22% across all soil types and temperatures. Incubation at higher temperatures (20 °C and 30 °C) impacted Net N_{min} of FM fertilizer in the Tifton series soil. Temperature and soil type had a relatively minor impact on the potentially mineralizable N of the PPL and MIX fertilizers after 120 d of incubation; however, both factors impacted the rate of fertilizer release shortly after application, which could impact the synchronicity of N availability and plant uptake. Temperature-related differences in the mineralization of organic fertilizers may not be large enough to influence a grower's decisions regarding N fertilizer inputs for vegetable crop production in the two soils. However, organic fertilizer source will likely play a significant role in N availability during the cropping season.

Keywords: Cecil clay loam; feather meal; mineralization; pelleted poultry litter; Tifton loamy sand

1. Introduction

Continued growth in certified organic vegetable production in the USA [1] has led to increased demand for alternatives to chemical fertilizers for plant nutritional needs. A range of organic materials are commonly used in agricultural systems to supply plant-available nitrogen (N), including cover crop residues, composts, manures and commercial organic fertilizers derived from various animal or plant sources [2,3]. However, unlike readily soluble synthetic fertilizers, the availability of nutrients from organic amendments and fertilizers can be highly variable [4]. Organic fertilizers may include some readily available inorganic N, ammonium (NH_4) and nitrate (NO_3), but the majority of N present is in the organic form and must undergo N mineralization to supply plant-available nitrogen during the growing season [5]. Failure to coordinate the timing of N release from organic fertilizers with the nutritional demands of crops may result in N loss and reduced productivity. Thus, it is important to accurately predict N mineralization rates from organic fertilizers and evaluate what factors may impact mineralization [6–8].

Organic fertilizers differ widely in physiochemical characteristics, and N mineralization rates can vary depending on these physiochemical properties. Total particle size, total N concentration, initial inorganic N levels and carbon to nitrogen (C:N) ratio can all affect the N mineralization of organic fertilizers [5–10]. In vitro incubation and field-based studies have shown that net N mineralization (Net N_{min}) among different organic fertilizers can vary between 20% to 93% of the total N applied [5,11–13]. However, the physiochemical characteristics of the materials themselves do not explain all the variability observed in N mineralization among fertilizers, and further information is needed to properly predict N release after application [14,15].

Nitrogen release from organic fertilizers has also been shown to be impacted by environmental conditions and the properties of the soils to which they are applied [14]. Microorganisms in the soil promote N mineralization through enzymatic reactions, which are primarily controlled by temperature and water content [16–18]. In general, the rate of mineralization in the soil increases with temperature up to a maximum and then declines [11,19,20]. Therefore, the season in which crops are planted and weather fluctuations during the season or across regions may impact the rate of N mineralization in the soil [17]. Several studies have evaluated the effects of temperature on N mineralization from soil organic matter [17,18,21,22]. However, research on N mineralization from organic fertilizers suggests that N transformations at different temperatures are likely to be dependent on a complex interaction between soil microorganisms and the mineralizable substrates in the amendment [11,23,24]. Factors such as the C:N ratio of the organic materials have been reported to be closely linked to the N mineralization of a range of organic amendments [3]. Temperature has also been shown to influence N availability, depending on material composition [14]. Studies suggest that N mineralization of high-N-containing organic materials are primarily influenced by temperature during the initial days after incorporation, whereas low-N-containing materials are more susceptible to temperature variations over an extended period [11,12,14,25].

In addition to environmental conditions, soil physical and/or chemical properties can alter the N mineralization of organic materials due to changes in the soil microbiota and the enzymatic reactions that drive the N release [10,14,26]. Further, soil type can significantly affect N mineralization [27–29]. In general, fine-textured soils with high clay contents have been reported to have lower N mineralization rates than coarse-textured, sandy soils, which is likely due to factors such as physical isolation of organic matter by clay particles or entrapment in small pores, where a substrate may be inaccessible to microorganisms [28,30]. Cassity-Duffey et al. [7] determined that, after 100 d of incubation, soil texture had an impact on the rate of N release but had minimal effects on total mineralizable N for two organic fertilizers, a feather meal (FM) and a pelleted fertilizer blend. Similarly, Lazicki et al. [14] found that soil texture and management history did not consistently affect total Net N_{min} from 22 different organic materials but may have influenced the rate of N release. Recently, de Jesus et al. [31] reported that location (soil type and weather conditions) impacted plant N accumulation for onions (*Allium cepa* L.) grown with a mixed organic (MIX) and pelletized poultry litter (PPL) fertilizers. In addition to N uptake by plants, residual soil inorganic N was greater at harvest when using a MIX fertilizer compared to a PPL product. Increased N uptake and soil accumulation were also linked to increased yields for onions grown with a MIX fertilizer in some locations and years. These results suggest that N mineralization may differ between organic fertilizer source, soil type and weather conditions, ultimately impacting the yield of the crop grown [31]. Further, mineralization of organic fertilizers at soil temperatures encountered during the production of onion during winter months in the Southeastern USA (<10 °C) have not been previously studied.

Understanding the factors affecting total mineralization and the rate of N release from organic fertilizers is important to determine the rate and timing of application [32]. Models have been proposed to simulate the N release from organic soil amendments in the field, but data are lacking specific to commonly used commercial organic fertilizers [11,23,33–35]. Although different kinetics may be utilized to predict N availability, first-order kinetics are often used to predict the N mineralization potential (N_0) from organic amendments, where the rate at which N mineralizes is assumed to be proportional to the amount of soil N available for mineralization [26,36]. An accurate estimate of N mineralization from organic fertilizers and the impact of temperature on the rate of mineralization would be a useful tool for organic vegetable growers. Thus, the objectives of this study were to determine the effects of four incubation temperatures and two soil textures on three commercially available organic fertilizers used for vegetable production in Georgia, USA through a 120 d laboratory incubation.

2. Materials and Methods

2.1. Soil and Organic Fertilizers

The soils used in this study were representative of two common agricultural soils in Georgia, USA. Soils were collected from US Department of Agriculture (USDA) certified organic land at the University of Georgia (UGA) Durham Horticulture Farm in Watkinsville, GA, USA (33°53′4.812″ N, 83°25′9.9876″ W) and at the UGA Horticulture Farm in Tifton, GA, USA (31°28′8.112″ N, 83°32′54.348″ W). The soil of the Watkinsville, GA, USA location is a Cecil sandy clay loam series (0% to 2% slope), while the Tifton, GA, USA location was a Tifton loamy sand series (2% to 5% slope) [37]. Approximately 20 kg of soil was collected from each location (0–15 cm depth) and passed through a 4-mm sieve. Soil samples were stored in 5-gallon buckets and kept aerated at room temperature prior to initiation of the incubation.

The maximum water holding capacity (WHC) of the soils was estimated through saturation and draining over a sand bath for 48 h [38,39], and was 0.25 g $H_2O \cdot g^{-1}$ soil and 0.30 g $H_2O \cdot g^{-1}$ soil for the Tifton and Cecil soils, respectively. Soils were analyzed for total N, total C, phosphorous (P), potassium (K), magnesium (Mg) and calcium (Ca) concentrations, pH and cation exchange capacity (CEC) at a commercial laboratory (Waters Agricultural Laboratories, Camilla, GA, USA) (Table 1).

Table 1. Chemical characteristics, including total nitrogen (N) and carbon (C), phosphorous (P), potassium (K), magnesium (Mg) and calcium (Ca) of the Cecil sandy clay loam (Cecil) and Tifton loamy sand (Tifton) used in this study. Units expressed per unit dry soil.

Soil	Total N	Total C	P	K	Mg	Ca	pHw [1]	CEC [i]
	(%)		($mg \cdot kg^{-1}$)				($mEq \cdot 100\ g^{-1}$)	
Cecil	0.31	1.79	33.5	116.5	63.5	677	5.5	7.0
Tifton	0.21	1.1	39.5	38.0	30.5	320	5.4	4.3

[1] Water pH = pHw measured using a 1:1 water:soil mixture, [i] Cation Exchange Capacity = CEC.

Three commercial fertilizers, a feather meal (FM) (13N-0P-0K; Mason City By-Products, Mason City, IA, USA), a pelleted poultry litter (PPL) (5N-1.8P-2.5K; Harmony Organic Fertilizer; Environmental Products LLC, Roanoke, VA, USA) and a mixed organic fertilizer (MIX) (10N-0.88P-6.6K; All Season Organic Fertilizer; Nature Safe, Irving, TX, USA) composed of feather-meat-bone-blood meal were sourced from a commercial supply company (7 Springs Farm Supply, Check, VA, USA). The total N and C of the organic fertilizers were determined by combustion according to the AOAC 993.13 method, while NO_3 and NH_4 were determined by distillation using AOC 2.068 and AOC 2.065 methods, respectively, at a commercial laboratory (Waters Agricultural Laboratories, Camilla, GA, USA) (Table 2).

Table 2. Characteristics, including total nitrogen (N), nitrate N (NO_3-N), ammonium N (NH_4-N), total carbon (C) and carbon to nitrogen (C:N) ratio of feather meal (FM), pelleted poultry litter (PPL) and mixed source (MIX) organic fertilizers used.

Material	Inorganic N	NO3-N	NH4-N	Total N	Total C	C:N Ratio
			%			
FM	1.18	0.85	0.33	13.61	48.17	3.54
PPL	1.04	0.90	0.14	4.36	27.74	6.36
MIX	0.49	0.35	0.14	9.44	34.35	3.64

2.2. Laboratory Incubation Study

To determine the rate of mineralization from the three organic fertilizers, a soil incubation study was performed for 120 d. Soil was wetted to 50% of estimated maximum WHC (0.12 g $H_2O \cdot g^{-1}$ soil and 0.16 g $H_2O \cdot g^{-1}$ soil for the Tifton and Cecil series soils, respectively) and allowed to pre-incubate under aerobic conditions for 48 h to mitigate the initial flux mineralization that typically occurs during soil rewetting [40]. Organic fertilizers were applied to soils at a rate to supply 100 mg $N \cdot kg^{-1}$ soil, based on the fertilizer labels, and Net N_{min} was later determined using the N values determined for each fertilizer by combustion (Table 2). The organic fertilizers were added to 300 g dry equivalent soil in resealable polyethylene bags (1 Quart Ziploc Freezer Bag, SC Johnson, Racine, WI, USA) and mixed thoroughly. There were 96 individual bags representing each fertilizer treatment (three fertilizers and non-treated control), two soil types and four incubation temperatures with each combination replicated three times and arranged in a split plot randomized complete block design, with incubation temperature being the main plot and fertilizer and soil type the subplots. Bags were incubated at 4, 10, 20 and 30 °C for 120 d. These temperatures are reflective of the range of soil temperatures commonly encountered during year-round field production of vegetables in the Southeastern USA. The bags were aerated, and water content maintained gravimetrically every 2 to 3 d. To determine the rate of release of inorganic N, 5 g subsamples of soil were taken at 0, 2, 4, 7, 14, 35, 56, 85 and 120 d. Subsamples were extracted with 40 mL 1 M KCl, shaken for 30 min and passed through filter paper (Whatman 42, Maidstone, Kent, UK). Prior to subsampling, soils were thoroughly mixed to ensure that fertilizers were homogenously distributed. Inorganic N was determined colorimetrically using the soil KCl extract method [41,42].

2.3. Mineralization Kinetics and Statistics

Cumulative Net N mineralized was calculated for the control soils (unamended), where:

$$\text{Control Net } N_{min} = \text{Inorganic } N_{t=x} - \text{Inorganic N Control}_{t=0} \quad (1)$$

Cumulative Net N mineralized (Net N_{min}) from the materials was calculated, where:

$$\text{Net } N_{min} = \text{Inorganic } N_{t=x} - \text{Inorganic N Control}_{t=x} - \text{Inorganic N of Material}_{t=0} \quad (2)$$

where Net N_{min} from the fertilizers (mg inorganic $N \cdot kg^{-1}$ dry material) was calculated as a function of the inorganic N at measured each time (t) point, the initial ammonium and nitrate present at time 0, and the N mineralization measured from the control soil (control Net N_{min}). Net N_{min} was calculated as a percentage of total organic N applied from each fertilizer. The Net N_{min} values for each time of incubation were analyzed using the ANOVA method by least-squares fit using JMP Pro 16.0 (SAS Institute Inc., Cary, NC, USA). Soil type, fertilizer treatment and temperature were treated as fixed effects, and replication was treated as a random effect. When statistically significant differences existed according to ANOVA ($p < 0.05$), mean separation was performed with Tukey's test at $\alpha = 0.05$.

To determine the N mineralization kinetics the Net N_{min} (mg inorganic N·kg^{-1} dry fertilizer) was fit to first-order kinetics using a non-linear model:

$$\text{Net } N_{min} = N_0\left[1 - e^{-kt}\right] \quad (3)$$

where N_0 is the potentially mineralizable pool of N predicted from the model, k is the rate coefficient of net N mineralization (mg inorganic N·kg^{-1}·d^{-1}) and t is time (days) [5,36]. The iterations for N_0 and k calculations and curve fitting of equations were carried out using a non-linear method with the G-Newton iteration using JMP Pro 16.0 (SAS Institute Inc., Cary, NC, USA). The performance of the fit of Net N_{min} modeled versus measured data was evaluated using the root mean squared error (RMSE), ratio of the RMSE to the standard deviation of measured data (RSR), Nash–Sutcliffe efficiency (NSE) and Percent bias (PBias). The RMSE is a commonly used error index statistics that quantifies the average magnitude of errors between predicted values and observed values [43–45]. A lower RMSE indicates a better model fit:

$$\text{RMSE} = \sqrt{\frac{\sum_{i=1}^{n}(O_i - P_i)^2}{n}} \quad (4)$$

The RSR is another way to assess the goodness of fit. The RSR standardizes RMSE by considering standard deviation of the observed data [43].

The NSE assesses the deviation between model predictions and observed values relative to the scattering of the observed data [43]. It ranges from negative infinity to 1, where an NSE of 1 indicates a perfect fit:

$$\text{NSE} = 1 - \left[\frac{\sum_{i=1}^{n}(O_i - P_i)^2}{\sum_{i=1}^{n}(O_i - \overline{O})^2}\right] \quad (5)$$

The PBias quantifies the systematic error or bias in a model's predictions by calculating the average difference between predicted and observed values. If the model underestimates, on average, the PBias is positive; conversely, if the model overestimates, on the average, the PBias is negative:

$$\text{PBIAS} = \left[\frac{\sum_{i=1}^{n}(O_i - P_i)^2 \times (100)}{\sum_{i=1}^{n}(O_i)}\right] \quad (6)$$

3. Results and Discussion

3.1. Laboratory Incubation

There was a linear correlation between incubation time and Net N_{min} in control (unamended) soils (Figure 1). Net N_{min} in the soil alone was significantly higher in the Cecil series soil compared to the Tifton series soil, which is likely to be related to the higher total C and total N content in the Cecil series compared to the Tifton soil (Table 1). In both soils, the Net N_{min} was significantly higher when incubated at 30 °C, while minimal differences were observed in Net N_{min} at 4 and 10 °C.

For the organic fertilizers, there was a trend of consistent significant differences for Net N_{min} among treatment combinations beginning at 35 d of incubation. Therefore, data from selected time points during the incubation (2, 4, 14, 35 and 120 d) for both Tifton and Cecil soils are presented. Complete results for Net N_{min} are available (Supplemental Tables S1 and S2). There were significant interactions between soil series, organic fertilizer and incubation temperature on Net N_{min} at several time points over the 120 d incubation (Tables 3 and 4). In the Cecil series soil, Net N_{min} differences between combinations of fertilizer and temperature were evident within the first 4 d of incubation (Table 3). In the Cecil soil, FM had a greater Net N_{min} compared to PPL and MIX when incubated at 20 and 30 °C from 4 to 35 d of incubation (Table 3). In contrast, minimal effects of temperatures were observed with PPL and MIX fertilizers in the first 35 d of incubation, with the exception of the PPL fertilizer at 14 d. In the Cecil soil, results suggest that high temperatures may have impacted the Net N_{min} immediately after application for the FM but less for the other fertilizers. Faster rates of mineralization soon after application might have favored the occurrence of alternative processes, such as ammonia (NH_3) volatilization [46,47]. At 120 d of incubation in the Cecil soil series, incubation temperature did not affect Net N_{min} of FM or PPL. However, the Net N_{min} of MIX fertilizer was greater at 20 °C than the other incubation temperatures at 120 d. These findings align with previous results reported by [14], which suggests that temperature has a larger impact during the first few days after incorporation of organic fertilizers, but the effects of temperature N mineralization decrease after a long period. Further, higher temperatures for some materials may also favor loss mechanisms such as NH_3 volatilization [46].

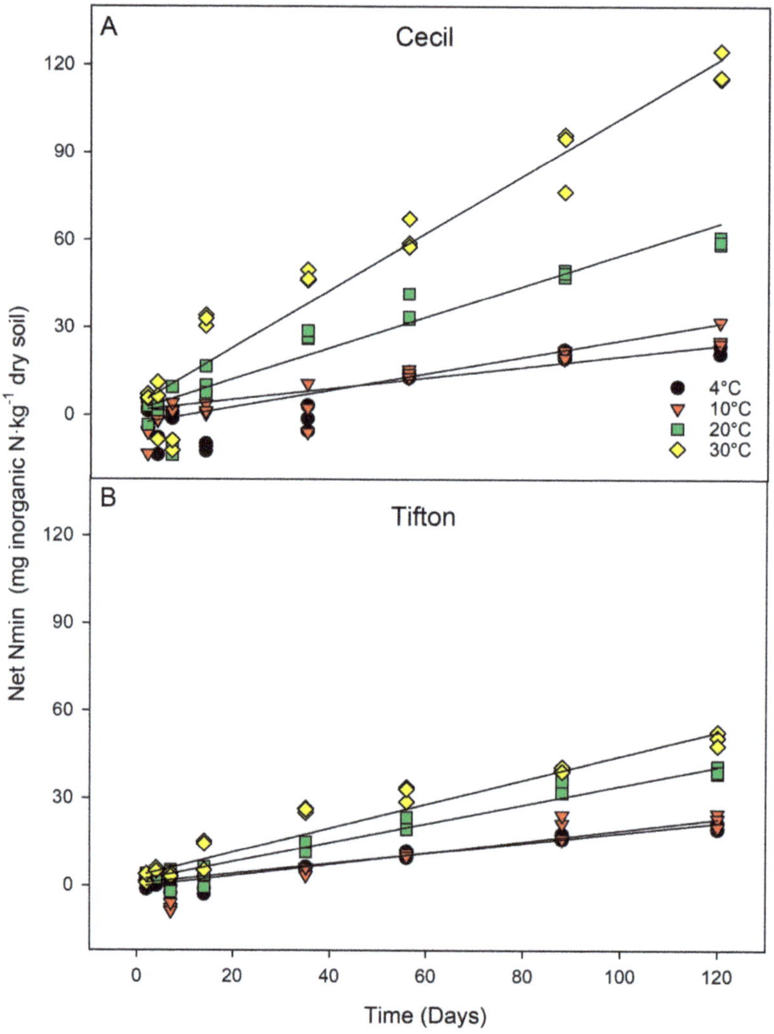

Figure 1. Net nitrogen mineralization (Net N_{min}) from control Cecil (**A**) and Tifton (**B**) series soils at 4, 10, 20 and 30 °C over 120 d of incubation.

Table 3. Net N mineralized (Net N_{min}[1]) from the feather meal (FM) and pelleted poultry litter (PPL), mixed source (MIX) fertilizers incubated at 4, 10, 20 and 30 °C for 120 d in a Cecil sandy clay loam soil. Differences in sampling time are noted by uppercase letters, while differences among temperatures are noted by lowercase letters.

Time of Incubation (Days)	Fertilizer	Temperature							
		4 °C		10 °C		20 °C		30 °C	
		Net N_{min} (mg Inorganic N·kg^{-1} Dry Material) [1]							
2	FM	0.56	A a [2]	3.72	A a	3.66	A a	0.00	A a
	PPL	0.00	A a	0.00	A a	4.20	A a	1.77	A a
	MIX	0.00	A a	4.61	A a	2.00	A a	0.00	A a
4	FM	12.31	AB ab	6.17	A b	22.02	A ab	34.09	A a
	PPL	16.91	A a	1.31	A a	10.17	B a	5.45	B a
	MIX	6.01	B ab	2.08	A b	3.08	C b	9.71	AB a
14	FM	31.84	A c	41.66	A b	53.71	A a	45.22	A ab
	PPL	15.70	A a	11.97	B a	11.06	C a	0.06	C b
	MIX	18.87	A a	--		26.69	B a	16.88	B a
35	FM	62.00	A b	58.90	A b	65.43	A ab	72.04	A a
	PPL	23.61	C a	20.72	B a	13.47	C a	11.53	C a
	MIX	32.67	B a	29.42	B a	30.20	B a	32.36	B a
120	FM	65.22	A a	60.53	A a	73.70	A a	72.14	A a
	PPL	17.15	C a	22.46	C a	18.48	B a	14.98	C a
	MIX	41.29	B ab	36.24	B b	58.85	A a	40.67	B ab

[1] Net N_{min} = Inorganic $N_{d=x}$ − Inorganic N $Control_{t=x}$ − Inorganic N of $Material_{t=0}$. [2] Different uppercase letters represent significant differences between fertilizers at each sampling time, and different lowercase letters represents significant differences among temperatures for each fertilizer and incubation time according to Tukey's honest significant difference test ($p < 0.05$).

Feather meal is derived from hydrolyzed, dried and ground poultry feathers. During mineralization of FM, long chains of keratin molecules are cleaved into smaller and more accessible components, which may enhance N release after relatively short incubation times [48,49]. Conversely, PPL and MIX have previously been shown to have slower mineralization rates than FM when incubated at 30 °C [5]. The PPL fertilizer had relatively low mineralization of the organic N (18 mg inorganic N·kg^{-1} dry material), which might be due to factors like the presence of bedding materials, moisture content and processing, which may affect the pool of mineralizable nitrogen [47]. The processing of the PPL can involve drying, grinding and compressing the material into uniform pellets, which can affect the availability and release of N [47,50,51]. It is important to note that the PPL fertilizer contained an initial 1.09% inorganic N (ammonium and nitrate) (Table 2), which was a relatively higher percentage of total N present compared to both FM and MIX sources. This inorganic N would be readily available for crop production and contribute to plant-available N in a field situation allowing for production of crops using PPL, despite having a relatively low rate of N mineralization. Several studies have reported the successful production of organic crops using PPL with a similar nutrient composition in the region where this study was conducted [31,52,53]. Total available N from the PPL fertilizer was notably greater than the organic mineralized N, suggesting that initial inorganic N from PPL may contribute to the total available N for plant uptake.

The effect of temperature on Net N_{min} of different fertilizers in the Tifton series soil was variable at different time points (Table 4). The FM fertilizer after 4 d of incubation had a significantly greater Net N_{min} at 20 and 30 °C (15 and 19 mg inorganic N kg^{-1} material, respectively) than at 4 and 10 °C (2 and 0 mg inorganic N kg^{-1} material, respectively). A similar trend in mineralization was observed with FM in the Cecil soil as well (Table 3). By 35 d of incubation Net N_{min} of FM had increased at lower temperatures. Temperature had a lesser impact on the Net N_{min} of MIX throughout the incubation period, with the greatest

Net N_{min} occurring at 20 °C after 120 d. The Net N_{min} of PPL at higher temperatures in the Tifton soil was low, with a complete absence of Net N_{min} at 30 °C starting from 14 d until 120 d, suggesting that higher temperatures favored the occurrence of other N processes in the PPL fertilizer instead of mineralization.

Table 4. Net N mineralized (Net N_{min} [1]) from the feather meal (FM), mixed source (MIX) and pelleted poultry litter (PPL) fertilizers incubated at 4, 10, 20 and 30 °C for 120 d in a Tifton loamy sand soil. Differences in sampling time are noted by uppercase letters, while differences among temperatures are noted by lowercase letters.

Time of Incubation (Days)	Fertilizer	Temperature							
		4 °C		10 °C		20 °C		30 °C	
		Net N_{min} (mg Inorganic N·kg^{-1} Dry Material) [1]							
2	FM	3.15	A a [2]	3.51	A a	3.38	A a	4.15	B a
	PPL	2.77	A ab	0.00	A b	4.94	A ab	11.26	A a
	MIX	0.24	A b	3.31	A ab	0.00	A b	4.02	B a
4	FM	1.86	A bc	0.00	B c	14.46	A ab	18.60	A a
	PPL	2.24	A b	9.65	A ab	13.26	A a	8.57	A ab
	MIX	1.15	A a	1.10	B a	0.91	B a	4.39	A a
14	FM	28.06	A a	28.68	A a	44.01	A a	33.82	A a
	PPL	15.21	AB a	7.95	B ab	6.97	B ab	0.00	B b
	MIX	6.04	B a	9.20	B a	21.71	AB a	17.88	AB a
35	FM	50.74	A ab	56.51	A a	48.93	A ab	43.39	A b
	PPL	11.04	C a	15.33	C a	3.08	B b	0.00	B b
	MIX	20.31	B a	28.73	B a	30.62	A a	35.51	A a
120	FM	61.91	A a	64.55	A a	50.09	A b	42.13	A b
	PPL	13.22	C a	9.74	C a	10.76	B a	0.00	B a
	MIX	28.43	B b	31.55	B b	49.49	A a	26.24	A b

[1] Net N_{min} = Inorganic $N_{d=x}$ − Inorganic N $Control_{t=x}$ − Inorganic N of $Material_{t=0}$. [2] Different uppercase letters represent significant differences within fertilizer type at each sampling point, and different lowercase letters represent significant differences among temperatures for each fertilizer and incubation time according to Tukey's honest significant difference test ($p < 0.05$).

Our results suggest that there was potential N loss from fertilizers during incubation, resulting from NH_3 volatilization or N immobilization, which may be favored by high temperatures [47]. Losses during early incubation periods from FM in the Cecil soil may be the result of a pattern of quick immobilization followed by gradual mineralization, reflecting the dynamic nature of microbial activity in response to availability of labile carbon compounds in the substrate [14,46,54]. It is also possible that NH_3 volatilization occurred in the PPL at 30 °C, resulting in a lack of measurable N mineralization. However, temperature is just one of several factors that influence NH_3 volatilization [46,55,56]. Soil properties such as moisture, pH, CEC and organic matter content also play a role [47,57]. In this context, NH_3 volatilization may be more likely to occur in the Tifton soil rather than the Cecil soil to due to its relatively higher pH, low organic matter and high sand content [58].

3.2. Mineralization Kinetics

A first-order model was fitted to the measured Net N_{min} during the 120 d incubation for each soil, fertilizer and temperature treatment (Table 5, Figure 2). This model allowed the determination of the rate of mineralization (k) and the predicted pool of mineralizable N (N_0) as a function of the temperature and the fertilizer applied. Goodness of fit of the first-order model were determined for each temperature, soil and fertilizer interaction and included RMSE, RSR, NSE and PBias.

Table 5. First order fit characteristics, root mean squared error (RMSE), ratio of the root mean square error to the standard deviation of measured data (RSR), Nash–Sutcliffe efficiency (NSE) and Percent bias (PBias) for net mineralized nitrogen (Net N_{min}) for feather meal (FM), pelleted poultry litter (PPL) and mixed source (MIX) organic fertilizers incubated in Cecil sandy clay loam (Cecil) and Tifton loamy sand (Tifton) soils at several temperatures.

Soil	Fertilizer	Temperature	Net N_{min} at 120 d		N_0 [3]	First Order Rate Coef. k [4]	RMSE	RSR	NSE	PBIAS
		°C	% [1]	mg Inorganic N·kg^{-1} Dry Soil [2]		d^{-1}				
Cecil	FM	4	68.22	65.22	64.54	0.04	4.64	0.26	0.93	1.42
		10	63.31	60.53	62.75	0.07	4.39	0.26	0.93	−5.52
		20	77.08	73.70	65.12	0.10	13.46	0.72	0.49	−15.47
		30	75.45	72.14	79.89	0.08	11.10	0.50	0.75	−4.83
	PPL	4	25.83	17.15	17.17	0.26	5.14	0.85	0.28	−11.74
		10	33.83	22.46	19.21	0.09	4.46	0.73	0.46	−24.97
		20	27.83	18.48	8.38	0.84	4.15	0.97	0.06	35.48
		30	22.55	14.98	15.04	0.03	8.21	0.82	0.33	−1.55
	MIX	4	46.14	41.29	38.01	0.04	4.59	0.36	0.87	−8.84
		10	40.49	36.24	37.41	0.03	4.70	0.41	0.83	−12.30
		20	65.76	58.85	55.57	0.03	6.14	0.35	0.88	−0.70
		30	45.44	40.67	46.31	0.04	6.23	0.39	0.85	−6.91
Tifton	FM	4	64.75	61.91	64.54	0.04	5.56	0.22	0.95	−2.98
		10	67.52	64.55	62.55	0.04	5.21	0.25	0.94	−6.00
		20	52.39	50.09	48.03	0.11	5.55	0.33	0.89	−2.06
		30	44.07	42.13	41.85	0.10	7.14	0.51	0.74	−3.77
	PPL	4	19.91	13.22	12.28	0.29	5.60	0.87	0.25	−2.32
		10	14.66	9.74	10.56	0.22	4.13	0.80	0.36	−5.90
		20	16.21	10.76	8.38	0.84	8.95	1.00	0.01	−0.27
		30	0	0	-	-	-	-	-	-
	MIX	4	31.76	28.43	32.72	0.02	3.25	0.26	0.93	−5.42
		10	35.26	31.55	30.61	0.03	4.82	0.38	0.86	−4.79
		20	55.29	49.49	48.93	0.03	7.85	0.41	0.83	−1.90
		30	29.32	26.24	30.15	0.09	8.39	0.64	0.59	−1.23

[1] Net mineralized N after 120 d of incubation as a percentage of organic N in the fertilizer applied. [2] Net measured mineralized N after 120 d of incubation. [3] N_0 is the amount of net mineralized N predicted from a first order kinetics model. [4] k is the rate coefficient of the first order model.

The goodness of fit for the first-order kinetic model varied with different treatment combinations. The model demonstrated good efficiency in predicting the N mineralization of the FM and MIX fertilizers under most temperature and soil conditions, with NSE values ranging from 0.74 to 0.95 and RSR values between 0.22 and 0.51. There was an exception for model fit of the MIX fertilizer in Cecil soil at 20 °C, where the NSE value dropped below 0.50 and RSR exceeded 0.70, suggesting a weaker correlation between observed and predicted values (Table 5; Figure 2). In this specific fit, there was a tendency for the model to overestimate N mineralization, resulting in a negative PBias of −15.47. In contrast, the model efficiency for predicting the N mineralization of MIX was satisfactory and consistent across different soil and temperature combinations, with NSE values ranging from 0.59 to 0.93 and RSR values between 0.26 and 0.64.

The model's performance was unsatisfactory for predicting the N mineralization of PPL fertilizer. In all first-order kinetic fits, RSR values exceeded 0.70, which may be attributed to the considerable variability in Net N_{min} observed among the samples of this material during the 120 d incubation (Tables 4 and 5). This variability negatively affected the correlation between observed and predicted values, which also resulted in NSE values between 0.06 and 0.46. Furthermore, the PBIAS of the models for the mineralization of PPL ranged from a positive value of 35.48 to negative values as low as −11.74, which indicates a tendency of both under- and overestimation of the model. The high initial inorganic N present in the PPL combined with low mineralization from the organic N pool (Figure 2, Table 5) also likely contributed to the poor fit of the first-order mineralization model.

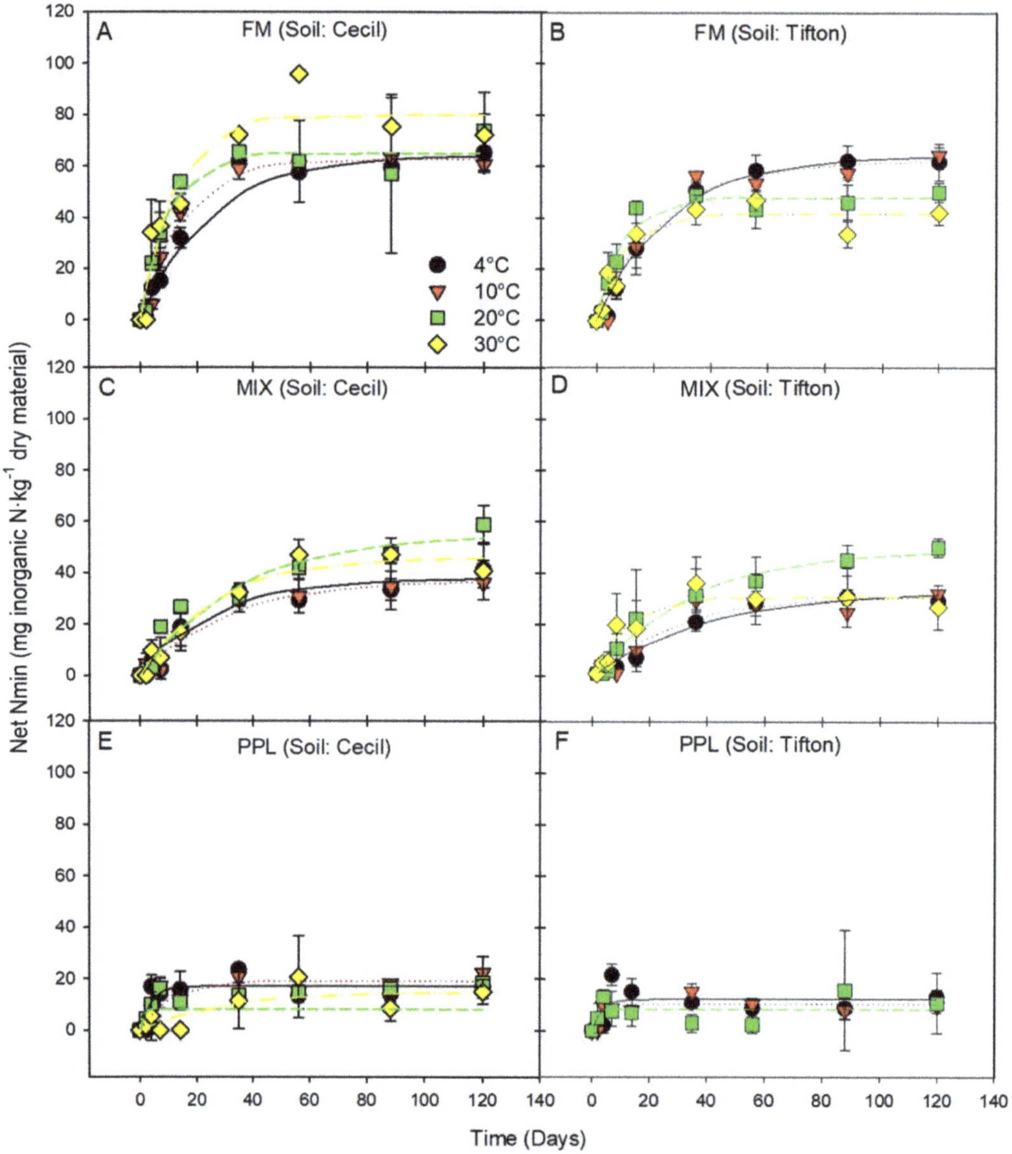

Figure 2. Nitrogen mineralization from the feather meal (FM) (**A**,**B**), mixed source (MIX) (**C**,**D**) and pelleted poultry litter (PPL) (**E**,**F**) fertilizers incubated at 4, 10, 20 and 30 °C for 120 d in Cecil and Tifton series soils. Regression lines show predicted first order regressions for N mineralization (N_0) and symbols show actual mineralized N (Net N_{min}) (error bars indicate standard deviation).

The interactions between soil and temperature are necessary for understanding the rate of N mineralization of different organic fertilizers. In a previous study, Lazicki et al. [14]

observed variability for potential plant-available N among 22 different organic materials trialed under warm and moist conditions (23 °C and 60% water holding capacity), with some undergoing N immobilization while others released up to 90% of applied N. The authors suggested, while the Net N_{min} was well correlated with the C:N ratio of each material, the soil texture and management history of the soils influenced the timing of N release [14]. Different soil attributes, such as pH, texture, organic matter and cation exchange capacity (CEC), create distinct physical and chemical environments for microorganisms, which may influence when and how quickly microorganisms decompose organic N compounds [14]. Furthermore, the degree of which the temperature can influence microbial growth and, consequently, the N mineralization rate is likely to vary across different environmental systems [17].

3.2.1. FM Fertilizer Mineralization Kinetics

The N_0 at 120 d for FM ranged from 62.75 to 79.89 mg inorganic N·kg^{-1} fertilizer applied in the Cecil soil (Table 5), and from 41.85 to 64.54 mg inorganic N·kg^{-1} fertilizer applied in the Tifton soil (Table 5). Similar N_0 values have been previously reported for FM [5,14,15]. In the present study, N mineralization of FM over the 120 d incubation in the Cecil soil showed two distinct phases, starting with a rapid release during the first 14 d of incubation, followed by a slow-release phase and a plateau at approximately 80 d (Figure 2A). In the Tifton soil, a similar pattern of rapid release occurred during the early stages of incubation followed by a steady and slow release at the temperatures of 4 and 10 °C. At 20 and 30 °C, the rapid phase was followed by a plateau at approximately 35 d (Figure 2B). The rapid mineralization observed from FM fertilizer in the 14 d of incubation (Figure 2B) corresponds to previous reports [5,12], and may be attributed to the presence of readily available N-containing compounds, such as urea, and simple organic forms of N that can be easily broken down by soil enzymes, even at lower temperatures. In contrast, the slower N mineralization after 14 d indicates the predominance of microbial degradation of complex organic N forms [12]. A model proposed by Geisseler et al. [15] predicted that 61% of total N from FM would be in the mineral form after 100 d of incubation, with half of that being in the mineral form after only 5 d of incorporation.

For FM, in the Cecil series soil models, the k and N_0 values were not significantly different among temperatures, with constant k values ranging between 0.04 and 0.10 and N_0 values between 62.75 and 79.89 mg inorganic N·kg^{-1} fertilizer applied (Table 5). In contrast, in the Tifton soil, N_0 was greater at the lower temperatures (4 °C and 10 °C), suggesting that temperature might affect N mineralization of FM in the Tifton soil to a greater degree compared with Cecil soil. This agrees with results reported by Dessurealt-Rompré et al. [16], who indicated that N mineralization rates in sandy loam soils were affected by temperature changes to a greater degree than predominately clay soils. Further, k and N_0 values were not significantly different between soils, with the only exception with the highest temperature of 30 °C, suggesting that overall differences for the rate of mineralization and potentially mineralizable N of FM in different mineral soils may be minimal at lower temperatures.

3.2.2. PPL Fertilizer Mineralization Kinetics

The N_0 for the PPL fertilizer after 120 d was less than 20 mg inorganic N·kg^{-1} fertilizer applied, regardless of soil type or temperature (Figure 2E,F). The model fit for both soils and temperature combinations showed most of the N release was during the first few days of incubation, followed by a plateau after the second week. On average, 10% of total N applied is predicted to be available from PPL after 7 d of incubation. Previous studies predicting N release from PPL reported higher values, with potential available N ranging between 25% to 50% of applied N [5,47], with most of the N released during the initial 7 d of incubation.

Model fits of the PPL product had a wide variability of k rates, with values ranging between 0.03 and 0.84 among treatments of soil and temperatures (Table 5). Further, N_0

values ranged between 8.38 and 19.21 and no significant differences were observed among treatments of soils and temperatures. This agrees with other studies evaluating the N mineralization of organic fertilizers, which suggest that the temperature affects k rate more than the N_0 [14,59]. Thus, changes in the N mineralization of PPL in response to environmental conditions are likely to be minimal.

3.2.3. MIX Fertilizer Mineralization Kinetics

The N_0 for the MIX fertilizer in the Cecil and Tifton soils after 120 d ranged between 37.41 and 55.57 mg inorganic N·kg^{-1} fertilizer applied and 30.15 and 48.93 mg inorganic N·kg^{-1} fertilizer applied, respectively (Table 5; Figure 2C,D). The N release throughout the 120 d incubation steadily increased, but at a lower rate after 35 d, except when incubated at 30 °C in the Tifton series soil, in which a plateau phase was observed starting at 35 d (Figure 2D). In a previous study, Cassity-Duffey et al. [7] investigated the N mineralization of organic fertilizers across four distinct soils over a 100 d incubation period. Their results suggested that most of N mineralized from a pelletized MIX fertilizer, composed of animal-based proteins, occurred within the initial 2 weeks of incubation, with approximately 53% to 58% of the initially applied N mineralized after 14 d.

The first order model fits for MIX fertilizer had rate k values between 0.02 and 0.09, while N_0 values ranged between 30.61 and 55.57 in both soils (Table 5). Further, k and N_0 values for the MIX product were not significantly different between soil types or incubation temperatures, suggesting that we may expect minimal differences for the N release of MIX fertilizer between the two soils. Cassity-Duffey et al. [7] also reported that there were no significant effects of soil texture between Cecil and Tifton series soils on the Net N_{min} of a MIX-type fertilizer. Interestingly, their research suggested that pH levels may have a more influential role than soil clay content in determining the potential N mineralization [7].

4. Conclusions

This study investigated N mineralization from three organic fertilizers in two soil types and soil temperatures that would be encountered during field vegetable production throughout the year in Georgia, USA. The Net N_{min} varied with soil, fertilizer type and temperature, indicating that accurately predicting N mineralization from organic fertilizers may be specific for individual farms and fertilizer sources. In general, a first-order kinetic model demonstrated good efficiency in predicting the N mineralization of FM and MIX fertilizers but was inefficient for predicting N mineralization of PPL. Temperature and soil type had a relatively minor impact on the potentially mineralizable N of the MIX and PPL fertilizers at 120 d. However, incubation at higher temperatures (20 °C and 30 °C) impacted the Net N_{min} of FM fertilizer in the Tifton series soil. Temperature-related differences may not be substantial enough to influence grower decisions regarding N fertilizer inputs for vegetable crop production in the two soils. However, organic fertilizer source will likely play a significant role in N availability during the cropping season. Further research is needed to determine how temperature may impact NH_3 volatilization and other loss mechanisms, as these processes may play a role in plant-available N at high soil temperatures.

Supplementary Materials: The following supporting information can be downloaded at: https://www.mdpi.com/article/10.3390/nitrogen5010004/s1, Table S1: Net N mineralization in a Cecil sandy clay loam soil. Table S2: Net N mineralization in a Tifton loamy sand soil.

Author Contributions: Conceptualization, K.C.-D. and T.C.; methodology, K.C.-D.; software, H.I.d.J.; formal analysis, H.I.d.J. and K.C.-D.; resources, H.I.d.J. and T.C.; data curation H.I.d.J. and K.C.-D.; writing—original draft preparation, H.I.d.J., K.C.-D. and T.C.; writing—review and editing, A.L.B.R.d.S., B.D., H.I.d.J., K.C.-D. and T.C.; project administration, H.I.d.J.; funding acquisition, B.D. and T.C. All authors have read and agreed to the published version of the manuscript.

Funding: This research was funded by the US Department of Agriculture, National Institute of Food and Agriculture Project Number GEOW-2019-03518; Evaluation of integrated bacterial disease management options for organic onion production in the Southeastern and Northcentral United States.

Data Availability Statement: The data presented in this study are available on request from the corresponding author.

Conflicts of Interest: The authors declare no conflicts of interest.

References

1. US Department of Agriculture. Results from the 2021 Organic Survey. 2022. Available online: https://www.nass.usda.gov/Publications/Highlights/2022/2022_Organic_Highlights.pdf (accessed on 11 November 2023).
2. Bergstrand, K.J. Organic fertilizers in greenhouse production systems—A review. *Sci. Hortic.* **2022**, *295*, 110855. [CrossRef]
3. Gale, E.S.; Sullivan, D.M.; Cogger, C.G.; Bary, A.I.; Hemphill, D.D.; Myhre, E.A. Estimating plant-available nitrogen release from manures, composts, and specialty products. *J. Environ. Qual.* **2006**, *35*, 2321–2332. [CrossRef] [PubMed]
4. Wortman, S.E.; Holmes, A.A.; Miernicki, E.; Knoche, K.; Pittelkow, C.M. First-season crop yield response to organic soil amendments: A meta-analysis. *Agron. J.* **2017**, *109*, 1210–1217. [CrossRef]
5. Cassity-Duffey, K.; Cabrera, M.; Gaskin, J.; Franklin, D.; Kissel, D.; Saha, U. Nitrogen mineralization from organic materials and fertilizers: Predicting N release. *Soil Sci. Soc. Am. J.* **2020**, *84*, 522–533. [CrossRef]
6. Cabrera, M.L.; Kissel, D.E.; Vigil, M.F. Nitrogen mineralization from organic residues: Research opportunities. *J. Environ. Qual.* **2005**, *34*, 75–79. [CrossRef] [PubMed]
7. Cassity-Duffey, K.; Cabrera, M.; Franklin, D.; Gaskin, J.; Kissel, D. Effect of soil texture on nitrogen mineralization from organic fertilizers in four common southeastern soils. *Soil Sci. Soc. Am. J.* **2020**, *84*, 534–542. [CrossRef]
8. Reganold, J.P.; Wachter, J.M. Organic agriculture in the twenty-first century. *Nat. Plants* **2016**, *2*, 1–8. [CrossRef]
9. Calderón, F.J.; McCarty, G.W.; Van Kessel, J.A.S.; Reeves III, J.B. Carbon and nitrogen dynamics during incubation of manured soil. *Soil Sci. Soc. Am. J.* **2004**, *68*, 1592–1599. [CrossRef]
10. Geisseler, D.; Horwath, W.R.; Joergensen, R.G.; Ludwig, B. Pathways of nitrogen utilization by soil microorganisms—A review. *Soil. Biol. Biochem.* **2010**, *42*, 2058–2067. [CrossRef]
11. Agehara, S.; Warncke, D.D. Soil moisture and temperature effects on nitrogen release from organic nitrogen sources. *Soil Sci. Soc. Am. J.* **2005**, *69*, 1844–1855. [CrossRef]
12. Hartz, T.K.; Johnstone, P.R. Nitrogen availability from high-nitrogen-containing organic fertilizers. *HortTechnology* **2006**, *16*, 39–42. [CrossRef]
13. Wild, P.L.; Van Kessel, C.; Lundberg, J.; Linquist, B.A. Nitrogen availability from poultry litter and pelletized organic amendments for organic rice production. *Agron. J.* **2011**, *103*, 1284–1291. [CrossRef]
14. Lazicki, P.; Geisseler, D.; Lloyd, M. Nitrogen mineralization from organic amendments is variable but predictable. *J. Environ. Qual.* **2020**, *49*, 483–495. [CrossRef] [PubMed]
15. Geisseler, D.; Smith, R.; Cahn, M.; Muramoto, J. Nitrogen mineralization from organic fertilizers and composts: Literature survey and model fitting. *J. Environ. Qual.* **2021**, *50*, 1325–1338. [CrossRef] [PubMed]
16. Dessureault-Rompré, J.; Zebarth, B.J.; Georgallas, A.; Burton, D.L.; Grant, C.A.; Drury, C.F. Temperature dependence of soil nitrogen mineralization rate: Comparison of mathematical models, reference temperatures and origin of the soils. *Geoderma* **2010**, *157*, 97–108. [CrossRef]
17. Guntiñas, M.E.; Leirós, M.C.; Trasar-Cepeda, C.; Gil-Sotres, F. Effects of moisture and temperature on net soil nitrogen mineralization: A laboratory study. *Eur. J. Soil Biol.* **2012**, *48*, 73–80. [CrossRef]
18. Sierra, J. Temperature and soil moisture dependence of N mineralization in intact soil cores. *Soil Biol. Biochem.* **1997**, *29*, 1557–1563. [CrossRef]
19. Bowles, T.M.; Acosta-Martínez, V.; Calderón, F.; Jackson, L.E. Soil enzyme activities, microbial communities, and carbon and nitrogen availability in organic agroecosystems across an intensively managed agricultural landscape. *Soil Biol. Biochem.* **2014**, *68*, 252–262. [CrossRef]
20. Miller, K.S.; Geisseler, D. Temperature sensitivity of nitrogen mineralization in agricultural soils. *Biol. Fertil. Soils* **2018**, *54*, 853–860. [CrossRef]
21. Clivot, H.; Mary, B.; Valé, M.; Cohan, J.P.; Champolivier, L.; Piraux, F.; Justes, E. Quantifying in situ and modeling net nitrogen mineralization from soil organic matter in arable cropping systems. *Soil Biol. Biochem.* **2017**, *111*, 44–59. [CrossRef]
22. Knoepp, J.D.; Swank, W.T. Using soil temperature and moisture to predict forest soil nitrogen mineralization. *Biol. Fertil. Soils* **2002**, *36*, 177–182. [CrossRef]
23. Cassity-Duffey, K.B.; Moore, A.; Satterwhite, M.; Leytem, A. Nitrogen mineralization as affected by temperature in calcareous soils receiving repeated applications of dairy manure. *Soil Sci. Soc. Am. J.* **2018**, *82*, 235–242. [CrossRef]
24. Sims, J.T. Nitrogen transformations in a poultry manure amended soil: Temperature and moisture effects. *J. Environ. Qual.* **1986**, *15*, 59–63. [CrossRef]
25. Hartz, T.K.; Smith, R.; Gaskell, M. Nitrogen availability from liquid organic fertilizers. *HortTechnology* **2010**, *20*, 169–172. [CrossRef]
26. Stanford, G.; Smith, S.J. Nitrogen mineralization potentials of soils. *Soil Sci. Soc. Am. J.* **1972**, *36*, 465–472. [CrossRef]

27. Gordillo, R.M.; Cabrera, M.L. Mineralizable nitrogen in broiler litter: II. Effect of selected soil characteristics. *J. Environ. Qual.* **1997**, *26*, 1679–1686. [CrossRef]
28. Hassink, J.; Bouwman, L.A.; Zwart, K.B.; Bloem, J.; Brussaard, L. Relationships between soil texture, physical protection of organic matter, soil biota, and C and N mineralization in grassland soils. *Geoderma* **1993**, *57*, 105–128. [CrossRef]
29. Abdelhafez, A.A.; Abbas, M.H.; Attia, T.M.; El Bably, W.; Mahrous, S.E. Mineralization of organic carbon and nitrogen in semi-arid soils under organic and inorganic fertilization. *Environ. Technol. Innov.* **2018**, *9*, 243–253. [CrossRef]
30. Soinne, H.; Keskinen, R.; Räty, M.; Kanerva, S.; Turtola, E.; Kaseva, J.; Nuutinen, V.; Simojoki, A.; Salo, T. Soil organic carbon and clay content as deciding factors for net nitrogen mineralization and cereal yields in boreal mineral soils. *Eur. J. Soil Sci.* **2021**, *72*, 1497–1512. [CrossRef]
31. De Jesus, H.I.; Ribeiro da Silva, A.L.B.; Dutta, B.; Cassity-Duffey, K.; Coolong, T. Influence of Organic Fertilizer Sources and Application Rates on Onion Production in Georgia, USA. *HortTechnology* **2023**, *33*, 398–406. [CrossRef]
32. Masunga, R.H.; Uzokwe, V.N.; Mlay, P.D.; Odeh, I.; Singh, A.; Buchan, D.; De Neve, S. Nitrogen mineralization dynamics of different valuable organic amendments commonly used in agriculture. *Appl. Soil Ecol.* **2016**, *101*, 185–193. [CrossRef]
33. Honeycutt, C.W. Nitrogen mineralization from soil organic matter and crop residues: Field validation of laboratory predictions. *Soil Sci. Soc. Am. J.* **1999**, *63*, 134–141. [CrossRef]
34. Kaupa, P.; Rao, B.R. Nitrogen mineralization and efficiency from co-applied animal manures and mineral fertilizer in sweetpotato under humid tropical conditions. *Field Crops Res.* **2014**, *168*, 48–56. [CrossRef]
35. Lei, L.; McDonald, L.M. Soil moisture and temperature effects on nitrogen mineralization in a high tunnel farming system. *Commun. Soil Sci. Plant Anal.* **2019**, *50*, 2140–2150. [CrossRef]
36. Cabrera, M.; Molina, J.A.; Vigil, M. Modeling the nitrogen cycle. In *Nitrogen in Agricultural Systems*; Schepers, J.S., Raun, W., Eds.; ASA: Madison, WI, USA, 2008; pp. 695–730.
37. US Department of Agriculture. 2022b. Soil Survey. Web Soil Survey—Soil Survey of Oconee and Tift Counties, Georgia. Available online: https://websoilsurvey.sc.egov.usda.gov/App/WebSoilSurvey.aspx (accessed on 28 March 2023).
38. Priha, O.; Smolander, A. Nitrogen transformations in soil under Pinus sylvestris, Picea abies and Betula pendula at two forest sites. *Soil Biol. Biochem.* **1999**, *31*, 965–977. [CrossRef]
39. Wyngaard, N.; Cabrera, M.L.; Shober, A.; Kanwar, R. Fertilization strategy can affect the estimation of soil nitrogen mineralization potential with chemical methods. *Plant Soil* **2018**, *432*, 75–89. [CrossRef]
40. Cabrera, M.L. Modeling the flush of nitrogen mineralization caused by drying and rewetting soils. *Soil Sci. Soc. Am. J.* **1993**, *57*, 63–66. [CrossRef]
41. Crooke, W.M.; Simpson, W.E. Determination of ammonium in Kjeldahl digests of crops by an automated procedure. *J. Sci. Food Agric.* **1971**, *22*, 9–10. [CrossRef]
42. Keeney, D.R.; Nelson, D.W. Nitrogen—Inorganic forms. In *Methods of Soil Analysis: Part 2 Chemical and Microbiological Properties*; Page, A., Ed.; American Society of Agronomy, Soil Science Society of America: Madison, WI, USA, 1983; pp. 643–698.
43. Moriasi, D.N.; Arnold, J.G.; Van Liew, M.W.; Bingner, R.L.; Harmel, R.D.; Veith, T.L. Model evaluation guidelines for systematic quantification of accuracy in watershed simulations. *Trans. ASABE* **2007**, *50*, 885–900. [CrossRef]
44. Singh, J.; Knapp, H.V.; Arnold, J.G.; Demissie, M. Hydrological modeling of the Iroquois river watershed using HSPF and SWAT 1. *J. Am. Water Resour. Assoc.* **2005**, *41*, 343–360. [CrossRef]
45. Vazquez-Amábile, G.G.; Engel, B.A. Use of SWAT to compute groundwater table depth and streamflow in the Muscatatuck River watershed. *Trans. Am. Soc. Agric. Eng.* **2005**, *48*, 991–1003. [CrossRef]
46. Erwiha, G.M.; Ham, J.; Sukor, A.; Wickham, A.; Davis, J.G. Organic fertilizer source and application method impact ammonia volatilization. *Commun. Soil Sci. Plant Anal.* **2020**, *51*, 1469–1482. [CrossRef]
47. Hadas, A.; Bar-Yosef, B.; Davidov, S.; Sofer, M. Effect of pelleting, temperature, and soil type on mineral nitrogen release from poultry and dairy manures. *Soil Sci. Soc. Am. J.* **1983**, *47*, 1129–1133. [CrossRef]
48. Hadas, A.; Kautsky, L. Feather meal, a semi-slow-release nitrogen fertilizer for organic farming. *Fert. Res.* **1994**, *38*, 165–170. [CrossRef]
49. Jan, M.T.; Roberts, P.; Tonheim, S.K.; Jones, D.L. Protein breakdown represents a major bottleneck in nitrogen cycling in grassland soils. *Soil Biol. Biochem.* **2009**, *41*, 2272–2282. [CrossRef]
50. Mazeika, R.; Staugaitis, G.; Baltrusaitis, J. Engineered pelletized organo-mineral fertilizers (OMF) from poultry manure, diammonium phosphate and potassium chloride. *ACS Sustain. Chem. Eng.* **2016**, *4*, 2279–2285. [CrossRef]
51. Purnomo, C.W.; Indarti, S.; Wulandari, C.; Hinode, H.; Nakasaki, K. Slow-release fertilizer production from poultry manure. *Chem. Eng. Trans.* **2017**, *56*, 1531–1536. [CrossRef]
52. Boyhan, G.E.; Hicks, R.; Torrance, R.; Riner, C.M.; Hill, C.R. Evaluation of poultry litter and organic fertilizer rate and source for production of organic short-day onions. *HortTechnology* **2010**, *20*, 304–307. [CrossRef]
53. Díaz-Pérez, J.C.; Bautista, J.; Gunawan, G.; Bateman, A.; Riner, C.M. Sweet onion (*Allium cepa* L.) as influenced by organic fertilization rate: 2. Bulb yield and quality before and after storage. *HortScience* **2018**, *53*, 459–464. [CrossRef]
54. Quan, Z.; Huang, B.; Lu, C.; Su, C.; Song, L.; Zhao, X.; Fang, Y. Effects of ryegrass amendments on immobilization and mineralization of nitrogen in a plastic shed soil: A 15N tracer study. *Catena* **2021**, *203*, 105325. [CrossRef]
55. Cabrera, M.L.; Chiang, S.C.; Merka, W.C.; Thompson, S.A.; Pancorbo, O.C. Nitrogen transformations in surface-applied poultry litter: Effect of litter physical characteristics. *Soil Sci. Soc. Am. J.* **1993**, *57*, 1519–1525. [CrossRef]

56. Hargrove, W.L. Evaluation of ammonia volatilization in the field. *J. Prod. Agric.* **1988**, *1*, 104–111. [CrossRef]
57. Zhenghu, D.; Honglang, X. Effects of soil properties on ammonia volatilization. *Soil Sci. Plant Nutr.* **2020**, *46*, 845–852. [CrossRef]
58. Wang, F.L.; Alva, A.K. Ammonium adsorption and desorption in sandy soils. *Soil Sci. Soc. Am. J.* **2000**, *64*, 1669–1674. [CrossRef]
59. Griffin, T.S.; Honeycutt, C.W. Using growing degree days to predict nitrogen availability from livestock manures. *Soil Sci. Soc. Am. J.* **2000**, *64*, 1876–1882. [CrossRef]

Nitrogen

MDPI

Article

Effect of Methyl Jasmonate Plus Urea Foliar Application on the Polysaccharide and Monosaccharide Composition of Tempranillo Grapes and Wines and on the Wine's Quality

Miriam González-Lázaro, Leticia Martínez-Lapuente, Teresa Garde-Cerdán *, Mikel Landín Ross-Magahy, Lesly L. Torres-Díaz, Eva P. Pérez-Álvarez, Zenaida Guadalupe and Belén Ayestarán *

Instituto de Ciencias de la Vid y del Vino (Universidad de La Rioja, CSIC, Gobierno de La Rioja), Ctra. de Burgos, Km. 6, 26007 Logroño, Spain

* Correspondence: teresa.garde.cerdan@csic.es (T.G.-C.); belen.ayestaran@unirioja.es (B.A.)

Abstract: Polysaccharides are the main group of macromolecules in wines. Climate change is a major problem for viticulturists as it leads to the production of unbalanced grapes. This is attributed to a mismatch between the technological maturity and phenolic maturity of grapes, which can negatively impact the production of high quality wines. To mitigate this effect, biostimulants can be applied to grapevines. For the first time in the literature, this work studied the foliar application of methyl jasmonate plus urea (MeJ + Ur) on the vineyard and its effect on the monosaccharide and polysaccharide composition of Tempranillo grapes and wines over two consecutive seasons. To achieve this, the extraction and precipitation of polysaccharides was conducted, and the identification and quantitation of monosaccharides was performed via GC–MS. The effect of MeJ + Ur foliar treatment in both the grapes and wines was season-dependent. The MeJ + Ur treatment had a slight impact on the monosaccharide composition of the grapes and also demonstrated a small effect on the wines. Multifactor and discriminant analysis revealed that the season had a greater influence on the monosaccharide and polysaccharide composition of grapes and wines compared to the influence of MeJ + Ur treatment. Interestingly, the MeJ + Ur-treated wines exhibited a higher sensory evaluation than the control wines in the second vintage. To gain further insights into the effect of MeJ + Ur foliar application on the monosaccharide and polysaccharide composition of grapes and wines, further investigations should be conducted.

Keywords: methyl jasmonate; urea; polysaccharide; monosaccharide; tempranillo; grape; wine; wine quality

1. Introduction

The quality of wines depends on a multitude of metabolites extracted from grapes, including polyphenols, amino acids, volatile compounds, and polysaccharides [1]. Among these, polysaccharides are one of the most important groups of macromolecules in wines [2] since these compounds act as colloidal protectors. Polysaccharides have various impacts on wine characteristics; for example, they influence anthocyanin extractability [3], affect astringency by reducing salivary-protein interactions [4], and potentially modify wine aromas through interactions with volatile compounds, altering their volatility [5]. Polysaccharides originate from both the cell walls and microorganisms, including yeasts, involved in the winemaking process [6].

Different families of polysaccharides have been distinguished, including polysaccharides rich in arabinose and galactose (PRAGs), such as type II arabinogalactan-proteins (AGPs), arabinans, rhamnogalacturonans type I (RG-I) and type II (RG-II), and homogalacturonans (HLs), which originate from grapes. Additionally, yeast fermentation releases glucans (GLs), mannans, and mannoproteins (MPs) [7]. In summary, the presence of AGP/PRAG, RG-II and MP contributes to the specific wine matrix and influences the perceived quality of the wine by affecting the colloidal state of the red wine and interacting with phenolic and volatile compounds. The grape cell wall is composed of polysaccharides (celluloses, hemicelluloses, and pectins), phenolic compounds, and proteins [3,6]. The cell wall acts as a barrier for the diffusion of aroma and phenolic compounds, including grape polysaccharides (RG-II, PRAG, HL), into wines. Therefore, polysaccharides are released to red wines during maceration and alcoholic fermentation [6]. In recent years, climate change has emerged as a significant concern for winemakers since it modifies vine development and fruit maturation patterns [8]. Grapes reach technological maturity when their berries attain maximum sugar concentration, indicating that they are ready to be harvested. However, phenolic maturity, characterized by an increase in the content of phenolic compounds in grapes, including total phenols and especially anthocyanins, occurs after technological maturity. This imbalance in grape development is a result of climate change, which, consequently, affects the quality of wine [9,10]. One potential strategy to alleviate the effects of climate change on grape composition is the foliar application of biostimulants [11–13]. Among the biostimulants sprayed on grapevines, elicitors and nitrogen compounds are particularly noteworthy. Elicitors are compounds capable of triggering a defensive response in plants, leading to the increased production of certain secondary metabolites [14,15]. Methyl jasmonate is an elicitor thoroughly employed as a foliar spray on grapevines. Its effects on enhancing volatile compounds [16,17], phenolic compounds [18,19], and amino acids [20,21] in grapes and wines have been well-documented. However, it should be noted that the increase in grape phenolic content achieved by methyl jasmonate does not always result in wines with a higher phenolic content [18,22].

The application of elicitors to grapevines produces a tightening of the cell walls in grape skins, making it more challenging to extract polyphenols, especially anthocyanins, from the skins into the must [23]. Regarding the impact of methyl jasmonate on the polysaccharide content of grapes, Paladines-Quezada et al. [24] reported that the cellulose content in the cell walls of MeJ-treated grapes was lower compared to control grapes, whereas Apolinar-Valiente et al. [23] indicated that the cell wall of grapes was reinforced after MeJ treatment. Paladines-Quezada et al. [25] concluded that the application of MeJ did not affect the concentration of MPs in the resulting wine, although MeJ treatment did reduce the content of RG-II and total polysaccharides in the wine. However, in a recent study examining the foliar application of MeJ and nanoparticles doped with MeJ [26], it was observed that foliar treatment had only a slight impact on the content of monosaccharides and polysaccharides in grapes, prompting the authors to conclude that the elicitors did not reinforce the skin cell walls of Tempranillo grapes. These studies suggest that the extent of skin cell wall reinforcement depends on the composition and morphology of the cell wall material, which varies among grape varieties. The effect on wine is influenced by the type of polysaccharide family and the vintage climate. On the other hand, the foliar application of nitrogen compounds in vineyards has also been investigated. Urea foliar application offers advantages over other fertilizers as it is inexpensive, reduces soil fixation, requires lower fertilizer amounts, improves crop quality, and accelerates plant uptake, response, and assimilation. Indeed, urea foliar application at veraison does not lead to increased vegetative growth or grapevine vigor. However, it does improve the sensory quality of grapes and the content of yeast-assimilable nitrogen (YAN) [27]. Studies have reported an increase in nitrogen [28,29] and phenolic [12,30] and aromatic compounds [31,32] in grapes following urea foliar application. However, to the best of the authors' knowledge, there are no papers related to the effect of urea foliar treatment on the monosaccharide and polysaccharide content of grapes and wines, nor on the effects of methyl jasmonate and urea (combined) foliar treatment in vineyards.

Therefore, based on the information discussed above, this study aimed to investigate the effect of methyl jasmonate and urea foliar application on the composition of monosaccharides and polysaccharides in grapes and wine derived Tempranillo grapevines over two consecutive vintages, something which has not been reported in the literature until now.

2. Materials and Methods

2.1. Vineyard Site, Experimental Design, and Vinification

The experiment was conducted in two consecutive vintages (2019–2020) using grapes from Tempranillo (*Vitis vinifera* L.) grapevines grown in the experimental vineyard of Finca La Grajera. This vineyard is located in Logroño, La Rioja (Spain) (Lat: 42°26′25.36″ North; Long: 2°30′56.41″ West; 456 m above sea level). The study was performed on grapevines planted in 1997; these vines were trained to a vertical shoot positioned (VSP) trellis system with a grapevine spacing of 2.80 m × 1.25 m and grafted onto an R-110 rootstock. For this study, two foliar applications were carried out: (i) control (sprayed with aqueous solution of Tween© (Polyethylene Glycol Sorbitan Monooleate) 80 alone) (Sigma-Aldrich, Madrid, Spain) and (ii) treatment with methyl jasmonate plus urea (MeJ + Ur). The products employed to facilitate foliar application were dissolved in water in a concentration of 10 mM of methyl jasmonate (Sigma-Aldrich, Madrid, Spain) and a dose of 6 kg N/ha of urea (Sigma-Aldrich), according to a protocol outlined in previous works [20,27]. Tween© 80 (1 mL/L) was used as a wetting agent. Treatments were carried out at veraison and one week later. In all cases, treatment was carried out by spraying 200 mL of solution over the leaves, and this was performed in triplicate. Ten vines were sprayed for each replication and treatment (n = ten vines treated in each treatment × two treatments × three replicates per treatment). The experimental design was arranged in a complete randomized block.

Grapes were harvested at their optimum technological maturity, i.e., when the weight of 100 berries remained constant and the potential alcohol content reached 13 (% v/v). For each replicate and treatment, a random sample of 100 berries were collected and immediately frozen at −20 °C for the subsequent analysis of grape polysaccharides. At the winery, the complete grape clusters, weighing approximately 25 kg, were destemmed and crushed. Grapes from the control treatment and the MeJ + Ur treatment, along with their respective replicates, were processed separately. General parameters were then determined in the resulting must. The resulting paste–must was introduced into individual 30 L-tanks, resulting in a total of 6 tanks corresponding to the six fermentations performed (two treatments × three repetitions/treatment). Alcoholic fermentation (AF) was carried out by inoculating the musts with a commercial *Saccharomyces cerevisiae* strain (Safoeno SC22, Fermentis, Marcq-en-Barœul, France) at a dosage of 20 g/hL. AF was carried out under a controlled temperature of 20 +/−2 °C. Once AF was complete (residual sugar content in must below 2.5 g/L), a commercial *Oenococcus oeni* strain (Viniflora CiNe, CHR Hansen, Hørsholm, Denmark) was inoculated into the wine at a rate of 1 g/hL to initiate malolactic fermentation (MLF) at a temperature of 17 ± 1 °C. Once MLF was complete, aliquot samples were frozen and stored at −20 °C until it was time to analyze the monosaccharide and polysaccharide compounds and general parameters of the wine.

2.2. General Parameters of Must and Wines

The general parameters (°Brix, potential alcohol, pH, and total acidity) of must were determined using the official methods outlined by the OIV [33]. Glucose, fructose, and their sum malic acid, ammonium nitrogen, amino nitrogen, yeast-assimilable nitrogen, and total phenols were analyzed using enzymatic equipment (Miura One, TDI, Barcelona, Spain). After malolactic fermentation, the wines were analyzed to determine their alcoholic concentration, pH, total acidity, volatile acidity, color intensity (CI), and total polyphenol index (TPI) [33]. The total anthocyanin content was studied following the methodology described by Ribéreau-Gayon and Stonestreet [34].

2.3. Analysis of Soluble Polysaccharides from Grapes and Wine via GC–MS

2.3.1. Extraction of Soluble Polysaccharides from Grapes

The grapes were defrosted and then were homogenized using an Ultra-Turrax T-18 (IKA, Staufen, Germany) at 18,000–20,000 rpm in static conditions. Moreover, 1 g of homogenate was employed for the extraction. Extraction was carried out for 18 h in tartaric acid (2.5 g/L) with a pH = 1 and in a 1:4 solid/liquid ratio following the extraction method outlined by Canalejo et al. [35].

2.3.2. Precipitation of Total Soluble Polysaccharides from Grapes and Wines

A method of recovery via precipitation after ethanolic dehydration was employed with grape extracts and wines [36,37]. Precipitation was carried out in triplicate (n = 3).

2.3.3. Identification and Quantification of Monosaccharides by GC–MS in Grapes and Wines

The monosaccharide contents of the grape and wine samples were analyzed via gas chromatography–spectrometry (GC–MS) using an Agilent 7890A gas chromatograph (Agilent Technologies, Waldbronn, Germany) coupled to a 5975C VL quadrupole mass detector. The methodology we followed was that described by Martínez-Lapuente et al. [26]. Each sample was injected in triplicate. A capillary chromatographic column of Teknokroma fused silica (30 m × 0.25 mm × 0.25 μm) of phase 5% phenyl–95% methylpolysiloxane was employed. The oven was initially heated to 120 °C, and the temperature was increased at a rate of 1 °C/min to 145 °C, then to 180 °C. The rate of temperature increase was then adjusted to 0.9 °C/min and later to 40 °C/min until a temperature of 230 °C was reached. The temperature of the injector, equipped with a 3.4 mm I.D. in a split ratio of 1:20, was 250 °C. Helium (99.996%) was employed as carrier gas at a flow rate of 1 mL/min. The ionization voltage was 70 eV. The temperature for the MS Quad was 150 °C, the temperature for the MS Source was 230 °C, and the temperature for the transfer line was 250 °C. The monosaccharide composition was determined by examining the trimethylsilyl-ester O-methyl glycosyl residues obtained after acid methanolysis and derivatization via GC–MS [37]. The content of each family of polysaccharides was estimated from the concentration of individual glycosyl residues, which are characteristic of structurally identified must and wine polysaccharides [2,36].

2.4. Sensory Analysis of the Wines

A sensory evaluation of the wines was carried out by a panel of 12 specialized panelists one month after the completion of malolactic fermentation. The evaluation took place in a designated test room following the guidelines outlined by the International Organization for Standardization. Prior to the evaluation, the panelists underwent training to familiarize themselves with sensory analysis terminology. The wine samples were evaluated in a totally random order using a blind tasting and comparison system. A tasting sheet approved by the OIV [38] was employed, and evaluations were made on a scale of 40 (insufficient) to 100 (excellent). Quantitative assessments of the olfactory attributes (compote, red fruit, black fruit, floral, spicy, smoked, alcoholic, vegetal/herbaceous, balsamic, lactic, oxidized, and reduced) along with the gustatory characteristics (sweet, acid, bitter, alcohol, astringency, and overall balance) were also performed, using a scale ranging from 1 (low intensity) to 10 (high intensity).

2.5. Statistical Analysis

Statistical analysis was performed using SPSS Version 21.0 statistical package for Windows (SPSS, Chicago, IL, USA). General parameters, monosaccharide and polysaccharide composition, and sensory analysis data were assessed using analysis of variance (ANOVA). The differences between means were compared using the Duncan test ($p \leq 0.05$). Multifactor analysis of the general parameters, monosaccharide, polysaccharide, and sensory data was carried out to study the effects of treatment, the season, and the interaction between these two factors. To classify samples, a discriminant analysis carried out by using the monosaccharide data of wines was conducted.

3. Results and Discussion

3.1. General Parameters of Must and Wines

Tables 1 and 2 show the general parameters of the control and MeJ + Urea (MeJ + Ur) grapes and wines for the two seasons studied.

Table 1. General parameters of control and MeJ + Urea (MeJ + Ur)-treated Tempranillo grapes from a vineyard located in La Rioja, Spain, in 2019 and 2020.

	2019		2020	
	Control	**MeJ + Ur**	**Control**	**MeJ + Ur**
Weight of 100 berries (g)	113.68 ± 11.07	131.52 ± 25.19	199.57 ± 7.27	222.83 ± 25.25
°Brix	24.7 ± 0.7 b	23.0 ± 0.6 a	22.3 ± 0.9 a	22.8 ± 0.7 a
Potential alcohol (% *v*/*v*)	14.63 ± 0.49 b	13.48 ± 0.42 a	12.97 ± 0.63	13.29 ± 0.51
pH	3.83 ± 0.05	3.80 ± 0.04	3.76 ± 0.01	3.71 ± 0.03
Total acidity (g/L) *	4.61 ± 0.11	5.11 ± 0.36	4.12 ± 0.33	3.83 ± 0.13
Glu (g/L)	120.18 ± 5.13 b	107.43 ± 3.65 a	107.31 ± 4.54	113.11 ± 6.85
Fru (g/L)	129.68 ± 4.84 b	119.25 ± 2.52 a	109.11 ± 6.53	115.75 ± 3.49
Malic acid (g/L)	2.24 ± 0.24	2.45 ± 0.46	1.21 ± 0.08 a	1.42 ± 0.05 b
Ammonium nitrogen (mg N/L)	78.00 ± 8.22 a	118.30 ± 6.54 b	121.16 ± 3.52	109.72 ± 8.59
Amino nitrogen (mg N/L)	118.51 ± 14.33 a	237.60 ± 30.51 b	152.53 ± 14.33	149.89 ± 7.06
YAN (mg N/L)	196.51 ± 21.18 a	355.90 ± 31.59 b	273.69 ± 17.69	259.61 ± 13.65
Total phenols (mg/L)	1185.33 ± 72.31 a	1351.83 ± 29.05 b	541.60 ± 64.02	578.17 ± 82.64

* As g/L of tartaric acid. All parameters are listed with their standard deviation (n = 3). For each season and parameter, different letters indicate significant differences between the samples ($p \leq 0.05$). Glu: glucose; Fru: fructose; YAN: yeast assimilable nitrogen.

Table 2. General parameters of control and MeJ + Urea (MeJ + Ur)-treated wines derived from Tempranillo grapes in 2019 and 2020.

	2019		2020	
	Control	**MeJ + Ur**	**Control**	**MeJ + Ur**
Alcohol (% *v*/*v*)	13.97 ± 0.31 b	12.80 ± 0.40 a	12.47 ± 0.70	12.53 ± 0.81
pH	3.96 ± 0.07	3.94 ± 0.13	3.66 ± 0.08	3.73 ± 0.13
Total acidity (g/L) *	4.27 ± 0.10 b	3.92 ± 0.06 a	4.43 ± 0.59	4.02 ± 0.23
V A [1] (g/L) **	0.23 ± 0.02	0.20 ± 0.01	0.22 ± 0.02	0.22 ± 0.03
Lactic acid (g/L)	1.32 ± 0.10	1.28 ± 0.12	0.86 ± 0.07 a	1.05 ± 0.09 b
YAN [2] (mg N/L)	18.06 ± 2.08 a	67.89 ± 8.90 b	30.36 ± 0.54	39.34 ± 10.65
T P [3] (mg/L)	2440.83 ± 123.16	2460.73 ± 124.74	1116.63 ± 106.69	1333.47 ± 153.38
T A [4] (mg/L)	1117.33 ± 69.97	1289.67 ± 102.00	130.99 ± 20.13	168.00 ± 18.68
CI [5]	18.27 ± 1.03	19.01 ± 1.14	6.05 ± 0.55 a	8.62 ± 1.10 b
TPI [6]	70.83 ± 3.47	73.32 ± 5.00	36.82 ± 4.05	44.73 ± 5.62

[1] V A: Volatile acidity, [2] YAN: yeast assimilable nitrogen, [3] T P: Total phenols, [4] T A: Total anthocyanins, [5] CI: Color index, [6] TPI: Total polyphenol index. * As g/L tartaric acid. ** As g/L acetic acid. All parameters are listed with their standard deviation (n = 3). For each season and compound, different letters indicate significant differences between the samples ($p \leq 0.05$).

Regarding the general parameters of grapes in 2019, MeJ + Ur grapes showed a lower °Brix and potential alcohol concentration in comparison with the control grapes. The glucose and fructose content was also lower in the MeJ + Ur-treated grapes than in the control grapes. This finding can be attributed to the delay in grape ripening caused by the MeJ + Ur foliar treatment during this season [24]. On the other hand, the MeJ + Ur grapes showed a higher content of ammonium nitrogen, amino nitrogen, YAN, and total phenols compared to the control grapes. Therefore, MeJ + Ur foliar treatment seems to improve the biosynthesis of the phenolic compounds in grapes. This effect has been previously described in the literature, albeit with respect to applying MeJ and Ur to Tempranillo grapes individually (not in tandem) [12,19]. Likewise, an increase in nitrogen content has also been previously observed in Tempranillo grapes after the individual foliar application of

Ur [27,28]. However, in the second vintage, different results were observed. Treatment with MeJ + Ur did not seem to affect grape maturation, as the only difference observed between the control and MeJ + Ur-treated grapes was in regard to the concentration of malic acid, which was higher in the MeJ + Ur-treated grapes. In addition, in contrast to the findings for 2019, MeJ + Ur did not improve the phenolic concentration of grapes in the second season studied (2020). Therefore, the effectiveness of MeJ + Ur foliar treatment appears to be dependent on the specific season, which, in the case of MeJ, has been previously established in both Tempranillo grapes and other grape varieties [24,39]. Notable differences in the weight of 100 berries and the nitrogen content of grapes were observed between the seasons, probably due to the higher pre-harvest rainfalls recorded in 2020 than in the 2019 vintage season (11.5 L/m^2 in 2019 versus 32.9 L/m^2 in 2020). In addition, there was a change in the time interval between the initial application of MeJ + Ur and the harvest date between the two vintages. In the 2019 season, one month elapsed between the initial application and the harvest date, while in 2020, this period was longer, lasting 1 month and 20 days. Turning to the general parameters of the wines (Table 2), it was observed that, in the first season, the control wines exhibited a higher alcohol content than MeJ + Ur wines, which is consistent with the results observed in the grapes. This effect is of particular interest in mitigating the effects of climate change, which, as previously stated, poses ongoing challenges to viticulture. However, no differences in the alcohol content of the wines from the second vintage were observed. Therefore, further studies should be conducted to elucidate this effect. In agreement with the higher concentration observed in MeJ + Ur-treated grapes in the 2019 vintage season, the MeJ + Ur-treated wines exhibited a higher YAN concentration. In the second vintage season, slight differences were noted between the control and MeJ + Ur-treated wines. The MeJ + Ur-treated wines displayed a higher CI compared to the control wines, which could have a significant effect on the quality of the wines. However, the MeJ + Ur-treated wines showed a higher lactic acid concentration than the control wines.

Therefore, treatment with MeJ + Ur had a limited effect on the general parameters of the wines in both of the seasons studied.

3.2. Monosaccharide Composition and the Polysaccharide Families of Tempranillo Grapes

Table 3 presents the glycosyl composition of the monosaccharide and the polysaccharide concentrations in both sets of grapes. According to Martínez-Lapuente et al. [26], glucose is the main component of various grape cell walls' polysaccharides, such as arabinoglucans, cellulose, and mannans. However, during the 2019 season and in both the control and MeJ samples, galactose was the predominant monosaccharide, followed by glucose, galacturonic acid, and arabinose. In contrast, in 2020, galactose, galacturonic acid, arabinose, and glucose were the main monosaccharides, as observed in Tempranillo grapes by Martínez-Lapuente et al. [26]. The concentration of monosaccharides varied significantly between seasons, which can be attributed to factors such as ripening and the structural properties of the skin cell walls [6]. In the first vintage season, the MeJ + Ur-treated grapes showed a higher concentration of 2-O-methyl-fucose, arabinose, and fucose while displaying a lower concentration of Kdo compared to the control grapes. No differences were recorded concerning the other monosaccharide compounds; thus, no variations in total monosaccharide content were observed.

In the first vintage season studied, Martínez-Lapuente et al. [26] found a lower concentration of Kdo in MeJ-treated grapes compared to control grapes.

However, in the 2020 season, no differences in inmonosaccharide content were observed between the control and MeJ + Ur-treated grapes, indicating that the foliar treatment did not affect the monosaccharide composition of grapes during this season. The different effects of MeJ + Ur foliar treatment on the monosaccharide concentration of Tempranillo grapes can be attributed to the season-dependent nature of elicitor effects [24,39]. The absence of differences in glucose concentration between the control and MeJ + Ur-treated grapes can be explained by the insignificant cell wall remodeling induced by the elicitor in the two seasons studied, as observed by Martínez-Lapuente et al. [26] following Tempranillo grapevine foliar treatment with MeJ. When plants are exposed to elicitors, they initiate a response that involves the accumulation of phenolic compounds, the depositing of callose in the cell wall, or the formation of lignin polymers to reinforce the skin cell wall [14]. Callose contains a high proportion of glucose [26]. However, the in MeJ + Ur-treated samples an increase in glucose content was not observed in either of the seasons studied. Hemicellulosic fractions are mainly composed of xyloglucans, mannans, and xylans. Xylose residues come from xyloglucans, mannose is derived from mannans, and hemicelluloses are present in the pericarp of grapes [40,41]. As was the case with glucose content, there were no differences in mannose and xylose content between the MeJ + Ur-treated samples and the control samples. Martínez-Lapuente et al. [26] observed that foliar treatment with MeJ did not induce any changes in the content of the major pectic monosaccharides in Tempranillo cell walls. Overall, foliar treatment with MeJ + Ur did not affect the concentration of monosaccharides in the grapes, which is consistent with the results reported in the literature for Tempranillo grapes following MeJ treatment [7]. Therefore, MeJ + Ur did not exhibit a synergetic effect on the biosynthesis of grape monosaccharides. Additionally, MeJ + Ur foliar treatment did not modify the polysaccharide composition of grapes. However, total polysaccharide concentration was approximately three times higher in the 2020 grapes compared to the 2019 grapes, which can be attributed to climatic differences. Across all seasons and samples, the polysaccharides rich in arabinose and galactose (PRAG) constituted the main family (64–74%), followed by HL (16–28%), mannans (4–6%), and RGII (3–4%). This result also aligns with the findings of Martínez-Lapuente et al. [26]. None of the polysaccharide families displayed differences between treatments in grape composition. These results indicate that MeJ + Ur foliar treatment did not reinforce the strength of the grape skin cell wall, which is antithetical to what has been described in the literature for MeJ-treated Monastrell grapes [23,25].

Table 3. Monosaccharide and polysaccharide composition of both the control and treated (MeJ + Ur) grapes derived from Tempranillo grapevines (mg/L) in 2019 and 2020 in a vineyard located in La Rioja, Spain.

	2019		**2020**	
	Control	**MeJ + Ur**	**Control**	**MeJ + Ur**
* 2-Omefu	0.04 ± 0.00 a	0.06 ± 0.00 b	0.14 ± 0.02	0.12 ± 0.03
* 2-OmeXyl	0.02 ± 0.00	0.03 ± 0.00	0.08 ± 0.00	0.06 ± 0.01
Apiose	0.01 ± 0.00	0.01 ± 0.00	0.04 ± 0.00	0.03 ± 0.01
Arabinose	1.59 ± 0.01 a	2.45 ± 0.45 b	4.74 ± 0.35	4.60 ± 0.81
Rhamnose	0.60 ± 0.08	0.69 ± 0.10	1.66 ± 0.14	1.33 ± 0.27
Fucose	0.02 ± 0.00 a	0.03 ± 0.00 b	0.05 ± 0.00	0.05 ± 0.01
Xylose	0.36 ± 0.08	0.29 ± 0.02	0.84 ± 0.07	0.72 ± 0.06
Mannose	0.72 ± 0.00	0.73 ± 0.05	1.46 ± 0.23	1.63 ± 0.37
Galactose	5.72 ± 0.24	6.31 ± 1.71	16.34 ± 0.90	16.01 ± 2.71
Galacturonic acid	2.81 ± 0.04	2.66 ± 0.52	11.32 ± 0.59	9.17 ± 2.04
Glucose	5.91 ± 1.10	4.92 ± 1.76	3.93 ± 0.75	3.09 ± 0.66
Glucuronic acid	0.53 ± 0.09	0.61 ± 0.03	1.52 ± 0.16	1.52 ± 0.19
Kdo	0.03 ± 0.00 b	0.02 ± 0.01 a	0.04 ± 0.00	0.05 ± 0.02
* TOTAL ms	18.38 ± 1.14	18.81 ± 2.55	42.16 ± 1.40	38.37 ± 3.59
* RGII	0.44 ± 0.02	0.46 ± 0.03	1.23 ± 0.07	1.03 ± 0.15
* Mannans	0.72 ± 0.00	0.73 ± 0.05	1.46 ± 0.23	1.63 ± 0.37
* PRAG	8.10 ± 0.44	9.59 ± 1.97	23.09 ± 1.11	22.51 ± 3.25
* HL	2.44 ± 0.15	2.14 ± 0.51	10.02 ± 0.62	8.11 ± 1.96
* PST	11.70 ± 0.47	12.92 ± 2.03	35.80 ± 1.29	33.28 ± 3.82

* 2-Omefu: 2-O-methyl-fucose; 2-OmeXyl: 2-O-methyl-xylose; Kdo: 3-deoxy-D-manooctulosonic acid; Total ms: total monosaccharides; RGII: rhamnogalacturonans-II; PRAG: polysaccharide rich in arabinose and galactose; HL: homogalacturonans; PST: total polysaccharides. For each compounds different letter indicate significant differences among treatments according to ANOVA test ($p \leq 0.05$). Absence of letters indicate no differences ($p > 0.05$).

3.3. Monosaccharide Composition and the Polysaccharide Families of Tempranillo Wines

Table 4 presents the monosaccharide and polysaccharide composition of the Tempranillo-derived wines. In the 2019 season, the wines exhibited higher concentrations of galactose, followed by galacturonic acid, mannose, arabinose, and glucose. This is in agreement with the results reported by Paladines-Quezada et al. [25], which showed that the main macromolecules in wines were PRAGs from berry cell walls and MP from yeast cell walls (in accordance with the monosaccharide composition of wines). Both wines contained rare sugars, namely, 2-O-methyl-fucose and 2-O-methyl-xylose, which serve as markers for RG-II [25,42]. Xylose was also present in the wines, indicating the solubilization of hemicellulose from grape cell walls [2]. No differences in the inmonosaccharide and polysaccharide concentrations were observed between the control and MeJ + Ur-treated wines. However, slight differences were observed in the grapes (Table 1). The absence of differences in the wines could be due to the fermentation process, which may have minimized the potential for differences [18]. As described by Martinez-Lapuente et al. [26] in their study on the effect of MeJ foliar treatment on Tempranillo vines, the total monosaccharide content was higher in the wines than in the grapes during both seasons studied.

Table 4. Monosaccharide and polysaccharide compositions of the control and treated (MeJ + Ur) Tempranillo wines (mg/L) in the 2019 and 2020 vintage seasons.

	2019		2020	
	Control	**MeJ + Ur**	**Control**	**MeJ + Ur**
Aceric acid	0.01 ± 0.00	0.23 ± 0.31	N.D	N.D.
* 2-Omefu	19.37 ± 5.68	15.87 ± 3.09	5.38 ± 0.56 b	1.46 ± 0.07 a
* 2-OmeXyl	9.29 ± 4.09	8.33 ± 1.77	3.28 ± 0.53 b	0.71 ± 0.28 a
Apiose	3.88 ± 2.29	4.25 ± 0.26	1.51 ± 0.32 b	0.58 ± 0.01 a
Arabinose	323.71 ± 116.87	326.83 ± 62.11	206.14 ± 45.21	170.41 ± 2.78
Rhamnose	161.22 ± 73.57	149.11 ± 19.02	43.00 ± 2.59	33.68 ± 12.39
Fucose	7.56 ± 2.28	7.05 ± 1.29	1.84 ± 0.29 b	0.93 ± 0.15 a
Xylose	22.77 ± 7.24	24.72 ± 4.56	9.84 ± 2.06	17.14 ± 4.83
Mannose	542.37 ± 171.67	653.40 ± 154.28	582.66 ± 108.83	510.12 ± 75.20
Galactose	1103.55 ± 427.71	1176.07 ± 219.70	623.81 ± 75.30	746.23 ± 59.81
Galacturonic acid	641.24 ± 73.45	695.02 ± 100.69	68.28 ± 4.30	55.60 ± 12.37
Glucose	178.68 ± 53.51	203.25 ± 14.38	78.06 ± 13.99	75.66 ± 27.47
Glucuronic acid	35.15 ± 17.42	56.76 ± 16.31	24.96 ± 7.64	23.97 ± 6.13
Kdo	11.61 ± 5.02	9.32 ± 5.00	1.42 ± 0.03 b	0.45 ± 0.08 a
* TOTAL ms	3060.41 ± 487.14	3330.21 ± 294.54	1650.18 ± 140.87	1636.94 ± 101.80
* RGII	176.64 ± 35.68	152.01 ± 24.61	46.34 ± 1.31 b	12.78 ± 1.40 a
* MP	542.37 ± 171.67	653.40 ± 154.28	582.66 ± 108.83	510.12 ± 75.20
* PRAG	1468.67 ± 466.28	1581.82 ± 247.49	854.90 ± 29.56	962.60 ± 68.95
* HL	466.91 ± 91.56	552.20 ± 93.73	19.89 ± 0.78 a	42.44 ± 11.94 b
* PST	2654.58 ± 506.50	2939.43 ± 307.31	1503.79 ± 79.80	1527.94 ± 147.76

* 2-Omefu: 2-O-methyl-fucose; 2-OmeXyl: 2-O-methyl-xylose; Kdo: 3-deoxy-D-manooctulosonic acid; Total ms: total monosaccharides; RGII: rhamnogalacturonans-II; MP: mannoproteins; PRAG: polysaccharide rich in arabinose and galactose; HL: homogalacturonans; PST: total polysaccharides. For each compounds different letter indicate significant differences among treatments according to ANOVA test ($p \leq 0.05$). Absence of letters indicate no differences ($p > 0.05$). N.D: Not detected.

In control wines RG-II accounted for 6.65%, MP for 20.43%, PRAG for 55.33%, and HL for 17.59% of the polysaccharide composition. Similar percentages were observed in the MeJ + Ur-treated wines, with RG-II accounting for 5.17%, MP accounting for 22.23%, PRAG accounting for 53.81% and HL accounting for 18.79%. Polysaccharides from Tempranillo grapes (RG-II, PRAG, and HL) represented 79.57% of the total of polysaccharides in the control wines and 77.77% of the total of polysaccharides in the MeJ + Ur-treated wines, which is in agreement with the figures described for Tempranillo wines in the literature [36]. Regarding the content of MP, no differences were observed between the control and MeJ + Ur-treated wine samples in both seasons (Table 4). This result is noteworthy as MP are primarily released by yeast, and the yeast strain used in the winemaking was the same for both the control group and MeJ + Ur-treated group. Martínez-Lapuente et al. [26] also described the effect of just MeJ treatment on the release of MP in wines. Likewise, they did not observe any differences in the PRAG and HL concentration between control and MeJ-treated wines in both seasons studied. To summarize, no differences in total polysaccharide content were found between the control and MeJ + Ur-treated wines in the 2019 season. In 2020, there was a slight variation in the distribution of monosaccharides in wines compared to the 2019 season. Galactose remained the monosaccharide with the highest concentration in both wines, followed by mannose, arabinose, glucose, and galacturonic acid. The control wines showed a higher concentration of 2-O-methyl-fucose, 2-O-methyl-xylose, apiose, fucose, and Kdo than the MeJ + Ur-treated wines. These results agree with those observed by Martínez-Lapuente et al. [26] in MeJ-treated wines in their second season studied, except for Kdo, which did not show any differences in the control wines. These differences in the monosaccharide concentrations of wines were relatively small, suggesting that the foliar application of MeJ + Ur did not significantly strengthen the skin cell wall of the Tempranillo grapes; therefore, the extraction of monosaccharides from grapes to wines was minimally affected. However, other studies have reported a strengthening of Monastrell grape cell walls due to MeJ treatment [24], resulting in greater difficulty in extracting saccharides from the skin cell wall [23,25]. Although the wines showed slight differences in the concentration of certain monosaccharides, the total monosaccharide contents of the control and MeJ + Ur-treated wines did not significantly differ, which is in agreement with what Martínez-Lapuente et al. [26] observed in their first vintage season studied for MeJ-treated wines. In addition, in 2020, the grapes did not show any differences concerning the monosaccharide concentration of control and MeJ + Ur treated grapes from MeJ + Ur (Table 3). Regarding the polysaccharide families, the distribution in control wines was as follows: RG-II represented 3.08% of the polysaccharides, MP represented 38.75%, PRAG represented 56.85%, and HL represented 1.32%. In the MeJ + Ur-treated wines, the percentages were as follows: RG-II accounted for 0.84%, MP for 33.39%, PRAG for 62.99%, and HL for 2.78%. The distribution of polysaccharide families varied significantly between vintages. In 2020, the percentage of MP was higher compared to 2019, whereas RG-II and HL content was lower in 2020 than in 2019. Polysaccharides from Tempranillo grapes represented 61.25% of the total polysaccharides in the control wines, whereas, in the MeJ + Ur-treated wines, this figure was 66.61%. The control wines showed a higher concentration of RG-II, which could be related to the higher concentration of 2-O-methyl-fucose, 2-O-methyl-xylose, and fucose in the control wines. Paladines-Quezada et al. [25] proposed that an increase in the rigidity of the Monastrell grape cell walls could hinder the extraction of RG-II. Consequently, MeJ + Ur treatment might have led to greater rigidity in the skin cell walls of the Tempranillo grapes in this vintage, as evidenced by the lower extraction of RG-II in the MeJ + Ur-treated wines compared to the control wines. The PRAG content in the control and MeJ + Ur-treated wines did not differ, which is consistent with previous observations in Monastrell wines obtained from grapevines treated solely with with MeJ [25], and Martínez-Lapuente et al. [26] also reported a lack of differences regarding polysaccharide families (except for RG-II) between control and MeJ-treated Tempranillo wines. Conversely, the MeJ + Ur-treated wines were characterized by a higher concentration of HL compared to the control wines. No differences were observed in MP, PRAG, and total polysaccharide content between the control and MeJ + Ur-treated wines. Therefore, a synergistic effect between MeJ and Ur regarding wine monosaccharides and polysaccharides was not observed.

The total monosaccharide and polysaccharide content in wines was higher in 2019 than in 2020. However, in grapes, the total monosaccharide and polysaccharide content was higher in 2020 compared to 2019. This can be attributed to skin cell wall polysaccharides contributing more to the total monosaccharides in the 2019 wines than those from the pulp [26]. This phenomenon occurs when berries are small, which can be explained by the higher recorded pre-harvest rainfall in 2020, which resulted in larger grape sizes compared to the 2019 vintage.

3.4. Multifactor Analysis of Variance of Monosaccharide Composition and the Polysaccharide Families of Tempranillo Grapes and Wines and Discriminant Analysis of Monosaccharide Compounds in Wines

A multifactor analysis was carried out the using general parameters data of grapes to study the impacts of treatment, season, and the interaction between these two factors (Table 5). It is noteworthy that the season effect was significant for all general parameters except for the YAN content in grapes. However, treatment had an effect on total phenols, ammonium and amino nitrogen, and the YAN concentration in grapes. This result is likely attributable to MeJ + Ur foliar treatment enhancing the biosynthesis of phenolic compounds and the accumulation of nitrogen in the grapes. The interaction between season and treatment significantly affected all general parameters studied in grapes, except for the weight of 100 berries, pH, malic acid, and total phenols, which are more influenced by the season.

Table 5. Percentage of variance attributable to season, treatment, and the interaction of both (season × treatment) regarding each general parameter studied in the Tempranillo grapes obtained from a vineyard in La Rioja, Spain.

	Grapes			
	Season (%)	**Treatment (%)**	**S × T (%)**	**Residual (%)**
Weight of 100 berries (g)	84.91 ***	4.57	0.08	10.44
°Brix	37.14 *	7.52	23.77 *	31.57
Potential alcohol (% *v*/*v*)	37.19 *	7.54	23.87 *	31.40
pH	59.44 **	10.32	0.93	29.31
Total acidity	69.25 ***	0.98	13.82 *	15.96
Glu	7.15	6.69	47.61 *	38,55
Fru	52.17 **	1.30	26.24 *	20.29
Malic acid	82.20 ***	3.42	0.00	14.38
Total phenols	95.10 ***	1.95 *	0.80	2.15
Ammonium nitrogen (mg N/L)	22.86 **	15.93 **	51.19 ***	10.02
Amino nitrogen (mg N/L)	8.37 *	39.41 ***	43.06 ***	9.16
YAN (mg N/L)	0.64	37.20 ***	53.02 ***	9.14

Glu: glucose; Fru: fructose; YAN: yeast assimilable nitrogen. Statistically significant at: * $p \leq 0.05$, ** $p \leq 0.01$, *** $p \leq 0.001$; the absence of * indicates that the factor had no significant effect ($p > 0.05$).

Multifactor analysis of variance was carried out using the general parameter data of the wines to study the effect of treatment, season, and the interaction between these two factors (Table 6). The season had an effect on all of the general parameters of the wines, except for total acidity, volatile acidity, and YAN concentration. Nevertheless, the treatment only affected the YAN content in wines, as well as the total anthocyanins and color intensity of wines. Therefore, the foliar treatment of MeJ + Ur appears to have a slight influence on the wine's quality. The interaction of these two factors was only significant for the YAN content of the wines.

Multifactor analysis of variance was conducted using monosaccharide and polysaccharide data to examine the impact of treatment, season, and the interaction between these two factors (Table 7). In grapes, the season had an effect on the concentration of all monosaccharides and polyssacharides. However, treatment only affected the concentration of aceric acid and xylose in grapes. The interaction between these factors had a significant impact on the contents of aceric acid, 2-O-methyl-fucose, and 2-O-methyl-xylose in grapes. In wines, the season also affected all of the monosaccharide compounds and polysaccharide families, except for aceric acid, mannose, and MP. Treatment only significantly impacted RG-II content. There was no significant interaction between these two factors for any compound. Therefore, as previously observed in the analytical data (Tables 3 and 4), MeJ + Ur foliar treatment had a limited effect on the monosaccharide and polysaccharide concentration in Tempranillo grapes and wines.

Table 6. Percentage of variance attributable to season, treatment, and the interaction of both (season × treatment) regarding each general parameter studied in the Tempranillo wines.

	Wines			
	Season (%)	**Treatment (%)**	**S × T (%)**	**Residual (%)**
Alcohol % *v/v*	32.56 *	12.62	15.87	38.95
pH	67.95 **	0.71	2.19	29.15
Total acidity	3.64	32.79	0.24	63.32
V.A. [1]	1.96	17.65	5.45	74.95
Lactic acid	73.46 ***	3.17	8.16	15.21
YAN [2]	4.46	58.45 ***	28.20 ***	8.88
TP [3]	95.70 ***	0.89	0.62	2.79
TA [4]	97.69 ***	0.96 *	0.40	0.94
CI [5]	95.39 ***	2.04 *	0.63	1.93
TPI [6]	91.49 ***	2.53	0.69	5.29

[1] V A: Volatile acidity, [2] YAN: yeast assimilable nitrogen, [3] T P: Total phenols, [4] T A: Total anthocyanins, [5] CI: Color index, [6] TPI: Total polyphenol index. Statistically significant at: * $p \leq 0.05$, ** $p \leq 0.01$, *** $p \leq 0.001$; the absence of * indicates that the factor had no significant effect ($p > 0.05$).

Table 7. Percentage of variance attributable to season, treatment, and the interaction of both (season × treatment) regarding the monosaccharide and polysaccharide content of Tempranillo grapes and wines.

	Grapes			Wines		
	Season (%)	**Treatment (%)**	**S × T (%)**	**Season (%)**	**Treatment (%)**	**S × T (%)**
Aceric acid	83.31 ***	5.17 *	5.17 *	14.11	11.52	11.52
* 2-Omefu	84.47 ***	0.37	6.17 *	82.79 ***	5.64	0.02
* 2-OmeXyl	88.62 ***	1.97	4.39 *	72.93 ***	4.89	1.02
Apiose	85.32 ***	4.11	3.09	68.81 **	0.60	3.25
Arabinose	87.09 ***	1.61	3.12	57.83 **	0.82	1.16
Rhamnose	84.45 ***	1.66	5.17	77.02 ***	0.65	0.01
Fucose	80.39 ***	0.33	5.78	87.08 ***	1.27	0.10
Xylose	91.31 ***	3.83 *	0.31	52.35 **	10.66	3.56
Mannose	82.45 ***	0.94	0.80	4.52	0.63	14.37
Galactose	93.08 ***	0.02	0.19	54.83 *	2.52	0.16
Galacturonic acid	91.09 ***	2.13	1.62	96.84 ***	0.11	0.29
Glucose	45.25 ***	10.51	0.08	81.35 ***	0.77	1.14
Glucuronic acid	94.64 ***	0.20	0.18	40.54 *	9.34	11.21
Kdo	45.73 ***	2.64	11.25	71.30 **	2.08	0.34
* TOTAL ms	95.45 ***	0.57	0.91	89.83 ***	0.61	0.75
* RGII	92.26 ***	1.53	2.46	89.54 ***	4.17 *	0.10
* Mannans in grapes MP in wines	82.45 ***	0.94	0.80	4.52	0.63	14.37
* PRAG	94.25 ***	0.10	0.52	65.33 *	2.10	0.00
* HL	90.43 ***	2.38	1.27	93.68 ***	1.19	0.40
* PST	96.55 ***	0.08	0.68	84.82 ***	1.23	0.88

* 2-Omefu: 2-O-methyl-fucose; 2-OmeXyl: 2-O-methyl-xylose; Kdo: 3-deoxy-D-manooctulosonic acid; Total ms: total monosaccharides; RGII: rhamnogalacturonans-II; MP: mannoproteins; PRAG: polysaccharide rich in arabinose and galactose; HL: homogalacturonans; PST: total polysaccharides. Statistically significant at: * $p \leq 0.05$, ** $p \leq 0.01$, *** $p \leq 0.001$; the absence of * indicates that the factor had no significant effect ($p > 0.05$).

To classify the different samples, a discriminant analysis was performed on monosaccharide compound data from both the control and MeJ + Ur-treated wines in the two consecutive vintages (Figure 1). Function 1 accounted for 98.9% of the variance, while function 2 accounted for 1.0% (total variance = 99.9%). The variables that contributed the most to the discriminant model were galacturonic acid, Kdo, and aceric acid in function 1, whereas in function 2, Kdo, 2-O-methyl-fucose, and aceric acid contributed the most. Wine samples were separated along function 1 according to a seasonal criterion, confirming the results observed in the multifactor analysis (Table 4). The control and MeJ +

Ur-treated samples were located close to each other within each vintage, as the differences in monosaccharide composition observed in wines were minor (Table 4).

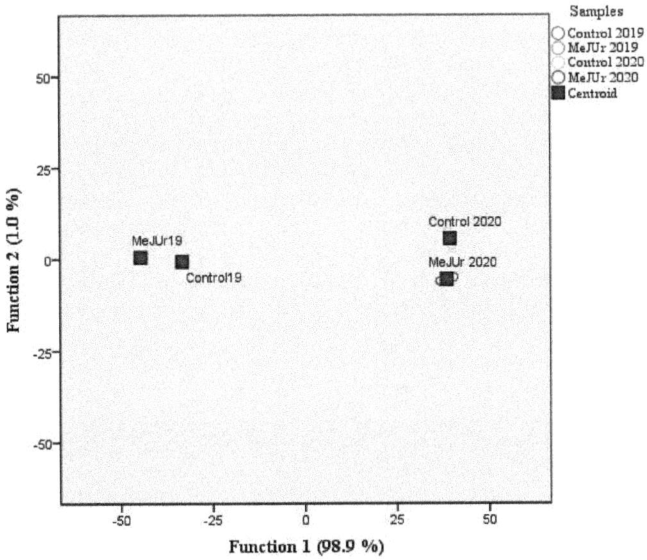

Figure 1. Discriminant analysis of the monosaccharide concentration of Tempranillo wines (control and MeJ + Ur) from both vintages studied.

3.5. Sensory Analysis of Tempranillo Wines

Table 8 presents a sensory evaluation of the control and MeJ + Ur wines. Wines from both vintages achieved a total evaluation above or near to 72 points, which, according to the scale used, means they can be classified as "good". In 2019, the wines only exhibited differences in persistence, with the control wines receiving higher scores than the MeJ + Ur-treated wines.

Table 8. Sensory evaluation of the control (C) and methyl jasmonate plus urea (MeJ + Ur)-treated Tempranillo wines one month later of the end of malolactic fermentation.

		2019		2020	
		Control	**MeJ + Ur**	**Control**	**MeJ + Ur**
Appearence	Cleanness	3.71 ± 0.74	3.63 ± 0.85	3.97 ± 0.82	4.13 ± 0.67
	Color	7.87 ± 1.26	7.67 ± 1.18	7.63 ± 1.54	7.87 ± 1.71
Aroma	Intensity	5.81 ± 1.40	6.33 ± 1.09	5.81 ± 1.18	6.32 ± 0.98
	Frankness	3.81 ± 0.95	4.13 ± 0.68	3.50 ± 0.88 a	4.10 ± 0.79 b
	Quality	11.74 ± 2.13	12.33 ± 1.30	11.25 ± 1.88	11.94 ± 1.59
Taste	Intensity	5.81 ± 1.17	6.07 ± 0.83	5.56 ± 1.44	5.81 ± 1.08
	Frankness	3.81 ± 0.60	4.07 ± 0.69	3.56 ± 0.76 a	4.03 ± 0.71 b
	Quality	15.29 ± 2.28	15.97 ± 1.77	14.44 ± 2.40 a	15.71 ± 2.10 b
	Persistence	6.13 ± 0.88 b	5.63 ± 0.81 a	5.75 ± 1.08	5.90 ± 0.91
Harmony		8.68 ± 0.70	8.97 ± 0.56	8.56 ± 0.80 a	8.97 ± 0.66 b
Total valuation		72.65 ± 8.66	74.80 ± 4.67	70.03 ± 8.41 a	74.77 ± 8.23 b

For each sensory attribute, different letters indicate significant differences between treatments ($p \leq 0.05$). The absence of letters indicates no differences ($p > 0.05$). The mean values ($n = 3$) are shown with their standard deviation.

However, in 2020, the sensory evaluation of the wines revealed more pronounced differences. The MeJ + Ur-treated wines were characterized by a higher evaluation in terms of the frankness of smell, frankness and quality of taste, harmony, and overall evaluation compared to the control wines. Therefore, the foliar treatment applied during the 2020 season resulted in wines that showed better results following the sensory analysis than the control wines. Interestingly, although the monosaccharide and polysaccharide concentrations of the wines did not show intense differences compared to the control wines, the CI of the MeJ + Ur-treated wines in 2020 was higher than that of the control wines (Table 2).

In the multifactorial analysis of the sensory evaluation (Table 9), it was found that treatment had a greater impact than the season. Treatment significantly influenced the intensity, frankness, and quality of the olfactory phase, as well as the frankness and quality of gustatory phase, with the MeJ + Ur-treated wines obtaining higher scores in these aspects. On the other hand, the season only had an effect on cleanness, and in any case, the interaction between these two factors did not significantly affect the sensory analysis.

Table 9. Multifactorial analysis of the sensory evaluation of the control (C) and methyl jasmonate plus urea (MeJ + Ur)-treated Tempranillo wines one month later of the end of malolactic fermentation.

		Treatment		**Season**		**Treatment xSeason**
		Control	**MeJ + Ur**	**2019**	**2020**	
View	Cleanness	3.84	3.88	3.67 a	4.05 b	N.S.
	Color	7.75	7.77	7.77	7.75	N.S.
Smell	Intensity	5.81 a	6.33 b	6.07	6.07	N.S.
	Frankness	3.65 a	4.12 b	3.97	3.80	N.S.
	Quality	11.50 a	12.13 b	12.04	11.59	N.S.
Taste	Intensity	5.68	5.94	5.94	5.68	N.S.
	Frankness	3.68 a	4.05 b	3.94	3.80	N.S.
	Quality	14.86 a	15.84 b	15.63	15.07	N.S.
	Persistence	5.94	5.77	5.88	5.83	N.S.
Harmony		8.62 a	8.97 b	8.82	8.77	N.S.
Total valuation		71.34 a	74.79 b	73.72	72.40	N.S.

For each sensory attribute different, letters indicate significant differences between samples ($p \leq 0.05$). The absence of letters indicates no differences ($p > 0.05$). N.S., not significant ($p > 0.05$).

4. Conclusions

The effect of MeJ + Ur foliar treatment was found to be season-dependent. In the first season, a reduction in the alcohol content of the MeJ + Ur-treated wines was observed, which is an interesting finding for mitigating the effects of climate change, a major global issue. This study, in addition to being novel, provides valuable information in addressing this problem. Additionally, in the first season, treatment with MeJ + Ur led to an increase in certain monosaccharides in the grapes, while no significant differences were observed in the MeJ + Ur-treated wines compared to the control wines. However, in the second season, the grapes from grapevines treated with MeJ + Ur did not exhibit differences compared to the control grapes, whereas the MeJ + Ur-treated wines showed lower concentrations of certain monosaccharides and polysaccharides. Multifactor analysis of variance indicated that the season had the largest influence on the monosaccharide and polysaccharide content of the grapes and wines studied, whereas treatment had only a slight effect on these compounds. The discriminant analysis of wines showed that the separation of wines was based on the seasonal criterion rather than the treatment. Furthermore, all of the wines studied were classified as "good" in the sensory analysis, and the multifactor analysis of sensory data revealed that the MeJ + Ur-treated wines had higher overall evaluations than the control wines, but this difference was only significant in the second vintage. Further studies should be carried out to elucidate the effects of MeJ + Ur foliar treatment on the monosaccharide and polysaccharide content of grapes and wines.

Author Contributions: Conceptualization, T.G.-C., E.P.P.-Á., and B.A.; methodology, T.G.-C., E.P.P.-Á., B.A., and L.M.-L.; formal analysis, L.M.-L. and M.L.R.-M.; investigation, M.G.-L. and L.L.T.-D.; resources, T.G.-C., E.P.P.-Á., and L.M.-L.; data curation, M.G.-L.; writing—original draft preparation, M.G.-L.; writing—review and editing, L.M.-L., T.G.-C., M.L.R.-M., L.L.T.-D., E.P.P.-Á., Z.G., and B.A.; funding acquisition, T.G.-C., E.P.P.-Á., B.A., and L.M.-L. All authors have read and agreed to the published version of the manuscript.

Funding: This work was made possible by funding from the Ministerio de Ciencia, Innovación y Universidades through the Project RTI2018-096549-B-I00.

Institutional Review Board Statement: Not applicable.

Informed Consent Statement: Not applicable.

Data Availability Statement: All data generated or analyzed during this study are included in the published paper.

Acknowledgments: M.G.-L. thanks to the Universidad de La Rioja for her Margarita Salas contract funding by the Ministerio de Universidades and the European Union (Financed by the European Union-Next GenerationEU). E.P.P.-Á. thanks the Ministerio de Ciencia, Innovación y Universidades for her Juan de la Cierva-Incorporación contract.

Conflicts of Interest: The authors declare no conflict of interest.

References

1. Unterkofler, J.; Muhlack, R.A.; Jeffery, D.W. Processes and purposes of extraction of grape components during winemaking: Current state and perspectives. *Appl. Microbiol. Biotechnol.* **2020**, *104*, 4737–4755. [CrossRef]
2. Doco, T.; Quellec, N.; Moutounet, M.; Pellerin, P. Polysaccharide patterns during the aging of Carignan noir red wines. *Am. J. Enol. Vitic.* **1999**, *50*, 25–32. [CrossRef]
3. Ortega-Regules, A.; Romero-Cascales, I.; Ros-García, J.M.; López-Roca, J.M.; Gómez-Plaza, E. A first approach towards the relationship between grape skin cell-wall composition and anthocyanin extractability. *Anal. Chim. Acta* **2006**, *563*, 26–32. [CrossRef]
4. Brandão, E.; Silva, M.S.; García-Estévez, I.; Williams, P.; Mateus, N.; Doco, T.; de Freitas, V.; Soares, S. The role of wine polysaccharides on salivary protein-tannin interaction: A molecular approach. *Carbohydr. Polym.* **2017**, *177*, 77–85. [CrossRef]
5. Jouquand, C.; Ducruet, V.; Giampaoli, P. Partition coefficients of aroma compounds in polysaccharide solutions by the phase ratio variation method. *Food Chem.* **2004**, *85*, 467–474. [CrossRef]
6. Martínez-Lapuente, L.; Guadalupe, Z.; Ayestarán, B. Properties of Wine Polysaccharides. In *Pectins—Extraction, Purification, Characterization and Applications*; IntechOpen: Rijeka, Croatia, 2020. [CrossRef]
7. Martínez-Lapuente, L.; Guadalupe, Z.; Ayestarán, B.; Ortega-Heras, M.; Pérez-Magariño, S. Changes in polysaccharide composition during sparkling wine making and aging. *J. Agric. Food Chem.* **2013**, *61*, 12362–12373. [CrossRef] [PubMed]
8. Mira de Orduña, R. Climate change associated effects on grape and wine quality and production. *Food Res. Int.* **2010**, *43*, 1844–1855. [CrossRef]
9. Gutiérrez, N.; López-De-silanes, L.; Escott, C.; Loira, I.; del Fresno, J.M.; Suárez-Lepe, J.A.; Morata, A. The effect of elicitors and canopy management in the chemical composition of vitis vinifera red varieties in warm and hot areas in spain. *Agronomy* **2021**, *11*, 1192. [CrossRef]
10. van Leeuwen, C.; Destrac-Irvine, A. Modified grape composition under climate change conditions requires adaptations in the vineyard. *Oeno One* **2017**, *51*, 147–154. [CrossRef]
11. Frioni, T.; VanderWeide, J.; Palliotti, A.; Tombesi, S.; Poni, S.; Sabbatini, P. Foliar vs. soil application of Ascophyllum nodosum extracts to improve grapevine water stress tolerance. *Sci. Hortic.* **2021**, *277*, 109807. [CrossRef]
12. Portu, J.; López-Alfaro, I.; Gómez-Alonso, S.; López, R.; Garde-Cerdán, T. Changes on grape phenolic composition induced by grapevine foliar applications of phenylalanine and urea. *Food Chem.* **2015**, *180*, 171–180. [CrossRef]
13. Portu, J.; López, R.; Santamaría, P.; Winterhalter, P.; Garde-Cerdán, T. Methyl jasmonate effect on Tempranillo (*Vitis vinifera* L.) grape phenolic content: A 2-year study. *Acta Hortic.* **2017**, *1188*, 127–134. [CrossRef]
14. Benhamou, N. Elicitor-induced plant defence pathways. *Trends Plant Sci.* **1996**, *1*, 233–240. [CrossRef]
15. Ruiz-García, Y.; Gómez-Plaza, E. Elicitors: A tool for improving fruit phenolic content. *Agriculture* **2013**, *3*, 33–52. [CrossRef]
16. D'Onofrio, C.; Matarese, F.; Cuzzola, A. Effect of methyl jasmonate on the aroma of Sangiovese grapes and wines. *Food Chem.* **2018**, *242*, 352–361. [CrossRef] [PubMed]

17. Gómez-Plaza, E.; Mestre-Ortuño, L.; Ruiz-García, Y.; Fernández-Fernández, J.I.; López-Roca, J.M. Effect of benzothiadiazole and methyl jasmonate on the volatile compound composition of *Vitis vinifera* L. Monastrell grapes and wines. *Am. J. Enol. Vitic.* **2012**, *63*, 394–401. [CrossRef]
18. Portu, J.; López, R.; Santamaría, P.; Garde-Cerdán, T. Methyl jasmonate treatment to increase grape and wine phenolic content in Tempranillo and Graciano varieties during two growing seasons. *Sci. Hortic.* **2018**, *240*, 378–386. [CrossRef]
19. Portu, J.; Santamaría, P.; López-Alfaro, I.; López, R.; Garde-Cerdán, T. Methyl jasmonate foliar application to tempranillo vineyard improved grape and wine phenolic content. *J. Agric. Food Chem.* **2015**, *63*, 2328–2337. [CrossRef] [PubMed]
20. Garde-Cerdán, T.; Portu, J.; López, R.; Santamaría, P. Effect of methyl jasmonate application to grapevine leaves on grape amino acid content. *Food Chem.* **2016**, *203*, 536–539. [CrossRef]
21. Gil-Muñoz, R.; Giménez-Bañón, M.J.; Moreno-Olivares, J.D.; Paladines-Quezada, D.F.; Bleda-Sánchez, J.A.; Fernández-Fernández, J.I.; Parra-Torrejón, B.; Ramírez-Rodríguez, G.B.; Delgado-López, J.M. Effect of methyl jasmonate doped nanoparticles on nitrogen composition of monastrell grapes and wines. *Biomolecules* **2021**, *11*, 1631. [CrossRef]
22. Gil-Muñoz, R.; Bautista-Ortín, A.B.; Ruiz-García, Y.; Fernández-Fernández, J.I.; Gómez-Plaza, E. Improving phenolic and chromatic characteristics of monastrell, merlot and syrah wines by using methyl jasmonate and benzothiadiazole. *Oeno One* **2017**, *51*, 17–27. [CrossRef]
23. Apolinar-Valiente, R.; Ruiz-García, Y.; Williams, P.; Gil-Muñoz, R.; Gómez-Plaza, E.; Doco, T. Preharvest Application of Elicitors to Monastrell Grapes: Impact on Wine Polysaccharide and Oligosaccharide Composition. *J. Agric. Food Chem.* **2018**, *66*, 11151–11157. [CrossRef] [PubMed]
24. Paladines-Quezada, D.F.; Moreno-Olivares, J.D.; Fernández-Fernández, J.I.; Bautista-Ortín, A.B.; Gil-Muñoz, R. Influence of methyl jasmonate and benzothiadiazole on the composition of grape skin cell walls and wines. *Food Chem.* **2019**, *277*, 691–697. [CrossRef] [PubMed]
25. Paladines-Quezada, D.F.; Gil-Muñoz, R.; Apolinar-Valiente, R.; Williams, P.; Fernández-Fernández, J.I.; Doco, T. Effect of applying elicitors to Vitis vinifera L. cv. Monastrell at different ripening times on the complex carbohydrates of the resulting wines. *Eur. Food Res. Technol.* **2022**, *248*, 2369–2381. [CrossRef]
26. Martínez-Lapuente, L.; Garde-Cerdan, T.; Pérez-Álvarez, E.P.; Guadalupe, Z.; Parra-Torrejón, B.; Ayestarán, B. Foliar applications to vines of methyl jasmonate and nanoparticles doped with methyl jasmonate: Impact on grape and wine polysaccharide composition. *OENO One* **2023**, *57*, 1–14. [CrossRef]
27. Pérez-Álvarez, E.P.; Ramírez-Rodríguez, G.B.; Carmona, F.J.; Martínez-Vidaurre, J.M.; Masciocchi, N.; Guagliardi, A.; Garde-Cerdán, T.; Delgado-López, J.M. Towards a more sustainable viticulture: Foliar application of N-doped calcium phosphate nanoparticles on Tempranillo grapes. *J. Sci. Food Agric.* **2021**, *101*, 1307–1313. [CrossRef]
28. Garde-Cerdán, T.; López, R.; Portu, J.; González-Arenzana, L.; López-Alfaro, I.; Santamaría, P. Study of the effects of proline, phenylalanine, and urea foliar application to Tempranillo vineyards on grape amino acid content. Comparison with commercial nitrogen fertilisers. *Food Chem.* **2014**, *163*, 136–141. [CrossRef]
29. Pérez-Álvarez, E.P.; Garde-Cerdán, T.; García-Escudero, E.; Martínez-Vidaurre, J.M. Effect of two doses of urea foliar application on leaves and grape nitrogen composition during two vintages. *J. Sci. Food Agric.* **2017**, *97*, 2524–2532. [CrossRef]
30. Garde-Cerdán, T.; Portu, J.; López, R.; Santamaría, P. Effect of foliar applications of proline, phenylalanine, urea, and commercial nitrogen fertilizers on stilbene concentrations in tempranillo musts and wines. *Am. J. Enol. Vitic.* **2015**, *66*, 542–547. [CrossRef]
31. Garde-Cerdán, T.; Gutiérrez-Gamboa, G.; López, R.; Rubio-Bretón, P.; Pérez-Álvarez, E.P. Influence of foliar application of phenylalanine and urea at two doses to vineyards on grape volatile composition and amino acids content. *Vitis J. Grapvine Res.* **2018**, *57*, 137–141. [CrossRef]
32. Garde-Cerdán, T.; Santamaría, P.; Rubio-Bretón, P.; González-Arenzana, L.; López-Alfaro, I.; López, R. Foliar application of proline, phenylalanine, and urea to Tempranillo vines: Effect on grape volatile composition and comparison with the use of commercial nitrogen fertilizers. *LWT-Food Sci. Technol.* **2015**, *60*, 684–689. [CrossRef]
33. OIV. *Compendium of Internationals Methods of Wine and Must Analysis*; OIV: Paris, France, 2009.
34. Ribéreau-Gayon, P.; Stonestreet, E. Determination of anthocyanins in red wine. *Bull. Soc. Chim. Fr.* **1965**, *9*, 2649–2652.
35. Canalejo, D.; Guadalupe, Z.; Martínez-Lapuente, L.; Ayestarán, B.; Pérez-Magariño, S. Optimization of a method to extract polysaccharides from white grape pomace by-products. *Food Chem.* **2021**, *365*, 130445. [CrossRef]
36. Ayestarán, B.; Guadalupe, Z.; León, D. Quantification of major grape polysaccharides (*Tempranillo* v.) released by maceration enzymes during the fermentation process. *Anal. Chim. Act.* **2004**, *513*, 29–39. [CrossRef]
37. Guadalupe, Z.; Martínez-Pinilla, O.; Garrido, A.; Carrillo, J.D.; Ayestarán, B. Quantitative determination of wine polysaccharides by gas chromatography-mass spectrometry (GC-MS) and size exclusion chromatography (SEC). *Food Chem.* **2012**, *131*, 367–374. [CrossRef]
38. OIV. OIV Standard for International Wine and Spirituous Beverages of Vitivinicultural Origin Competitions. Annex 3.1. Score Sheet. 2009 In RESOLUTION OIV/CONCOURS 332A/2009. Available online: https://www.oiv.int/public/medias/4661/oiv-concours-332a-2009-en.pdf (accessed on 31 January 2023).
39. Garde-Cerdán, T.; Gutiérrez-Gamboa, G.; Baroja, E.; Rubio-Bretón, P.; Pérez-Álvarez, E.P. Influence of methyl jasmonate foliar application to vineyard on grape volatile composition over three consecutive vintages. *Food Res. Int.* **2018**, *112*, 274–283. [CrossRef]

40. Arnous, A.; Meyer, A.S. Quantitative prediction of cell wall polysaccharide composition in grape (*Vitis vinifera* L.) and apple (*Malus domestica*) skins from acid hydrolysis monosaccharide profiles. *J. Agric. Food Chem.* **2009**, *57*, 3611–3619. [CrossRef] [PubMed]
41. Minjares-Fuentes, R.; Femenia, A.; Garau, M.C.; Candelas-Cadillo, M.G.; Simal, S.; Rossell'o, C. Ultrasound-assisted extraction of hemicelluloses from grape pomace using response surface methodology. *Carbohydr. Polym.* **2016**, *138*, 180–191. [CrossRef]
42. Pellerin, P.; Doco, T.; Vidal, S.; Williams, P.; Brillouet, J.M.; O'Neill, M.A. Structural characterization of red wine rhamnogalacturonan II. *Carbohydr. Res.* **1996**, *290*, 183–197. [CrossRef]

Nitrogen

MDPI

Communication

"Alperujo" Compost Improves Nodulation and Symbiotic Nitrogen Fixation of Soybean Inoculated with *Bradyrhizobium diazoefficiens*

Germán Tortosa [1,*], Socorro Mesa [1], María J. Delgado [1] and Carol V. Amaya-Gómez [2,*]

[1] Department of Soil and Plant Microbiology, Estación Experimental del Zaidín (EEZ-CSIC), Profesor Albareda, 1, 18008 Granada, Spain

[2] Corporación Colombiana de Investigación Agropecuaria-Agrosavia, Centro de Investigación La Libertad, Villavicencio 500009, Colombia

* Correspondence: german.tortosa@eez.csic.es (G.T.); camaya@agrosavia.co (C.V.A.-G.)

Abstract: The utilization of compost to enhance plant productivity and symbiotic nitrogen fixation (SNF) has been recognized as an effective alternative to synthetic nitrogen fertilizers. This environmentally sustainable method is readily accessible to farmers. This study investigated the effect of olive pomace ("alperujo", AL) compost on the nodulation and SNF of soybeans (*Glycine max* L.) and their natural symbiont (*Bradyrhizobium diazoefficiens*). For that, soybean plants were subjected to several doses of AL compost under controlled greenhouse conditions. At the end of the experiment, the dry weight of plant biomass (aerial part and roots), the number and fresh weight of nodules, and nitrogen and leghaemoglobin contents were analyzed. The application of AL compost significantly improved soybean growth, as demonstrated by an increase in both plant biomass and height. Furthermore, nodular leghaemoglobin content and nitrogen content were found to be enhanced by the addition of AL compost (7 and 40%, respectively), indicating an increase in nodule effectiveness and symbiotic efficiency. Our results provide clear evidence of the synergetic effect of AL compost on the soybean-*B. diazoefficiens* association, probably due to AL-compost improved soybean roots development, and rhizospheric organic matter and nutrients assimilation by rhizobia.

Keywords: compost; leghaemoglobin; legume; nitrogen fixation; nodules; rhizobia; symbiotic effectiveness

1. Introduction

Legumes are the main source of vegetable proteins for animal and human food and are also one of the main strategies for the sustainability of agricultural ecosystems [1]. An example is soybean (*Glycine max*), which accounts for about 50% of the total cultivated area and 68% of the world production of legumes [2]. On the other hand, legumes establish a mutual symbiosis with certain soil microorganisms known as rhizobia. The rhizobia-legume interaction encompasses a tight regulation process that only allows specific rhizobia strains to induce the formation of nodules in certain legumes. Inside the nodules, rhizobia carry out symbiotic nitrogen fixation (SNF), a process where the nitrogenase enzyme reduces atmospheric nitrogen (N_2) to ammonia (NH_4^+). This way, non-bioavailable nitrogen is converted into nitrogen compounds that plants can use for their growth and development [3].

Composting is a biotechnology that transforms organic waste into humic materials that can be utilized as fertilizers in agriculture. An example can be found in the two-phase olive pomace or "alperujo" (AL), which comes from the production of the olive oil and it is one of the most important organic wastes produced in Spain. It has been recently demonstrated that composting is a viable technology for its treatment and also as a feasible strategy for the production of solid and liquid organic commercial fertilizers [4,5].

In addition to rhizobia, other microbial strains that facilitate the acquisition of nutrients or produce phytohormones or compounds with antimicrobial activity are present [6,7]. Given the enhancement of nutrients and the mitigation of environmental stresses, compost improves the formation of legume nodules and, therefore, SNF. For instance, compost application to soils led to an increase in the number of nodules compared to soybeans grown without compost [8–10]. Considering all these benefits, compost has become a viable technology for the nutrient enrichment of soils, and it is also considered a feasible strategy to produce solid and liquid organic commercial fertilizers [4,5].

Compost is a very important source of organic matter, humic substances and plant nutrients, as well as a reservoir of beneficial microorganisms that can promote plant growth. Thus, the main objective of this work was to study the feasibility of AL compost to improve the SNF process and its efficiency in legume crops, specifically in soybean plants inoculated with *Bradyrhizobium diazoefficiens*, its natural endosymbiont.

2. Materials and Methods

2.1. AL Compost

The compost used in this study was previously prepared by mixing an equivalent amount of AL and sheep manure, as described in earlier work [4]. In short, a trapezoidal pile of 10 t was made and handled with a backhoe loader. A total of 7 turning operations were performed, which were more frequent during the bioxidative phase, and more time wise as the composting progressed. The pile humidity was maintained at values close to 40% thanks to a sprinkler irrigation system. The process lasted for a total of 22 weeks. The obtained compost had a slightly alkaline pH (7.67 ± 0.23) and low salinity (EC of 1.11 ± 0.08 dS m^{-1}). Its organic matter content was 54.5 ± 1.8%, with a significant lignocellulosic fraction and a remarkable humic content, with 78.30 ± 3.39% of humic acids. The total nitrogen content was 1.50 ± 0.30%, essentially organic due to its low mineral content (ammonium, nitrate and nitrite). The AL compost did not show any phytotoxic properties, showing values of 93 ± 1% in the germination index (Zucconi test).

2.2. Cultivation of *B. diazoefficiens*

In this experiment, the parental strain (USDA 110) of *B. diazoefficiens* (U.S. Department of Agriculture, Beltsville, MD, USA) was used as soybean endosymbiont. This bacterium was routinely grown on PSY (peptone-salts-yeast) medium at 28 °C supplemented with chloramphenicol (15 μg L^{-1}) and arabinose (0.1% *w*/*v*), as previously described [11]. The preparation of the inoculum was made by growing *B. diazoefficiens* to stationary phase and then centrifuging it at 12,000× *g* at 4 °C for 10 min. Finally, the pellet was resuspended in sterile saline (NaCl, 0.9%) and achieved an optical density (600 nm) of 0.8, equivalent to about 10^8 cells per mL.

2.3. Soybean Inoculation and Growth

Soybean (*Glycine max* L. Merr., cv. Williams) plants were inoculated with *B. diazoefficiens* USDA 110 as previously described [12]. Seeds were surface sterilized with ethanol (96%, *v*/*v*) and H_2O_2 (30%, *v*/*v*) and germinated in Petri dishes (1% agar, *w*/*v*) for 3 days at 28 °C. Seeds were transplanted into sand-filled pots (0.25 L) and inoculated with 1 mL of *B. diazoefficiens* inoculum. Plants were periodically watered (twice a week) with 40 mL of a nitrogen-free nutrient solution and grown under controlled light and temperature conditions in the Greenhouse and Growth Chamber Service of the Estación Experimental del Zaidín (EEZ-CSIC; https://www.eez.csic.es/en/green-houses; accessed on 15 May 2020) for 28 days [12] until the vegetative stage of V3-V4 (Figure 1).

Figure 1. Soybeans before harvesting. Plants were grown with different doses of AL compost (D0, D1, D2 and D3). Non-inoculated plants grown in the absence of compost were used as the control.

2.4. Physiological Determinations

Shoot dry weight (SDW), root dry weight (RDW), nodule number (NN) and nodule fresh weight (NDW) were measured after harvesting (Figure 2). Nodule functionality and fitness were evaluated by fluorimetrically analyzing leghaemoglobin concentration (Lb) after the acidic reaction of the nodular fraction, as it is described in [12] (Figure 2). Furthermore, the nitrogen content in dry shoots (N_{shoots}), roots (N_{roots}), nodules ($N_{nodules}$) and seeds (N_{seeds}) were determined by the Dumas method using a LECO TruSpec CN Elemental Analyzer (USA) available at the Elemental Analysis Service of the EEZ-CSIC) [11]. Total nitrogen accumulated (T_{NA}) was estimated according to the following formula: $T_{NA} = (N_{shoots} + N_{roots} + N_{nodules}) - N_{seeds}$. Data were expressed as mg N plant^{-1} (dry weight).

Figure 2. Soybean nodules after harvesting. Inoculated soybean plants were grown with different doses of AL compost: D0, D1, D2 and D3.

2.5. Experimental Design and Statistical Analysis

Five treatments were tested according to AL compost doses applied: Non-inoculated soybeans with no compost added (Control), inoculated soybean with *B. diazoefficiens* without compost (D0), inoculated soybean with *B. diazoefficiens* with compost (2.2 g per pot or 28 t ha^{-1}) (D1), inoculated soybean with *B. diazoefficiens* with compost (4.4 g per pot or 56 t ha^{-1}) (D2) and inoculated soybean with *B. diazoefficiens* with compost (6.6 g per pot or 84 t ha^{-1}) (D3).

The number of replicates per treatment was 9 and the assay was performed three times during one year of experimentation. Statistical treatment of the data consisted of the calculation of some descriptive statistics such as arithmetic mean and standard deviation, as well as an inferential calculation based on the one-way analysis of variance (ANOVA) and Tukey tests ($p < 0.05$).

3. Results

The effect of AL compost on the growth of soybean plants was positive for all the parameters analyzed, especially plant biomass (Table 1). Both SDW and RDW increased by about 5 and 45%, respectively, compared to the D0 treatment. Improvements in plant height and trifoliate leaves were also observed, but without being statistically significant. Furthermore, the shoot to root ratio (SDW/RDW) decreased, which meant a differentiation in plant development produced by the different doses of AL compost added during growing.

Table 1. Effect of AL compost doses (D0, D1, D2 and D3) on plant height, trifoliate leaves, shoot (SDW) and root dry weight (RDW) and shoot to root ratio (SDW/RDW) of soybean plants. A set of non-inoculated plants (Control) grown without compost was included in the assay.

Treatment	Plant Height (cm)	Trifoliate Leaves	SDW (g Plant^{-1})	RDW (g Plant^{-1})	SDW/RDW (Plant^{-1})
Control	15 b	2 b	1.23 c	0.32 c	3.84 a
D0	24 a	3 a	1.42 b	0.45 b	3.16 b
D1	28 a	4 a	1.51 a	0.61 a	2.48 c
D2	27 a	4 a	1.49 a	0.67 a	2.22 c
D3	26 a	4 a	1.48 a	0.62 a	2.39 c

Values in a column followed by the same letter are not statistically different according to the Tukey test ($p < 0.05$).

Regarding soybean-*B. diazoefficiens* symbiosis, the addition of AL compost significantly improved NN by 7 to 20% compared to the D0 treatment (Table 2). In addition, a slight increase in nodular biomass was observed, especially in D2, which was 0.66 g NFW per plant. This data represented 8% more than the value obtained in D0. In terms of symbiotic efficiency, which was measured as Lb content, it was observed that AL compost produced a significant improvement depending on the dose applied. Lb concentration boosted from 6.82 in D0 to 7.12, 7.25 and 7.32 mg NFW^{-1} in D1, D2 and D3 treatments, respectively.

Table 2. Effect of AL compost doses (D0, D1, D2 and D3) on nodule number (NN), nodules fresh weight (NFW), fresh weight per nodule (NFW/NN) and nodular leghaemoglobin content (Lb) of soybean plants. Non-inoculated plants grown without compost were used as a reference for the comparisons (Control).

Treatment	NN (Plant^{-1})	NFW (g Plant^{-1})	NFW/NN (mg Nodule^{-1})	Lb (mg NFW^{-1})
Control	-	-	-	-
D0	65 b	0.61 b	6.82 c	6.82 c
D1	70 ab	0.63 b	7.12 b	7.12 b
D2	76 a	0.66 a	7.25 a	7.25 a
D3	78 a	0.62 b	7.32 a	7.32 a

Values in a column followed by the same letter are not statistically different according to the Tukey test ($p < 0.05$).

Finally, the nitrogen accumulation in soybean biomass was also determined. AL compost addition significantly increased nitrogen content in shoots, roots and also nodules of soybean plants inoculated with *B. diazoefficiens* (Table 3). N_{shoot} rose 1.2–1.3 fold with AL compost compared to D0 treatment. Similarly, N_{root} was around 1.4 and 1.6 fold, and $N_{nodules}$ was close to 1.3 and 1.5 fold. Particularly, D2 was the best of the AL compost treatments for N acquisition in soybean plants, with 38.60 ± 2.1 mg plant^{-1} of T_{NA} (Figure 3).

Table 3. Effect of AL compost doses (D0, D1, D2 and D3) on nitrogen content in soybean shoots (N_{shoot}), roots (N_{roots}) and nodules ($N_{nodules}$).

Treatment	N_{shoot} (mg Plant^{-1})	N_{root} (mg Plant^{-1})	$N_{nodules}$ (mg Plant^{-1})
D0	21.47 b	4.31 c	2.31 c
D1	26.21 a	5.89 b	3.12 b
D2	28.26 a	6.78 a	3.56 a
D3	26.31 a	6.04 ab	2.91 b

Values in a column followed by the same letter are not statistically different according to Tukey test ($p < 0.05$).

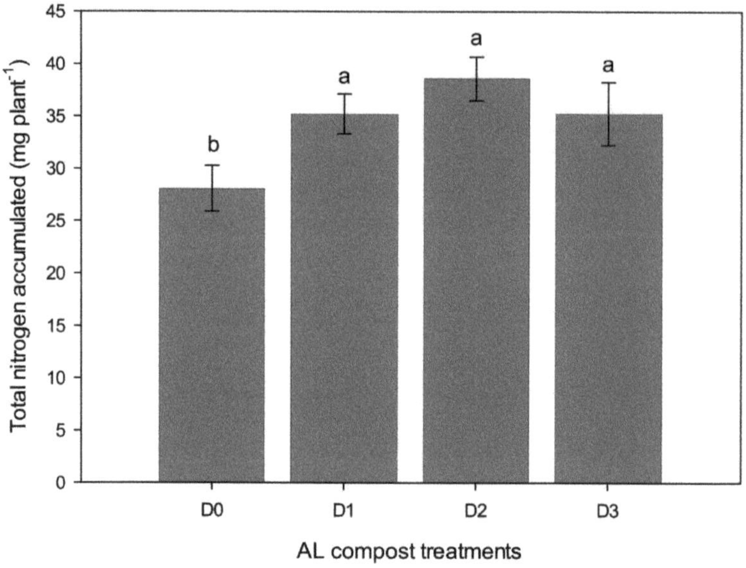

Figure 3. Total nitrogen accumulated (T_{NA}) in soybean plants grown with different doses of AL compost (D0, D1, D2 and D3). Similar letters are not statistically different according to the Tukey test ($p < 0.05$).

4. Discussion

Compost is an important source of organic matter, plant nutrients and plant growth-promoting microorganisms [13,14]. It has been demonstrated that compost is a valuable material for the production of commercial organic amendments and fertilizers [4,5], and also, an efficient alternative to peat as a carrier for rhizobia inoculants [15]. The AL compost used in our experiments presented an important organic matter content, but did not contain any rhizobia able to nodulate soybean plants (data not shown). However, as shown by the presence of nodules in the D1, D2 and D3 treatments, the inoculation of soybean with *B. diazoefficiens* helped to overcome the lack of the specific resident rhizobia in the AL compost where plants were grown. The absence of the proper rhizobia strain can be an obstacle for nodule formation, and so for the SNF. For this reason, recent studies have evaluated the improvement in nodule formation by the addition of the rhizobia strain able to nodule the legume of interest. Ulzen et al. [8] demonstrated that soybean plants inoculated with *Bradyrhizobium* strain USDA 110 and grown in soil fertilized with fertisoil increased the number of nodules by 65% compared to non-inoculated plants. Moreover, the inoculation of *Medicago sativa* (cv. Siriver) plants with the *Rhizobium* strain (RhOL1) grown in soil amended with quack grass compost doubled the number of nodules compared to plants that were not inoculated [16]. These results highlight the requirement for the addition of inoculants containing the rhizobia strains to improve the effect of compost over nodulation efficiency.

In addition to the increment in nodule number, higher total nitrogen content was observed in treatments where compost was added (D1–D3). Similar results were found in faba bean grown in calcareous soil fertilized with traditional compost and vermicompost [17]. The soil residual nitrogen, and the nitrogen content in seeds and straws in plants fertilized with the two types of compost, were two to three times higher than in the control plants [17]. The effect of compost in plant growth has also been well documented. Compost promotes plant biomass, nutrient assimilation and crop yields. Specifically, compost improves legume crops such as common beans, soybeans, etc. [18,19]. Despite these advantages, compost presented a disadvantage related to nitrogen assimilation due to it being mostly under its organic form. Compost cannot be compared with mineral nitrogen fertilization during plant growth, but when both factors are combined (e.g., compost and nitrate), a synergetic effect is usually observed [14]. That means that the supplementation of compost with an easily-nitrogen form is required for relaying optimal crop yields, especially under greenhouse conditions. According to that, SNF could be a reliable strategy to achieve this goal.

Compost can improve SNF in soybeans through several mechanisms. Compost improves soil structure by increasing aggregation and promoting the formation of stable soil aggregates [20]. This allows for better aeration and water infiltration, which in turn can enhance the growth and activity of nitrogen-fixing bacteria, as well as the formation of root nodules. Compost increases the organic matter content of the soil, which can help to enhance the soil's water-holding capacity and provide a favorable environment for nitrogen-fixing bacteria [4]. Organic matter also serves as a food source for microorganisms, further supporting their growth and activity. Compost can suppress soilborne diseases and pests by promoting the growth of beneficial microorganisms that compete with or antagonize pathogens [13]. This can lead to healthier soybean plants with a better ability to establish symbiotic relationships with nitrogen-fixing bacteria. Finally, compost can stimulate root growth in soybean plants, increasing the surface area for nutrient uptake and nodule formation [8,15,16]. A more extensive root system can provide more opportunities for nitrogen-fixing bacteria to colonize and establish a symbiotic relationship with the plant.

Our data demonstrate, for the first time, that AL compost can improve soybean plant growth, as well as their nodulation capacity. Nitrogen accumulation was also increased by compost addition compared to the control treatment (without compost). As far as we know, the contribution of AL compost to the nitrogen fertilization of plants grown under greenhouse is scarce [14]. Therefore, we can conclude that most of the nitrogen of the soybeans derived from SNF, and that AL compost contributes to the improvement of SNF of the soybean-*B. diazoefficiens* symbiosis.

5. Conclusions

In this study, the effect of AL compost on the nodulation and SNF of soybean—*B. diazoeffciens* symbiosis was evaluated. The addition of AL compost enhanced the growth of soybean plants as it was revealed by the plant biomass augmentation and plant height. Furthermore, the nodule effectiveness and the symbiotic efficiency, measured by nodular leghemoglobin content and T_{NA}, respectively, were increased by AL compost. Despite this, the study was carried out under greenhouse conditions (microcosms), the results showed clear evidence of the synergetic effect of AL compost on soybean SNF. Further research needs to be carried out in order to confirm its viability under real conditions.

Author Contributions: Conceptualization, methodology, investigation, writing original draft, visualization, G.T.; Resources, writing review and editing, Funding acquisition, S.M. and M.J.D.; Conceptualization, writing review and editing, C.V.A.-G. All authors have read and agreed to the published version of the manuscript.

Funding: This work was funded by FEDER-co-financed grants AGL2017-85676-R and PID2021-1240070B to M.J.D. and grant PID2020-114330GB-100 to S.M. (Ministerio de Ciencia e Innovación [MICINN], Spain). Grants P18-RT-1401 to M.J.D., and S.M. from the Junta de Andalucía, Spain, and Tv22 to C.V.A.-G. funded by the Ministry of Agriculture and Rural Development of Colombia and Corporación Colombiana de Investigación Agropecuaria—Agrosavia, are also acknowledged. Analysis Service of the EEZ-CSIC thanks to grant EQC2019-005472-P funded by MCIN/AEI/10.13039/501100011033 and, as appropriate, by "ERDF A way of making Europe".

Institutional Review Board Statement: Not applicable.

Informed Consent Statement: Not applicable.

Data Availability Statement: Not applicable.

Acknowledgments: G.T. thanks to Paloma Pizarro-Tobias her comments and English improvements.

Conflicts of Interest: The authors declare no conflict of interest. The funders had no role in the design of the study; in the collection, analyses, or interpretation of data; in the writing of the manuscript; or in the decision to publish the results.

References

1. Voisin, A.S.; Guéguen, J.; Huyghe, C.; Jeuffroy, M.H.; Magrini, M.B.; Meynard, J.M.; Mougel, C.; Pellerin, S.; Pelzer, E. Legumes for feed, food, biomaterials and bioenergy in Europe: A review. *Agron. Sustain. Dev.* **2014**, *34*, 361–380. [CrossRef]
2. Herridge, D.F.; Peoples, M.B.; Boddey, R.M. Global inputs of biological nitrogen fixation in agricultural systems. *Plant Soil* **2008**, *311*, 1–18. [CrossRef]
3. Poole, P.; Ramachandran, V.; Terpolilli, J. Rhizobia: From saprophytes to endosymbionts. *Nat. Rev. Microbiol.* **2018**, *16*, 291–303. [CrossRef] [PubMed]
4. Tortosa, G.; Alburquerque, J.A.; Ait-Baddi, G.; Cegarra, J. The production of commercial organic amendments and fertilisers by composting of two-phase olive mill waste ("alperujo"). *J. Clean. Prod.* **2012**, *26*, 48–55. [CrossRef]
5. Tortosa, G.; Alburquerque, J.A.; Bedmar, E.J.; Ait-Baddi, G.; Cegarra, J. Strategies to produce commercial liquid organic fertilisers from "alperujo" composts. *J. Clean. Prod.* **2014**, *82*, 37–44. [CrossRef]
6. Tondello, A.; Fasolo, A.; Marcato, S.; Treu, L.; Bonato, T.; Zanardi, W.; Concheri, G.; Squartini, A.; Baldan, B. Characterization of bacterial communities isolated from municipal waste compost and screening of their plant-interactive phenotypes. *Sci. Total Environ.* **2022**, *806*, 150592. [CrossRef] [PubMed]
7. Hayat, R.; Sheirdil, R.A.; Iftikhar-ul-Hassan, M.; Ahmed, I. Characterization and identification of compost bacteria based on 16S r-RNA gene sequencing. *Ann. Microbiol.* **2013**, *63*, 905–912. [CrossRef]
8. Ulzen, J.; Abaidoo, R.C.; Ewusi-Mensah, N.; Osei, O.; Masso, C.; Opoku, A. Organic manure improves soybean response to rhizobia inoculant and P-fertilizer in northern Ghana. *Front. Agron.* **2020**, *2*, 9. [CrossRef]
9. Vergara Cid, C.; Ferreyroa, G.V.; Pignata, M.L.; Rodriguez, J.H. Biosolid Compost with wood shavings and yard trimmings alleviates stress and improves grain quality in soybean grown in lead polluted soils. *Environ. Sci. Pollut. Res.* **2020**, *27*, 27786–27795. [CrossRef] [PubMed]
10. Santos, M.S.; Nogueira, M.A.; Hungria, M. Microbial inoculants: Reviewing the past, discussing the present and previewing an outstanding future for the use of beneficial bacteria in agriculture. *AMB Expr.* **2019**, *9*, 205. [CrossRef] [PubMed]
11. Tortosa, G.; Pacheco, P.J.; Hidalgo-García, A.; Granados, A.; Delgado, A.; Mesa, S.; Bedmar, E.J.; Delgado, M.J. Copper modulates nitrous oxide emissions from soybean root nodules. *Environ. Exp. Bot.* **2020**, *180*, 104262. [CrossRef]
12. Tortosa, G.; Parejo, S.; Cabrera, J.J.; Bedmar, E.J.; Mesa, S. Oxidative Stress Produced by Paraquat Reduces Nitrogen Fixation in Soybean-*Bradyrhizobium diazoefficiens* Symbiosis by Decreasing Nodule Functionality. *Nitrogen* **2021**, *2*, 30–40. [CrossRef]
13. Pascual, J.A.; Ceglie, F.; Tuzel, Y.; Koller, M.; Koren, A.; Hitchings, R.; Tittarelli, F. Organic substrate for transplant production in organic nurseries. A review. *Agron. Sustain. Dev.* **2018**, *38*, 35. [CrossRef]
14. Tortosa, G.; González-Gordo, S.; Ruiz, C.; Bedmar, E.; Palma, J. "Alperujo" compost improves the ascorbate (Vitamin C) content in pepper (*Capsicum annuum* L.) fruits and influences their oxidative metabolism. *Agronomy* **2018**, *8*, 82. [CrossRef]
15. Albareda, M.; Rodríguez-Navarro, D.N.; Camacho, M.; Temprano, F.J. Alternatives to peat as a carrier for rhizobia inoculants: Solid and liquid formulations. *Soil Biol. Biochem.* **2008**, *40*, 2771–2779. [CrossRef]
16. Ben-Laouane, R.; Ait-El-Mokhtar, M.; Anli, M.; Boutasknit, A.; Ait Rahou, Y.; Raklami, A.; Oufdou, K.; Wahbi, S.; Meddich, A. Green compost combined with mycorrhizae and rhizobia: A strategy for improving alfalfa growth and yield under field conditions. *Gesunde Pflanz.* **2021**, *73*, 193–207. [CrossRef]
17. Bezabeh, M.W.; Haile, M.; Sogn, T.A.; Eich-Greatorex, S. Yield, nutrient uptake, and economic return of faba bean (*Vicia faba* L.) in calcareous soil as affected by compost types. *J. Agric. Food Res.* **2021**, *6*, 100237. [CrossRef]
18. Abbott, L.K.; Macdonald, L.M.; Wong, M.T.F.; Webb, M.J.; Jenkins, S.N.; Farrell, M. Potential roles of biological amendments for profitable grain production: A review. *Agric. Ecosyst. Environ.* **2018**, *256*, 34–50. [CrossRef]

19. Rady, M.M.; Semida, W.M.; Hemida, K.A.; Abdelhamid, M.T. The effect of compost on growth and yield of *Phaseolus vulgaris* plants grown under saline soil. *Int. J. Recycl. Org. Waste Agric.* **2016**, *5*, 311–321. [CrossRef]
20. Gonzalez, R.F.; Cooperband, L.R. Compost effects on soil physical properties and field nursery production. *Compost Sci. Util.* **2002**, *10*, 226–237. [CrossRef]

Nitrogen

MDPI

Article

Environmental Impact Assessment for Animal Waste, Organic and Synthetic Fertilizers

Vassilis D. Litskas

Laboratory Ecology & Protection of Environment, School of Veterinary Medicine, Aristotle University of Thessaloniki, 54124 Thessaloniki, Greece; litskas@vet.auth.gr; Tel.: +30-2310-999967

Abstract: The use of fertilizers is of the utmost importance for food security on a global scale. However, fertilizer production and overuse may yield environmental issues. In this research, Life Cycle Assessment (LCA) was used to estimate eighteen environmental impact categories for six different fertilizer products: three synthetic (ammonium nitrate; calcium ammonium nitrate; and urea ammonium nitrate) and three organic (cattle manure; compost; and a mixture of compost and synthetic fertilizer). The processes for fertilizer production were obtained from the Agribalyse database. The system boundaries were from cradle to factory gate (or farm gate in the case of animal waste), and the impact indicators were calculated per kg of nitrogen (N). The data showed that the organo-mineral fertilizer (a mix of compost and synthetic fertilizer) had the highest environmental impact according to the results for most of the impact categories. The median values for this product regarding water consumption, fossil resource use and global warming potential were 322.5 L, 3.82 kg oil equivalent and 13.70 kg CO_2 equivalent, respectively, per kg of N. The respective values for cattle manure, for which the lowest environmental impact was observed, were 0.23 L of water, 0.002 kg oil-eq and 3.29 kg of CO_2-eq, respectively, per kg of N. Further research should focus on the determination of the impact from other stages of the life cycle (e.g., transportation and application to the field) which were not included in this work. This research could support the selection of N fertilizer in sustainable food production.

Keywords: food security; LCA; nitrogen; agribalyse; sustainability; climate change

1. Introduction

Ensuring food security for a projected worldwide population of nearly 10 billion people by 2050 is a significant problem, particularly in an era of global climate change and land degradation [1]. Fertilizers have been crucial in providing food, and a little less than half of the world's population is thought to be dependent on synthetic fertilizers [2]. On one hand, fertilizers can assist the environment because they can boost crop production [3], leading to less land being needed to produce food due to increased yields. However, they also harm the ecosystem [1,3]. Many nations overuse fertilizers, which causes nutrient runoff into ecosystems and water systems [4]. Negative environmental effects can result from fertilizer overuse, such as increasing environmental pollution [5], land degradation [6], chemical runoff [4], and biodiversity losses [7]. Through ammonia volatilization and/or biological processes such as nitrification and denitrification, a significant amount of fertilizer nitrogen (in some cases exceeding 50%) is ultimately lost to the environment [8], leading to environmental pollution and a low crop nitrogen use efficiency [9]. It is possible to reduce these losses and increase fertilizer efficiency with enhanced efficiency fertilizers (such as controlled release fertilizers) and urease and nitrification inhibitors [8], but their effectiveness varies depending on the soil type and climate [10,11], and some non-biodegradable polymers used in fertilizer coatings have negative environmental effects [9].

In Europe, as well as other parts of the world, nitrate-based fertilizers are the most widely used fertilizers [12]. The primary products are urea and urea ammonium nitrate

(UAN) aqueous solution, which are widely used in other areas of the world, and nitrate-based fertilizers such ammonium nitrate (AN) and calcium ammonium nitrate (CAN), which are well suited to most European soils and climatic conditions [12]. Ammonium sulphate and ammonium sulphate nitrate, calcium nitrate, sodium nitrate, Chilean nitrate, and anhydrous ammonia are also classified as nitrogen fertilizers. The main types of organic fertilizers are animal waste (e.g., slurries, manures), and crop residues [3]. They vary in nutritional significance, but they are typically found on farms, where the nutrients and organic carbon they contain are recycled. Animal waste, however, contains pollutants such as heavy metals and xenobiotics, as well as harmful microorganisms [13,14]. Various nutrient sources with varying physical characteristics and nutrient concentrations are covered by animal manures and slurries. Furthermore, depending on the species of cattle and the farm management technique, their nutrient content varies [9]. There are numerous so-called fertilizing products (e.g., organic fertilizers, organo-mineral fertilizers, mineral fertilizer growing mediums, and plant biostimulants). These can assist farmers in tailoring their fertilization practices to their environmental and farm conditions [3].

According to current estimates from the International Panel on Climate Change (IPCC), the manufacture of synthetic nitrogen fertilizers alone uses around 2% of the world's energy [15,16]. Nitrogen fertilizer use is a main source of anthropogenic non-CO_2 GHG emissions [15,16]. Nitrous oxide (N_2O), methane (CH_4), and carbon dioxide (CO_2) are the most common GHG emissions, although ammonia (NH_3) emissions are also significant. According to estimates [15], agriculture is responsible for more than 80% of anthropogenic N_2O emissions, and 70% of anthropogenic NH_3 emissions, which are primarily caused by the application of livestock manure and inorganic fertilizer. Relevant to livestock and organic fertilizers is 40% of anthropogenic CH_4 emissions which are primarily caused by enteric fermentation [15,16].

Even if the emission factors for synthetic fertilizers are known [9], other impact categories have not been studied. Life Cycle Assessment (LCA) is a useful approach to studying the environmental impact of fertilizers [9]. Defining similar system boundaries and functional units for different fertilizer types can lead to a holistic evaluation of the environmental impact of fertilizer production. This could support decision-making for reducing the GHG emissions from agriculture. In addition, the novelty of this research is related to the comparison of different fertilizers, in terms of their environmental impacts per kg of nitrogen added to the crops. Typically, the environmental impact is provided per kg of fertilizer, which is misleading as the nutrients are what the crops require.

This research aimed to evaluate the environmental impact of the production of six different fertilizers, organic and conventional. The objectives were (1) to apply LCA to determine key environmental impact indicators for each of the fertilizer types, and (2) to compare the environmental impacts due to fertilizer production. The results could be useful for GHG emission mitigation in the food production sector.

2. Materials and Methods

2.1. Fertilizers

The fertilizers used for this research were (1) ammonium nitrate (AN), (2) calcium ammonium nitrate (CAN), (3) urea ammonium nitrate (UAN), (4) cattle manure (MAN), (5) organic fertilizer (OF) and (6) organo-mineral fertilizer (OMF). Their basic properties are provided in Table 1.

Based on the UUID, these processes can be obtained via the Agribalyse database [17] and details about the production of these fertilizers can be found. The fertilizers were selected as proxies for globally used products (e.g., AN) and alternatives to synthetic fertilizers (e.g., cattle manure). Regarding the synthetic fertilizers, the processes used represented the production of 1 kg of ammonium nitrate (AN), calcium ammonium nitrate and urea ammonium nitrate, as nitrogen, at the factory gate. The organic and organo-mineral fertilizer processes that are available in Agribalyse are based on the work of Avadí [18,19]. The cattle manure process in Agribalyse represents the organic material produced after the fresh manure is stored in a pit, outside the farm facilities, for a typical period of six months before it is applied to the field.

Table 1. Characteristics of the fertilizers used for this research and identification code in the Agribalyse database.

Fertilizer	NPK	Agribalyse Process—UUID
Ammonium Nitrate (AN)	33.5-0-0	Ammonium Nitrate, as N (00b72b4a-0121-47f4-b0a6-8994ea3045cb)
Calcium Ammonium Nitrate (CAN)	26-0-0	Calcium Ammonium Nitrate, as N (951a335f-1630-4363-8248-552951786331)
Urea Ammonium Nitrate (UAN)	46-0-0	Urea Ammonium Nitrate, as N (c0e960a7-33bd-457d-9645-0f5baaa9412b)
Cattle manure (MAN)	0.77% N w.w.	Manure from cattle, stocked in surface or pit (ccbe7eec-2d09-464c-ab89-2fb771142a6d)
Organic fertilizer (OF)	3-2-3	Organic fertilizer, 3-2-3, granulate, packaged (fef98b5f-e133-4b75-916b-a27990b6f4b1)
Organo-mineral (OMF)	5-5-10	Organo-mineral fertilizer, 5-5-10, granulate, packaged (1178a3f6-cd65-4f27-bc03-f1deb4c437f7)

2.2. *Life Cycle Assessment*

A Life Cycle Assessment was performed to determine the environmental impact of the six fertilizer products (presented in Table 1). The system boundaries were from material production to the factory gate (or farm gate, in the case of cattle manure). The functional unit was 1 kg of nitrogen. The processes for the production of the six fertilizers were obtained from the Agribalyse database [17]. The OpenLCA software was used [20], and for the environmental impact assessment, the ReCiPe 2016 midpoint (H) method was employed [21]. The following impact categories were determined:

1. Fine particulate matter formation (kg $PM_{2.5}$ equivalent);
2. Fossil resource scarcity (kg oil-eq);
3. Freshwater ecotoxicity (kg 1,4-DCB);
4. Freshwater eutrophication (kg P equivalent);
5. Global warming (kg CO2 equivalent);
6. Human carcinogenic toxicity (kg 1,4-DCB);
7. Human non-carcinogenic toxicity (kg 1,4-DCB);
8. Ionizing radiation (kBq Co-60 equivalent);
9. Land use (m^2a crop equivalent);
10. Marine ecotoxicity (kg 1,4-DCB);
11. Marine eutrophication (kg N equivalent);
12. Mineral resource scarcity (kg Cu equivalent);
13. Ozone formation, Human health (kg NOx equivalent);
14. Ozone formation, Terrestrial ecosystems (kg NOx equivalent);
15. Stratospheric ozone depletion (kg CFC11 equivalent);
16. Terrestrial acidification kg (SO_2 equivalent);
17. Terrestrial ecotoxicity (kg 1,4-DCB);
18. Water consumption (m^3).

The above-presented indicators were expressed per kg of nitrogen, based on the nitrogen content of each of the six fertilizers: AN (33.5% N), CAN (26%), UAN (46%), OF (3%), OMF (5%) and MAN (0.77%). Therefore, the indicators provide the environmental impact per kg of N applied in the field.

2.3. Data Treatment and Statistical Analysis

Uncertainty calculation was performed using a Monte Carlo simulation (10,000 runs) in OpenLCA, according to Hsu [22]. All uncertainty distributions that were defined in the flows (e.g., inputs for machinery production) that were used in the LCA were considered for the simulation. Accordingly, for the impact categories reported for each of the 6 fertilizers, the median value and range were reported. Cluster analysis was used to group the 6 fertilizers according to the values obtained for the 18 impact categories. Euclidean distance was used for the grouping.

2.4. LCA Assumptions and Limitations

The models for fertilizer production (Table 1), as presented in the Agribalyse database, should be treated as proxies for fertilizer production. The main assumptions that were made in this work are: (1) The geographic location of the processes for the fertilizers was Central Europe (France). (2) Emissions or C storage due to soil processes, after fertilizer application, were not considered. (3) Emissions due to organic material degradation in the manure pit (storage) were not considered. (4) Nitrogen content in the cattle manure was 0.77% wet weight [23].

3. Results

In Figure 1, the impact categories—water consumption (L), fossil resource scarcity (kg oil equivalent) and global warming potential (kg CO_2-equivalent)—are presented for synthetic (a, c, e) and organic (b, d, f) fertilizers (per kg of N). In the case of water consumption and fossil resource scarcity, the lowest median values were observed for manure and the highest for organo-mineral fertilizer.

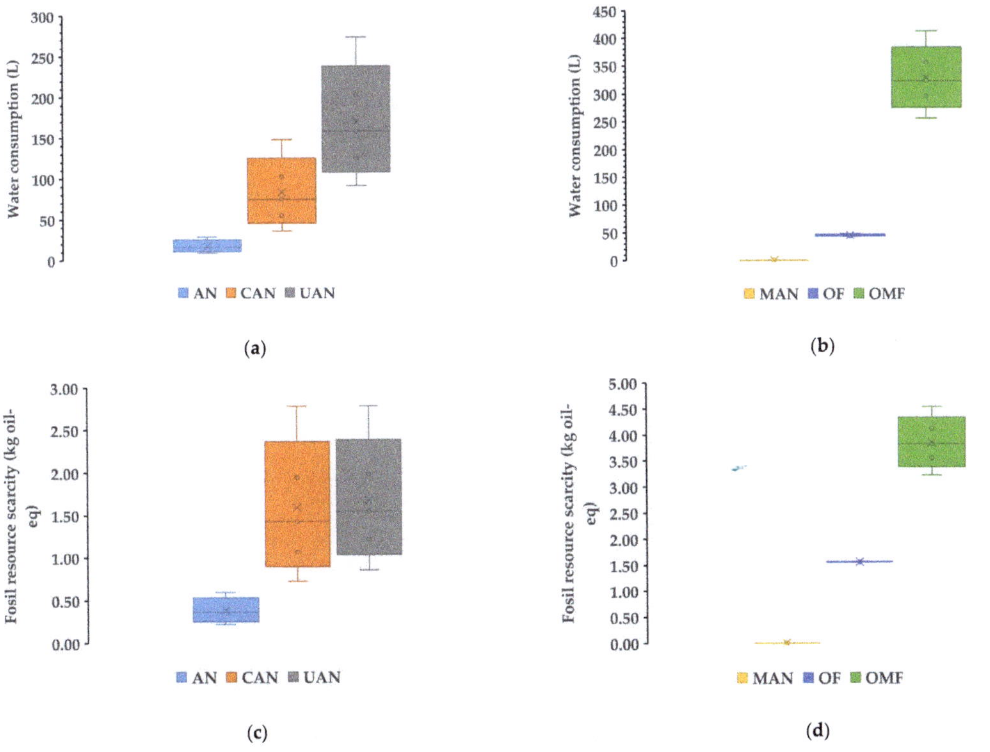

Figure 1. *Cont.*

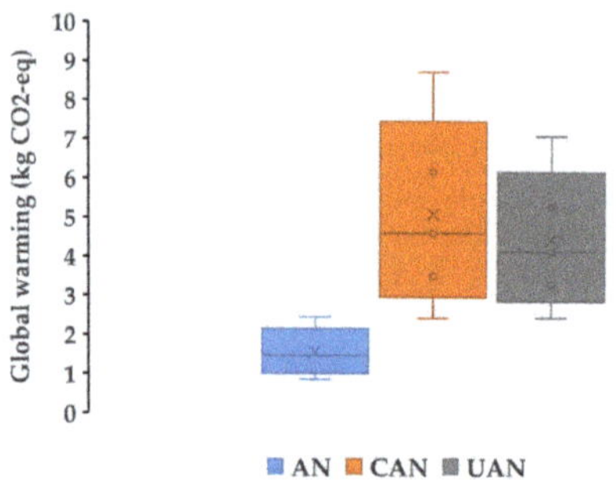

(e)

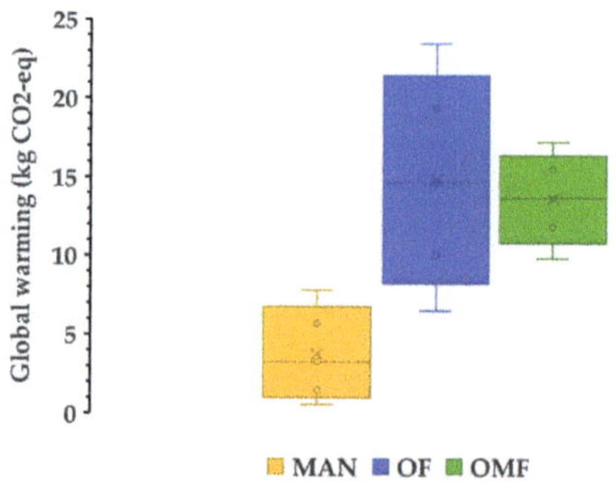

(f)

Figure 1. Water consumption (L), fossil resource scarcity (kg oil equivalent) and global warming potential (kg CO_2-eq) for synthetic (**a**,**c**,**e**) and organic (**b**,**d**,**f**) fertilizers. The boxplots provide the range, upper (5%) and lower (95%) percentile, median (line) and average (X) values.

The median values for water consumption (Figure 1a,b) due to the production of fertilizer were 16.37, 74.52, 158.58, 0.23, 45.01 and 322.54 L per 1 kg of N for AN, CAN, UAN, MAN, OF and OMF, respectively. Regarding fossil resource scarcity (Figure 1c,d) the respective median values were 0.36, 1.43, 1.55, 0.002, 1.56 and 3.82 kg oil equivalent per 1 kg of N for AN, CAN, UAN, MAN, OF and OMF, respectively. For global warming potential, median values for the six different fertilizers were 1.47, 4.59, 4.14, 3.29, 14.71 and 13.70 per 1 kg of N for AN, CAN, UAN, MAN, OF and OMF, respectively. For water consumption, the lowest median value was found for manure while the highest was observed for the organo-mineral fertilizer. In the case of resource scarcity, the same was observed. For the global warming potential (per kg of N), the lowest median value was observed for ammonium nitrate while the highest was for the organic fertilizer. In Table 2, the remaining (15) impact categories (median values; range) are presented for the six different fertilizer types.

Table 2. Median values (minimum and maximum) for impact categories (ReCiPe 2016 Midpoint H), for synthetic (AN, CAN, UAN) and organic fertilizers (MAN, OF, OMF).

Impact Category	AN	CAN	UAN	MAN	OF	OMF
Particulate matter (g $PM_{2.5}$ eq)	1.42 (0.88–2.49)	4.41 (2.29–8.53)	3.91 (2.26–6.88)	11.80 (0.93–20.80)	19.71 (7.40–31.46)	22.75 (16.41–29.17)
Freshwater ecotoxicity (g 1,4-DCB)	32.01 (17.95–63.93)	117.56 (57.44–233.47)	92.99 (49.65–173)	0.271 (0.269–0.275	126.63 (107.83–166.17)	393.82 (335.26–482.73
Freshwater eutrophication (g P eq)	0.38 (0.20–0.77)	0.95 (0.46–1.90)	0.76 (0.40–1.44)	0.003 (0.001–0.012)	1.026 (0.795–1.659)	5.164 (4.273–6.301)
Human carcinogenic toxicity (g 1,4-DCB)	39.57 (21.69–79.69)	99.12 (49.19–195.38)	78.81 (42.79–145.15)	1.112 (1.119–1.125)	111.6 (110.45–111.6)	386.61 (328.26–478.25)
Human non-carcinogenic toxicity (g 1,4-DCB)	1293.77 (725.7–2642.2)	3453.19 (1720.9–6790)	2737.64 (1497.1–5019.2)	8.13 (8.03–8.25)	5149.3 (3298.1–8778.1)	12,883 (10,884–16,186)

Table 2. *Cont.*

Impact Category	AN	CAN	UAN	MAN	OF	OMF
Ionizing radiation (Bq Co-60 eq)	65.41 (32.14–145.92)	208.62 95.62–423.25)	150.22 (71.94–303.87)	0.001 (0.0002–0.03)	2.29 (2.26–2.29)	1.71 (1.60–1.86)
Land use (m2a crop eq)	0.04 (0.02–0.11)	0.06 (0.03–0.11)	0.04 (0.02–0.08)	0.007 (0.001–0.012)	6.07 (4.77–8.31)	3.32 (2.77–4.14)
Marine ecotoxicity (g 1,4-DCB)	45.38 (25.87–89.72)	170.12 (83.35–337.33)	135.57 (72.45–252.33)	0.387 (0.383–0.398)	144.69 (138.33–157.88)	552.04 (466.7–679.1)
Marine eutrophication (mg N eq)	81.16 (45.01–142.87)	454.27 (219.65–907.02)	519.56 (276.79–965.38)	0.0001 (0.0001–0.005)	0.003 (0.002–0.007)	0.002 (0.001–0.003)
Mineral resource scarcity (g Cu eq)	2.96 (1.76–5.63)	14.31 (7.06–28.31)	11.39 (6.07–20.97)	0.090 0.089–0.091)	21.08 (15.58–35.18)	3966.34 (2645.5–6165.4)
Ozone formation, human health (g NOx eq)	3.09 (1.92–5.51)	5.86 (3.27–10.85)	5.38 (3.29–9.17)	0.036 (0.028–0.055)	18.55 (17.58–20.51)	23.92 (20.74–28.26)
Ozone formation, terrestrial ecosystems (g NOx eq)	3.13 (1.95–5.58)	5.98 (3.34–11.08)	5.52 (3.38–9.40)	0.037 (0.029–0.056)	18.94 (17.97–20.90)	24.44 (21.21–28.86)
Stratospheric ozone depletion (mg CFC11 eq)	11.88 (5.21–24.97)	26.33 (14.77–44.33)	26.511 (15.21–49.28)	48.88 (10.73–111.22)	211.25 (14.80–404.78)	84.25 (12.85–154.75)
Terrestrial acidification (g SO_2 eq)	4.90 (2.97–8.59)	19.13 (9.58–37.58)	15.30 (8.55–27.58)	96.28 (7.51–201.3)	127.6 (28.11–224.13)	94.5 (53.87–135.61)
Terrestrial ecotoxicity (kg 1,4-DCB)	3.62 (2.69–5.79)	12.21 (6.87–22.56)	10.35 (6.19–17.87)	0.026 (0.025–0.050)	22.19 (22.03–24.53)	42.47 (36.06–52.63)

In Figure 2, the comparison of the six fertilizers for the 18 impact categories is presented. Based on the values provided in Figure 1 and Table 2, the lowest median value is coloured green and the highest red. In this sense, fertilizers with a green colour for most of the impact categories are better in terms of environmental impact. It can be observed that in comparison to the other (5) fertilizer types, the organo-mineral fertilizer has the highest values (red colour) in the most impact categories. Therefore, this fertilizer has the highest environmental impact.

In Figure 3, cluster analysis was used to group the six fertilizers according to the median values for the 18 impact categories. Two groups were observed: one containing ammonium nitrate, calcium ammonium nitrate, manure, and urea, and the second containing organic and organo-mineral fertilizers. This means that these groups have similar values in respect of the indicators' values.

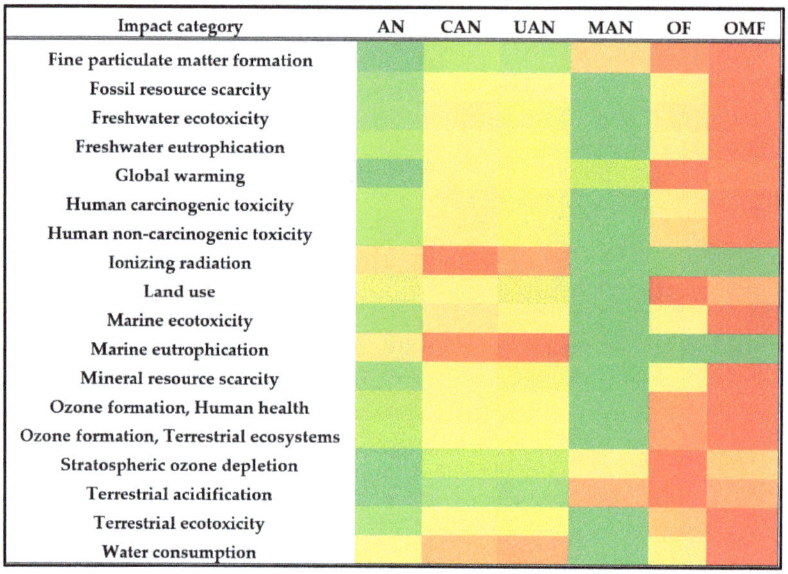

Figure 2. Comparison of the six fertilizers for 18 environmental impact categories. The lowest value per impact category is coloured green while the highest is red. AN: ammonium nitrate; CAN: calcium ammonium nitrate; UAN: urea ammonium nitrate MAN: cattle manure; OF: organic fertilizer; OMF: organo-mineral fertilizer.

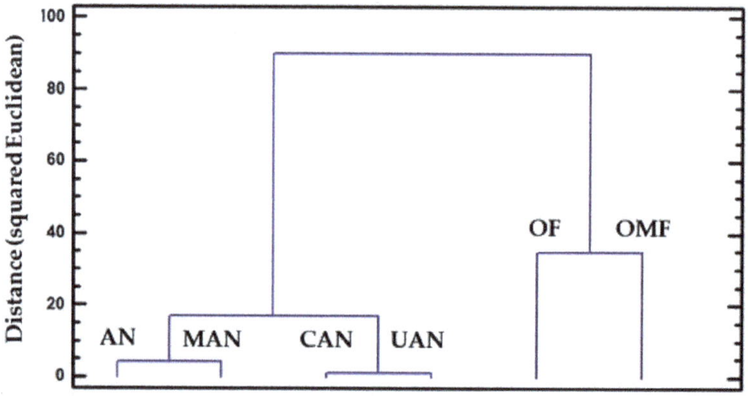

Figure 3. Cluster analysis for the grouping of six fertilizers, according to the values of the 18 impact categories. Two clusters appear, one containing AN, MAN, CAN and UAN and the other OF and OMF.

4. Discussion

Life Cycle Assessment was performed on six fertilizers typically used in global agriculture [9]. The system boundaries for the LCA were from cradle to factory (or farm) gate. The impact categories of the ReCiPe (2016 Midpoint H) method captured the effects on resource use (e.g., fossil resource scarcity; water consumption; land use), ecotoxicity (e.g., marine, and terrestrial), atmosphere (e.g., fine particulate matter formation; ionizing radiation, stratospheric ozone), human health (e.g., non-carcinogenic toxicity), climate change (e.g., global warming potential), eutrophication and acidification. The functional unit selected for the assessment was 1 kg of N. This means that the environmental impact of six different fertilizers (products) was compared based on providing 1 kg of nitrogen to the crop.

One of the most recognized environmental impact indicators is global warming potential (or carbon footprint), expressed as kg of CO_2 equivalent per functional unit [24]. The results showed that the range for the fertilizers was 1.47–14.71 kg CO_2-eq/kg N. The lowest value was observed for ammonium nitrate, while the highest values were observed for the organic (3-2-3) and organo-mineral fertilizers. Ammonium nitrate had a lower global warming potential even when compared to manure. This is due to the low N content of cattle manure [23]. It should be stressed that these values only account for fertilizer production. Fertilizer application in the field leads to greenhouse gas emissions (e.g., N_2O) [24] which were not taken into consideration in this research. It should also be stressed that C storage in the soil, due to organic material application, was not taken into consideration. The addition of organic material in the soil could decrease greenhouse gas emissions and benefit soil functions [25]. The results of this study, regarding the CO_2 emissions of synthetic N fertilizers, are per the review of Walling and Vaneeckhaute [9], which report values of 1.3–7.2 kg CO_2-eq/kg of N for urea and ammonium nitrate fertilizers produced in several countries/continents. However, this work provides the emissions of organic fertilizers per ton of material produced, thus making direct comparison difficult. Our results stress the importance of making similar comparisons among N fertilizers of different types before recommending several types to be broadly used. Additionally, LCA is a useful tool for holistic assessment of the environmental impact of fertilizers (Figures 2 and 3).

Besides the CO_2 emissions, data in the literature for the other impact categories (Table 2) remain scarce. In this study, manure, which is a byproduct of cattle-raising, showed a smaller environmental impact in comparison to the other five fertilizer types (Figure 1; Table 2). On the other hand, it was observed that per kg of nitrogen, the organic fertilizers, especially the organic and organo-mineral fertilizers, have higher impacts compared to the conventional in terms of particulate matter, ecotoxicity, eutrophication, human health, and stratospheric ozone depletion (Table 2; Figure 2). The higher impacts for many of the categories for the organic and organo-mineral fertilizers are related to the inputs used for their production and the low N content. This results in increased impact when data are expressed per kg of N and not per kg or ton of product, which is the typical functional unit of organic fertilizers in LCA studies [9,18]. For instance, electricity and fuel are used for the production and packaging of both organic and synthetic fertilizers. The lower N content of organic fertilizers (5% for organo-mineral and 3% for compost) compared to the respective content in synthetic fertilizers (e.g., 33.5% for ammonium nitrate), results in higher values of the indicators, such as particulate matter emissions, when the impact is expressed per kg of N.

Further research should include the other life cycle stages (e.g., transportation, application to the field, soil processes) to obtain a complete picture of fertilizer use impacts on the environment. Transportation over large distances (e.g., aeroplanes, trains, ships) increases the environmental impact of products due to fuel consumption [26–28]. To reduce the environmental impact, fertilizers produced and distributed at the national scale have a lower impact due to decreased values in almost all categories of the LCA [9].

The results of this work favour the use of ammonium nitrate and cattle manure (Figure 2) in comparison to the other four fertilizers, due to lower values in many of the impact categories of the LCA study. It should be highlighted that the nitrogen release from manure is much slower than from the ammonium nitrate fertilizer [29,30]. In the long term, however, the use of organic material supports soil health [25,30].

Finally, LCA for the determination of the environmental impact of fertilizers could be a useful tool for the selection of products that support food security with the lowest environmental impact. In this case, the impacts should be provided per kg of nutrient and not per kg of the final product.

5. Conclusions

The use of fertilizers is of the utmost importance for food security on a global scale. However, fertilizer production and overuse could yield environmental issues. This research shows that the fertilizers with the lowest environmental impacts were cattle manure, followed by ammonium nitrate. Our work stresses that the comparison for the values of the indicators should be carried out per kg of nutrient, not per kg of product, as is typically used in LCA research. According to this study, organic materials such as animal waste are promising to ensure food security in combination with low environmental impact. The use of packaged organic fertilizers could result in a higher environmental impact than that of simpler synthetic fertilizers, such as ammonium nitrate. Further work is required to consider LCA stages, such as transportation, soil functions, and N_2O release due to N application.

Funding: This research received no external funding.

Data Availability Statement: Not applicable.

Conflicts of Interest: The author declares no conflict of interest.

References

1. Wang, X. Managing Land Carrying Capacity: Key to Achieving Sustainable Production Systems for Food Security. *Land* **2022**, *11*, 484. [CrossRef]
2. Liang, G. Nitrogen Fertilization Mitigates Global Food Insecurity by Increasing Cereal Yield and Its Stability. *Glob. Food Secur.* **2022**, *34*, 100652. [CrossRef]
3. Bebber, D.P.; Richards, V.R. A Meta-Analysis of the Effect of Organic and Mineral Fertilizers on Soil Microbial Diversity. *Appl. Soil Ecol.* **2022**, *175*, 104450. [CrossRef]
4. Xu, Y.; Elbakidze, L.; Yen, H.; Arnold, J.G.; Gassman, P.W.; Hubbart, J.; Strager, M.P. Integrated Assessment of Nitrogen Runoff to the Gulf of Mexico. *Resour. Energy Econ.* **2022**, *67*, 101279. [CrossRef]
5. Badrzadeh, N.; Samani, J.M.V.; Mazaheri, M.; Kuriqi, A. Evaluation of Management Practices on Agricultural Nonpoint Source Pollution Discharges into the Rivers under Climate Change Effects. *Sci. Total Environ.* **2022**, *838*, 156643. [CrossRef]
6. Hossain, M.E.; Islam, M.S.; Sujan, M.H.K.; Tuhin, M.M.-U.-J.; Bekun, F.V. Towards a Clean Production by Exploring the Nexus between Agricultural Ecosystem and Environmental Degradation Using Novel Dynamic ARDL Simulations Approach. *Environ. Sci. Pollut. Res.* **2022**, *29*, 53768–53784. [CrossRef]
7. Wang, Y.; Lu, Y.; Yuan, J.; He, G. Evaluating the Risks of Nitrogen Fertilizer-Related Grain Production Processes to Ecosystem Health in China. *Resour. Conserv. Recycl.* **2022**, *177*, 105982. [CrossRef]
8. Besen, M.R.; Minato, E.A.; Cassim, B.M.A.R.; Macon, C.R.; de Oliveira Zampar, E.J.; Inoue, T.T.; Batista, M.A. The Efficiency of Stabilized Fertilizers in Reducing Ammonia Loss Is Influenced by Liming Method. *Commun. Soil Sci. Plant Anal.* **2022**, *53*, 1552–1564. [CrossRef]
9. Walling, E.; Vaneeckhaute, C. Greenhouse Gas Emissions from Inorganic and Organic Fertilizer Production and Use: A Review of Emission Factors and Their Variability. *J. Environ. Manag.* **2020**, *276*, 111211. [CrossRef]
10. Vangeli, S.; Posse, G.; Beget, M.E.; Otero Estrada, E.; Valdettaro, R.A.; Oricchio, P.; Kandus, M.; Di Bella, C.M. Effects of Fertilizer Type on Nitrous Oxide Emission and Ammonia Volatilization in Wheat and Maize Crops. *Soil Use Manag.* **2022**, *38*, 1519–1531. [CrossRef]
11. Fan, D.; He, W.; Smith, W.N.; Drury, C.F.; Jiang, R.; Grant, B.B.; Shi, Y.; Song, D.; Chen, Y.; Wang, X.; et al. Global Evaluation of Inhibitor Impacts on Ammonia and Nitrous Oxide Emissions from Agricultural Soils: A Meta-Analysis. *Glob. Change Biol.* **2022**, *28*, 5121–5141. [CrossRef]
12. Fertilizers Europe Types of Fertilizer. Available online: https://www.fertilizerseurope.com/fertilizers-in-europe/types-of-fertilizer/ (accessed on 22 November 2022).
13. Sui, Q.; Zhang, J.; Chen, M.; Wang, R.; Wang, Y.; Wei, Y. Fate of Microbial Pollutants and Evolution of Antibiotic Resistance in Three Types of Soil Amended with Swine Slurry. *Environ. Pollut.* **2019**, *245*, 353–362. [CrossRef]
14. Xu, Y.; Li, J.; Zhang, X.; Wang, L.; Xu, X.; Xu, L.; Gong, H.; Xie, H.; Li, F. Data Integration Analysis: Heavy Metal Pollution in China's Large-Scale Cattle Rearing and Reduction Potential in Manure Utilization. *J. Clean. Prod.* **2019**, *232*, 308–317. [CrossRef]
15. IPCC AR5 Climate Change 2014: Mitigation of Climate Change 2014. Available online: https://www.ipcc.ch/report/ar5/wg3/ (accessed on 29 November 2022).
16. IPCC Climate Change 2022: Mitigation of Climate Change. Available online: https://www.ipcc.ch/report/ar6/wg3/ (accessed on 29 November 2022).
17. Lemberger, P. Agribalyse 3.0.1—An Update! | OpenLCA.Org. Available online: https://www.openlca.org/agribalyse-3-0-1-update/ (accessed on 29 November 2022).

18. Avadí, A.; Aissani, L.; Pradel, M.; Wilfart, A. Life Cycle Inventory Data on French Organic Waste Treatments Yielding Organic Amendments and Fertilisers. *Data Brief* **2020**, *28*, 105000. [CrossRef]
19. Avadí, A. Screening LCA of French Organic Amendments and Fertilisers. *Int. J. Life Cycle Assess.* **2020**, *25*, 698–718. [CrossRef]
20. OpenLCA.Org | OpenLCA Is a Free, Professional Life Cycle Assessment (LCA) and Footprint Software with a Broad Range of Features and Many Available Databases, Created by GreenDelta since 2006. Available online: https://www.openlca.org/ (accessed on 20 November 2022).
21. Huijbregts, M.A.; Steinmann, Z.J.; Elshout, P.M.; Stam, G.; Verones, F.; Vieira, M.D.M.; Hollander, A.; Zijp, M.; van Zelm, R. *ReCiPe 2016: A Harmonized Life Cycle Impact Assessment Method at Midpoint and Endpoint Level Report I: Characterization*; Rijksinstituut voor Volksgezondheid en Milieu RIVM: Utrecht, The Netherlands, 2016.
22. Hsu, D.D. Life Cycle Assessment of Gasoline and Diesel Produced via Fast Pyrolysis and Hydroprocessing. *Biomass Bioenergy* **2012**, *45*, 41–47. [CrossRef]
23. Loyon, L. Overview of Animal Manure Management for Beef, Pig, and Poultry Farms in France. *Front. Sustain. Food Syst.* **2018**, *2*, 36. [CrossRef]
24. Litskas, V.; Mandoulaki, A.; Vogiatzakis, I.N.; Tzortzakis, N.; Stavrinides, M. Sustainable Viticulture: First Determination of the Environmental Footprint of Grapes. *Sustainability* **2020**, *12*, 8812. [CrossRef]
25. Ioannidou, S.C.; Litskas, V.D.; Stavrinides, M.C.; Vogiatzakis, I.N. Linking Management Practices and Soil Properties to Ecosystem Services in Mediterranean Mixed Orchards. *Ecosyst. Serv.* **2022**, *53*, 101378. [CrossRef]
26. Yao, X.; Shah, W.U.H.; Yasmeen, R.; Zhang, Y.; Kamal, M.A.; Khan, A. The Impact of Trade on Energy Efficiency in the Global Value Chain: A Simultaneous Equation Approach. *Sci. Total Environ.* **2021**, *765*, 142759. [CrossRef]
27. Adedoyin, F.F.; Alola, A.A.; Bekun, F.V. The Alternative Energy Utilization and Common Regional Trade Outlook in EU-27: Evidence from Common Correlated Effects. *Renew. Sustain. Energy Rev.* **2021**, *145*, 111092. [CrossRef]
28. Dou, G.; Choi, T.-M. Does Implementing Trade-in and Green Technology Together Benefit the Environment? *Eur. J. Oper. Res.* **2021**, *295*, 517–533. [CrossRef]
29. Lawrencia, D.; Wong, S.K.; Low, D.Y.S.; Goh, B.H.; Goh, J.K.; Ruktanonchai, U.R.; Soottitantawat, A.; Lee, L.H.; Tang, S.Y. Controlled Release Fertilizers: A Review on Coating Materials and Mechanism of Release. *Plants* **2021**, *10*, 238. [CrossRef]
30. Bergstrand, K.-J. Organic Fertilizers in Greenhouse Production Systems—A Review. *Sci. Hortic.* **2022**, *295*, 110855. [CrossRef]

Nitrogen

MDPI

Article

Using the Haney Soil Test to Predict Nitrogen Requirements in Winter Wheat (*Triticum aestivum* L.)

Debankur Sanyal [1] and Christopher Graham [2,*]

[1] Department of Environmental Science, College of Agriculture and Life Science, University of Arizona, Tucson, AZ 85721, USA; dsanyal@arizona.edu

[2] Department of Agronomy, Horticulture and Plant Science, South Dakota State University Extension West River Ag Center, Rapid City, SD 57703, USA

* Correspondence: christopher.graham@sdstate.edu

Abstract: Managing nitrogen (N) is one of the of the biggest challenges in achieving environmental and economic sustainability in the agroecosystem. As N fertilizer prices have increased significantly, farmers are considering a revised N recommendation to optimize crop production, while addressing negative environmental impacts of excess N in water bodies. This study analyzes the accuracy of using the Haney Soil Test (HST) to predict the N requirement (HSTNR) of winter wheat (*Triticum aestivum* L.) in a semi-arid climate. The accuracy of the HST to predict the economically optimum N rate (EONR) was dependent on in-season precipitation. In drought conditions, the HSTNR was 33 kg N ha^{-1} lower on average than the EONR. Conversely, in wetter years, the HSTNR was 35 kg N ha^{-1} higher than the EONR. Net return was approximately USD 19 ha^{-1} lower than that with the EONR under both precipitation scenarios. Similar differences were found for protein content. There was a strong correlation between soil respiration and the soil health calculation, within the HST, and the difference between the net return on yield from the HSTNR and the EONR yield. These indicators may serve as useful metrics for formulating soil health-based N recommendations in winter wheat. However, in drought-prone areas, the HSTNR may significantly underpredict the EONR in many years due to an overestimation of N mineralization.

Keywords: nitrogen recommendation; economically optimum nitrogen rate (EONR); Haney soil test nitrogen requirement (HSTNR); protein content; soil health

1. Introduction

Mitigating agricultural nitrogen (N) pollution is one of the major environmental concerns in the twenty-first century [1,2]. Additionally, as N fertilizer prices increase, growers are considering a revision of their current N recommendations to use an optimum N rate to enhance N use efficiency [3]. N fertilizer recommendations must bridge the shortfall between N provided by the soil and that required by the plant. To be most effective in improving current N recommendations, new methodologies need to accurately incorporate N provided by the soil and optimize economic profits. Some current N fertilizer recommendations incorporate economics to calculate an economically optimal N rate (EONR) by including the cost of fertilizer, the price of the commodity, and the response curve of the crop to N additions into a rate recommendation—independent of soil conditions [4,5].

Conversely, the Haney Soil Test (HST) was developed to include a suite of lab-based methods to predict the amount of N supplied to the crop from the soil, which could improve the accuracy of N fertilizer recommendations [6–8]. Traditional soil testing uses potassium chloride (KCl) to extract plant-available N from the soil, while the HST utilizes an extractant comprising organic acids designed to mimic the mechanisms used by plant roots to acquire nutrients at a solution pH similar to that observed in the soil [8]. Additionally, the HST approach assesses water-extractable organic carbon and organic N in the soil along with soil respiration as a 1-day CO_2 evolution to estimate the microbial contribution to plant-available N [7,9,10].

Initial evaluations of the HST have primarily focused on the southern and central United States, with a particular focus on corn (*Zea mays* L.) production [11,12]. It is less clear how well the HST predicts N supply in more northern climates with small grains. Wheat (*Triticum aestivum* L.) differs from many other commodity crops in that protein content is measured along with yield. Because wheat protein is well correlated with N fertility, one concern from producers is that reducing N fertilizer will reduce protein, even if yields remain adequate. The objective of the current study was to compare two approaches in estimating the N requirement of wheat: an economics-based approach that does not consider soil N mineralization with the HST method that specifically measures soil N supply. Thus, our study evaluated the accuracy of the HST for N recommendations in winter wheat in a semi-arid climate where N limitations often reduce yield and protein content due to slow rates of mineralization.

2. Materials and Methods

2.1. Experimental Design and Setup

This study took place over three growing seasons (2017–2019) at three locations (Sturgis, Vivian and Wall) in western South Dakota. However, the field sites where experimental plots were laid out were different for different years, but had similar soil types and were managed similarly. Initial soil test information is given in Table 1. Two site-years (Vivian, 2018 and Sturgis, 2019) were omitted from the study due to unresponsiveness to N and flooding, respectively. As excess precipitation coupled with soil texture created an anaerobic condition, it impacted the N dynamics; therefore, the research dataset could not be used to maintain a logical interpretation of the overall research outcomes. Hard red winter wheat (HRW) was planted in the Fall on 25 cm row spacing with a no-till grain drill (Model 750, John Deere Co., Moline, IL, USA) at a population of 297 pure live seeds m^{-2}. In 2017, urea granules were broadcast at rates of 0, 45, 90, 134, 180 and 224 kg N ha^{-1} at planting. In 2018 and 2019, urea was applied at the 'Feekes 4' growth stage of wheat, also known as 'green-up' when active tiller formation occurs [13]. These differences in N application dates are assumed to be negligible, as previous research suggests insignificant differences in wheat production due to early-season application timing in semi-arid climates [14].

Table 1. Initial values for soil organic matter (SOM), pH, electrical conductivity (EC), organic nitrogen released (ON release), total Haney available nitrogen (N), and soil health calculations (SHC) using Haney Soil Test protocols.

Site	SOM %	pH	EC	ON Release 0–15 cm	ON Release 15–60 cm	Total Haney Available N	SHC 0–15 cm	SHC 15–60 cm
			dS m^{-1}		Kg N ha^{-1}			
Sturgis 2017	3.1	6.8	0.5	22	-	33	16.5	-
Vivian 2017	4.4	7.3	0.8	7	-	33	5.5	-
Wall 2017	1.8	6.4	0.3	21	-	63	11.9	-
Sturgis 2018	3.6	6.1	0.3	30	25	67	15.4	10.7
Wall 2018	2.7	6.9	1.0	50	27	92	15.6	7.9
Vivian 2019	2.6	6.5	0.3	25	22	56	7.6	6.1
Wall 2019	5.2	7.8	0.7	36	35	116	22.4	20.4

All sites used the same randomized complete block design with four replications. Plots consisted of six planted rows with a total area of 14 m^2. At harvest maturity, the front and back of each plot were trimmed back and only the middle four rows were harvested to avoid edge effects (total harvested area = 11.6 m^2). Plots were harvested with a small plot combine (Wintersteiger, Salt Lake City, UT, USA), and yield was adjusted to 130 g H_2O kg-grain^{-1}. A subsample from each replication was taken for further protein analysis using whole grain near-infrared transmittance (Foss, Eden Prairie, MN, USA). Each protein value was a composite of the measured light transmittance through the grain from ten sub-samples.

2.2. Soil Sampling and Haney N Recommendations

Soil samples for N analysis were gathered in early spring prior to urea application. Each sample consisted of a composite of 8–12 samples per plot. In 2017, N recommendations were based on a 0–15 cm sample only per HST recommendations. In 2018 and 2019, samples were taken from 0–15 cm and 15–60 cm depths to align more closely with standard N testing for wheat in South Dakota [15].

Haney soil test protocols were conducted at the USDA-ARS Grassland Soil and Water Research Laboratory on air-dried samples previously passed through a 2-mm sieve. All analyses in the HST protocol were based on previously described methods, which together were combined to create a N recommendation (HSTNR) based on estimated plant-available N [6–8,16–18]. The final HSTNR was determined using guidelines established by the South Dakota State Experiment Station [15]:

$$\text{Haney Soil Test Fertilizer Recommendation (HSTNR)} = 2.5 \times \text{YG} - \text{HSTN}$$

where YG is the yield goal in bu ac^{-1} (where 1 bushel of wheat weighs 60 lbs.; therefore, 1 bu acre^{-1} is equal to 67.25 kg ha^{-1}; we converted YG to kg ha^{-1}), '2.5' is a numeric factor, derived from N rate studies in South Dakota [15]; HSTN is the estimated HST plant-available N (kg ha^{-1}) from either a 0–15 cm or 0–60 cm soil sample.

To determine how well the HSTNR approximates the EONR, the YG was determined as the yield corresponding to the EONR of each site and year. The maximum return to N (MRTN) approach (the N rate that maximizes the economic return to N application) was used to determine the EONR and corresponding yield [4,19,20]. This approach predicts N rates based on replicated N response trials, independent of starting soil N values. However, the maximum grain accumulation is the same with or without accounting for soil N, hence the validity of using this value as the YG. Based on the similarity of fit statistics but significantly different join point, both a quadratic plateau and a linear plateau curve were used to fit the yield response for each N trial conducted (Table 2). The linear plateau model is defined by Equations (1) and (2):

$$Y = a + bX \; if \; X < J \tag{1}$$

$$Y = P \; if \; X \geq J \tag{2}$$

where Y is the yield of grain (kg ha^{-1}) and X is the N application rate (kg ha^{-1}); a (intercept), b (linear coefficient), J (join point, occurring at the intersection of the linear and the plateau lines), and P (plateau yield) are constants obtained by fitting the model to the data. The quadratic-plus-plateau model is defined by Equations (3) and (4):

$$Y = a + bX + cX^2 \; if \; X < J \tag{3}$$

$$Y = P \; if \; X \geq J \tag{4}$$

where Y is the yield of grain (kg ha^{-1}) and X is the N application rate (kg ha^{-1}); a (intercept), b (linear coefficient), c (quadratic coefficient), J (join point, occurring at the intersection of the quadratic and the plateau lines), and P (plateau yield) are constants obtained by fitting the model to the data [21]. The join point (J) is considered to be the point at which increasing the fertilizer rate is no longer effective at increasing yield. The plateau (P) is the value at which yield is maximized for the site.

Table 2. Regression statistics for the linear plateau (LP) and quadratic plateau (QP) models and plateau N rate and corresponding grain yield at the hinge point, where R^2 is the coefficient of determination, RMSE is root mean square error, and AIC is Akaike information criterion.

Location	Year	LP R^2	QP R^2	LP RMSE	QP RMSE	LP AIC	QP AIC	LP Plateau N	QP Plateau N	LP Plateau Grain Yield	QP Plateau Grain Yield
									kg ha^{-1}		
Sturgis	2017	0.81	0.80	241	248	339	341	134	204	2481	2523
Sturgis	2018	0.60	0.60	424	425	366	367	72	109	3481	3487
Vivian	2017	0.81	0.81	322	324	353	354	101	155	3531	3558
Vivian	2019	0.80	0.81	218	212	335	333	106	159	4304	4318
Wall	2017	0.50	0.51	495	495	375	374	95	120	3278	3267
Wall	2018	0.70	0.71	414	410	336	335	111	162	5446	5458
Wall	2019	0.87	0.89	402	372	364	360	101	136	4641	4624
Average		0.73	0.73	359	355	353	352	103	149	3880	3891

Using the equation derived from fitting each model (i.e., site) (Figure 1), the net return can be calculated and plotted as the increase in yield multiplied by the grain price at a given N rate, minus the cost of that same amount of N. In this paper, we used a ratio of 5.55 for cost of fertilizer N: price of wheat grain (equivalent to USD 1.11 kg N^{-1} (USD 0.50 lb N^{-1}) and USD 0.202 kg grain^{-1} (USD 5.50 bu grain^{-1})). The EONR is the N rate at which this return is maximized.

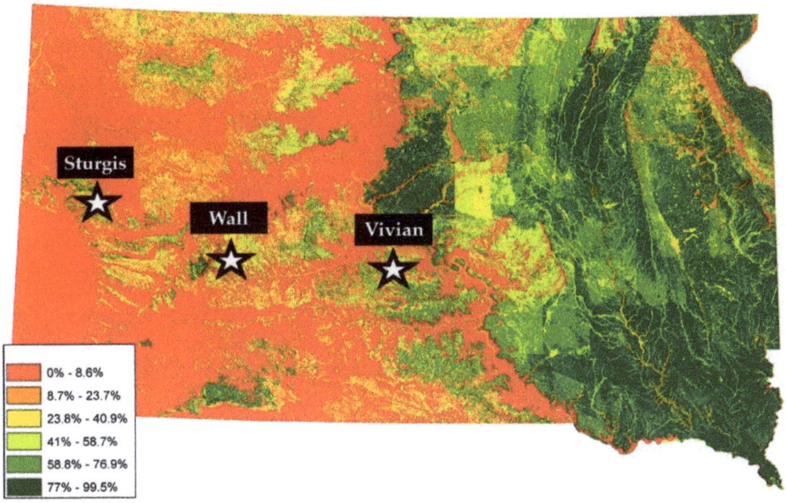

Figure 1. Study locations in South Dakota on a map depicting predicted cropland intensities by soil units in 2017 as proposed by the United States Department of Agriculture (USDA).

All statistical analyses were conducted in the R statistical package (R Core Development Team, 2014). Analysis examined both differences between the HSTNR and EONR using either the plant-available N estimated from 0–15 cm soil depth or as the total plant-available N from the 0–60 cm soil depth where available (2018 and 2019).

3. Results and Discussion

Precipitation during the three growing seasons strongly influenced the differences between HSTNR and MRTN (Figures 2 and 3). The 2017 growing season experienced severe drought, receiving 60–70% of the long-term average precipitation (https://climate.sdstate.edu/: accessed on 1 July 2021). During this year, the HSTNR was underestimated by an average of 35 kg ha^{-1} for the quadratic plateau yield curve and 30 kg ha^{-1} for the linear plateau estimation using a 0–15 cm soil sample. Conversely, in wetter years (2018 and 2019, https://climate.sdstate.edu/: accessed on 1 June 2021), the HSTNR was on average 29 kg ha^{-1} and 41 kg ha^{-1} higher than the MRTN for the quadratic plateau and linear plateau curves, respectively (Table 3). It is important to note that the HST is a lab-based method that measures the CO_2 flux following soil rewetting to predict the potential availability of N to the plant through mineralization [11]. This is due in large part to the strong correlation between soil respiration and water-soluble organic N [7]. Further, soils in more arid environments tend to exhibit a lower CO_2 burst upon rewetting, which is presumed to be due to a higher frequency of drying/rewetting cycles and an increasing trend of available carbon depletion [22]. However, where the rewetting cycle is limited (e.g., drought), so too is respiration, which presumes a decreased rate of N mineralization and an overprediction of N available to the plant.

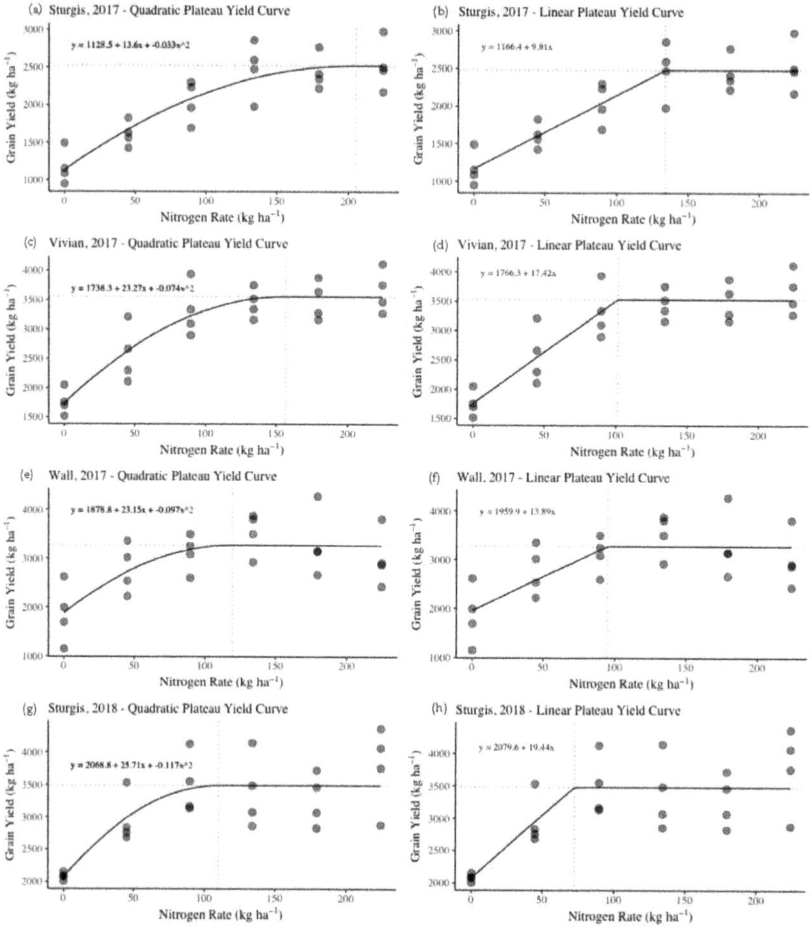

Figure 2. *Cont.*

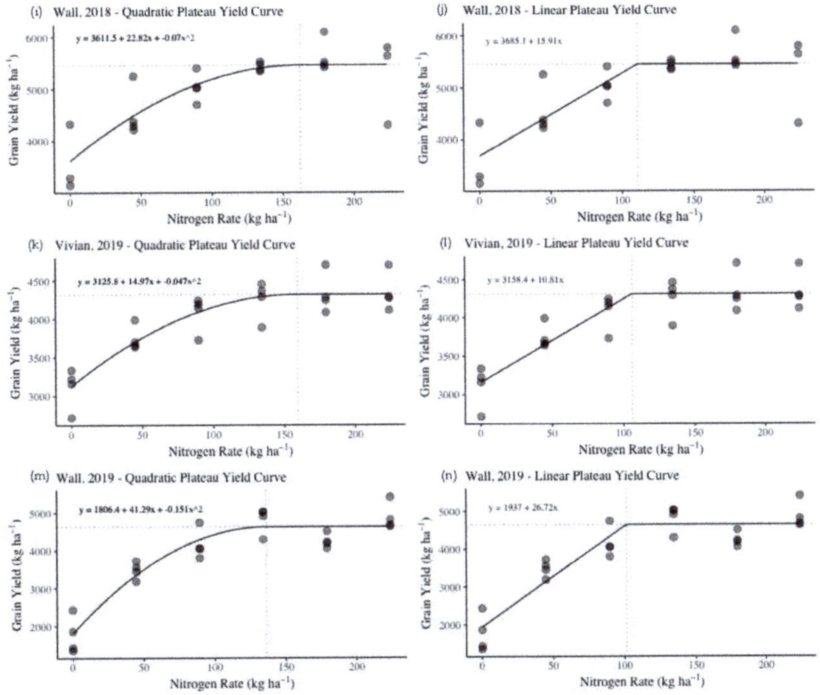

Figure 2. Graphical representation of quadratic plateau yield curve (left column) and linear plateau yield curve (right column) depicting relationships between grain yield (kg ha^{-1}) on the Y-axis and nitrogen rate (kg ha^{-1}) on the X-axis at the three study locations over three years between 2017–2019. The circles in each graph represent the yield measurements (replications) for each N rate. The dotted horizontal and vertical lines represent the join point (N rate) for each site/year and plateau (grain yield). Equations in each graph represent Equations (1) and (3) depicting each site's unique coefficients.

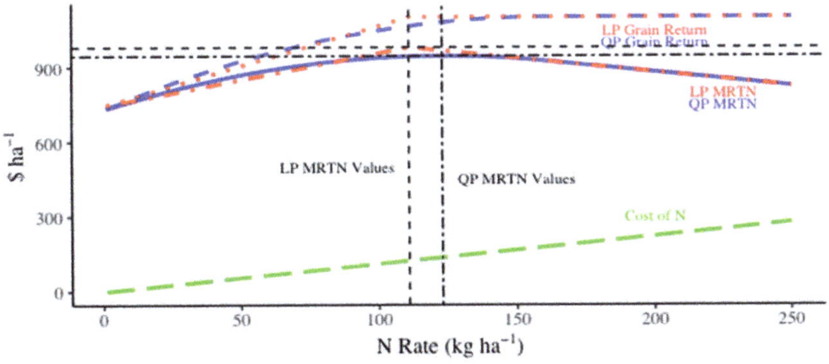

Figure 3. A graphical representation of the maximum return to N (MRTN) approach. The dotted red line and dashed blue line represent a typical linear plateau (LP) and quadratic plateau (QP) grain yield, respectively. The dot-dash red line and solid blue line represent the LP and QP MRTN. The MRTN curve

is derived from the difference between the respective yield curve and the cost of N fertilizer (represented as a green dashed line). The MRTN curve peaks, as represented by the dotted and dot-dash black lines, and then declines with increasing N rate due to the additional cost of N without additional return on grain. The corresponding peak values for each curve represent the most economically efficient N rate (i.e., MRTN).

Table 3. Estimated nitrogen rate (kg ha^{-1}) for the linear plateau (LP) and quadratic plateau (QP) models at the estimated Haney Soil Test (HST) and MRTN values.

Site	Estimated Available N (0–15 cm)	Estimated Available N (0–60 cm)	HSTNR QP (0–15 cm)	HSTNR LP (0–15 cm)	HSTNR QP (0–60 cm)	HSTNR LP (0–60 cm)	QP MRTN	LP MRTN
	kg N ha^{-1}							
Sturgis 2017	37	-	58	66	-	-	122	134
Vivian 2017	37	-	106	109	-	-	119	101
Wall 2017	71	-	61	65	-	-	91	95
Sturgis 2018	39	75	103	105	67	69	87	72
Wall 2018	31	63	191	195	159	163	123	111
Vivian 2019	62	130	110	117	42	49	101	106
Wall 2019	27	41	163	166	149	152	118	101
Average			113	118	104	108	109	103

Moreover, using a 0–60 cm soil sample will always decrease the HSTNR over a 0–15 cm soil sample because it incorporates a larger soil volume and accounts for more soil N. This means that differences between the HSTNR and MRTN will likely be exacerbated during dry years but may be closer to the optimum in wetter years/climates. Overall, incorrect estimation of N mineralization can be attributed to its dependency on climatic variables, specifically rainfall [23–25].

Because the HSTNR varied by weather, wheat yields were similarly affected. During the dry 2017 growing season, the estimated HST yield was 276 kg ha^{-1} lower than the MRTN using a quadratic plateau curve and was 361 kg ha^{-1} lower when a linear plateau curve was used (Table 4). This was largely due to two of the three sites having HSTNR values significantly lower than the MRTN rate (Table 3). Ironically, the Vivian site's 2017 HSTNR was very close to the MRTN rate, but this was due to the low soil health score from the Haney test, which resulted in a low predicted N mineralization (hence, higher predicted HSTNR). During more conducive growing conditions, the HST yield using a linear plateau estimation generally fell beyond the curve join point, resulting in N being applied without returning an increase in grain yield. Using a quadratic plateau also resulted in yields beyond the MRTN yield but to a lesser extent, averaging 66 kg ha^{-1} more than the MRTN (Table 4).

Table 4. Estimated grain yield (kg ha^{-1}) for the linear plateau (LP) and quadratic plateau (QP) models at the estimated Haney soil test (HST) and MRTN values.

Site	HST QP Yield (0–15 cm)	HST LP Yield (0–15 cm)	HST QP Yield (0–60 cm)	HST LP Yield (0–60 cm)	QP MRTN Yield	LP MRTN Yield
	kg grain ha^{-1}					
Sturgis 2017	1806	1814	-	-	2294	2481
Vivian 2017	3369	3531	-	-	3454	3531
Wall 2017	2932	2863	-	-	3186	3278
Sturgis 2018	3481	3481	3268	3421	3424	3481
Wall 2018	5458	5456	5450	5456	5352	5446
Vivian 2019	4204	4304	3672	3688	4158	4304
Wall 2019	4625	4641	4625	4641	4572	4641
Average	3696	3727	4254	4302	3777	3880

From an economic perspective, the quadratic plateau estimation of the MRTN was always lower than the linear plateau model (Table 5). By definition, the MRTN value will always equal the plateau/join point, whereas with the quadratic plateau model, the MRTN value is generally significantly lower than the plateau join point. As a result, net return is more volatile in the linear plateau estimation and, in general, the net HSTN return was closer to the MRTN in the quadratic model. On average, the quadratic plateau model HSTN return was USD 21.27 ha^{-1} less than the MRTN, whereas the linear plateau model HSTN was USD 45.34 less than the MRTN. These differences were roughly 97% and 93% of the MRTN for the quadratic plateau and linear plateau models, respectively (Table 5).

Table 5. Estimated returns (USD ha^{-1}) for the linear plateau (LP) and quadratic plateau (QP) model parameters at the estimated Haney soil test (HST) and MRTN values.

Site	HSTN QP Return (0–15 cm)	HSTN LP Return (0–15 cm)	HSTN QP Return (0–60 cm)	HSTN LP Return (0–60 cm)	QP MRTN Return	LP MRTN Return
			USD ha^{-1}			
Sturgis 2017	300.38	293.14	-	-	328.01	352.42
Vivian 2017	562.87	592.27	-	-	565.59	601.15
Wall 2017	524.49	506.13	-	-	542.57	556.71
Sturgis 2018	588.85	586.61	585.86	614.44	595.05	623.24
Wall 2018	890.51	885.66	933.28	921.18	944.54	978.90
Vivian 2019	727.03	739.54	695.04	690.6	727.84	751.75
Wall 2019	753.26	768.8	753.22	768.76	792.66	825.37
Average	621.06	624.59	741.85	748.75	642.32	669.93

Components of the HST have been shown to be moderately correlated to the EONR of corn [12]. While the data are limited, our study supports this finding in wheat as a strong correlation was found in the difference between net return from the HSTN and the EONR return and either the soil health calculation (SHC) or the 1-day CO_2 burst (Figure 4). The SHC is a relative measure of soil health, which incorporates the 1-day CO_2 burst (hence autocorrelation) along with measures of soil organic C and N. While not used explicitly in the HSTNR, it appears to offer a potential guide for estimating N recommendations along with CO_2 evolution. The negative correlation in average or above average precipitation growing seasons suggests that a higher SHC or CO_2 burst value indicates that the HSTNR was a more accurate approximation of the EONR in this study. Additionally, the positive correlation in the drier year corroborates our inference that the HST is overestimating the N supplying abilities of the soil with the implicit assumption that a high SHC (i.e., a healthier soil) equates to more N mineralization and a lower HSTNR. Another report from South Dakota also recommended CO_2 burst (soil respiration) as a potential tool for N recommendation in corn [26,27]. Further study should be directed toward the influence of rainfall on the estimation of the HSTNR to ensure that the effect is directly related to rainfall and not confounded by other site-specific factors such as soil texture and land use history.

The flush of CO_2 is a measure of microbial biomass and well-correlated with N mineralization, but this is a simulated measure under idealized conditions [17]. Our results suggest that the efficacy of the 'CO_2 burst' would be maximized by incorporating as much in-season weather as possible. Hence, the later the soil test is taken into the growing season, the closer it is likely to approximate the EONR, which may necessitate a split N application in wheat; similar reports are available for corn [26–28].

Likewise, protein content is a critical measure of wheat quality and is positively correlated with timing and concentration of plant-available N [29,30]. Hence, protein content at the HSTNR was not significantly different between the quadratic plateau and linear plateau models (Table 6). During the drier 2017 growing season, the HSTNR estimated protein trended much lower, which could be problematic in years where protein discounts and premiums are in effect. Winter wheat typically requires a protein content of 12%. Where nitrogen mineralization does not meet expectations (i.e., dry growing conditions), a protein shortfall is more likely. Therefore, a split application of N towards reproductive growth stages can provide additional benefit by improving wheat quality; a recent report indicated similar findings in wheat [14,31].

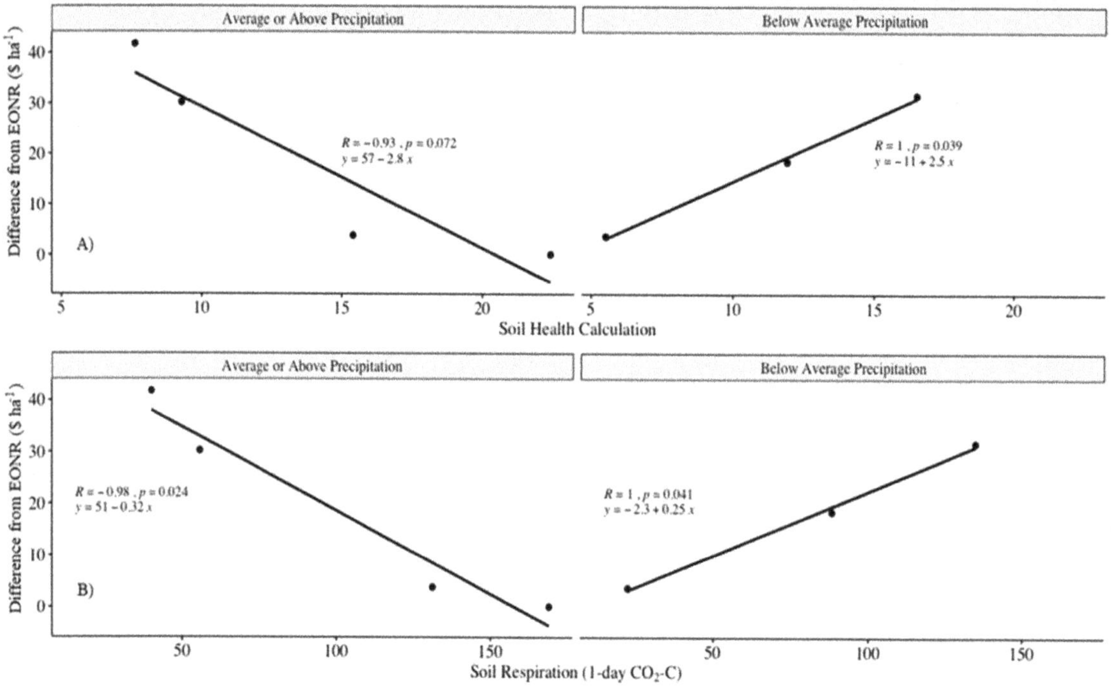

Figure 4. Regression of the difference in return from the HSTN and EONR (EONR − HSTN) on soil health calculation (**A**) and soil respiration (1-day CO_2 burst) (**B**) based on precipitation level.

Table 6. Estimated parameterized percent protein values based on the regression of applied fertilizer N on protein content for each site-year.

Site-year	Regression Equation	R^2	HSTN QP	HSTN LP	HSTN QP	HSTN LP	EONR QP	EONR LP
			0–15 cm		0–60 cm			
Sturgis 2017	y = 0.016x + 11.26	0.80	12.2	12.3	-	-	13.2	13.4
Vivian 2017	y = 0.017x + 11.15	0.78	13.0	13.0	-	-	13.2	12.9
Wall 2017	y = 0.024x + 9.55	0.87	11.0	11.1	-	-	11.7	11.8
Sturgis 2018	y = 0.013x + 12.13	0.61	13.5	13.5	13.0	13.0	13.3	13.1
Wall 2018	y = 0.021x + 9.05	0.92	13.1	13.1	12.4	12.5	11.6	11.4
Vivian 2019	y = 0.010x + 11.99	0.89	13.1	13.2	12.4	12.5	13.0	13.1
Wall 2019	y = 0.011x + 10.49	0.65	12.3	12.3	12.1	12.2	11.8	11.6
Average	y = 0.016x + 10.80	0.79	12.6	12.7	12.5	12.5	12.5	12.4

4. Conclusions

Devising efficient N recommendations is difficult because historically it has been largely based on available N (generally nitrate N) in soil before/at the time of planting and does not consider N mineralization during crop growth. The HST attempts to account for N mineralization in its recommended N rate, which offers the prospect of reducing overall N rates. The SHC and 1-day CO_2 burst provide useful indicators of the potential for the soil to supply N to the plant. However, in-season precipitation plays a critical role in the efficacy of this test, especially in dryland ecosystems, where plant-available water is primarily dependent on in-season precipitation, ultimately influencing N mineralization. In our study, the HST appeared to underestimate N mineralization under dry conditions. Since N mineralization is a microbially mediated process and water limitations severely reduce microbial metabolism and growth, drought effectively 'penalized' the soils with higher mineralization potential (i.e., 'healthy soils' with higher SOM, mineralizable N, soil protein content or microbial activities), where that potential was not realized in-season due to an assumed limitation on N mineralization during critical plant growth stages. In semi-arid climates, where adequate protein concentration is of concern, a 0–15 cm soil sample for HSTNR, rather than a 0–60 cm sample, likely approximates the EONR more closely with adequate protein in drought years or when N is applied as a single application. However, under wetter conditions, a 0–60 cm soil sample may perform similarly, particularly where a split application is utilized for protein content.

Designing N recommendations that incorporate N mineralization during crop growth is critical to improving N recommendations overall. Under dryland conditions in particular, attaining accuracy in estimating N mineralization is difficult. However, split N application provides an opportunity to incorporate more weather information during the wheat growing season to better optimize N use efficiency. Our study provides valuable information regarding the impact of in-season precipitation on HST-based N recommendation that necessitates a revision of the parameters used in the estimation of model parameters. In spite of other limitations in the scope of this study due to limited numbers of years and locations, it indicates an important knowledge gap to improve N recommendation. Future studies should incorporate more diverse locations in terms of soil characteristics, climate, etc. over multiple years to develop a more effective model for N recommendation to optimize N use efficiency.

Author Contributions: Conceptualization, C.G.; methodology, C.G.; formal analysis, C.G.; investigation, C.G. and D.S.; resources, C.G.; data curation, C.G. and D.S.; writing—original draft preparation, C.G.; writing—review and editing, D.S.; visualization, C.G. and D.S.; supervision, C.G.; project administration, C.G.; funding acquisition, C.G. All authors have read and agreed to the published version of the manuscript.

Funding: This research was supported by a Conservation Innovation Grant funded by USDA-NRCS.

Institutional Review Board Statement: Not available.

Informed Consent Statement: Not available.

Data Availability Statement: Research data can be made available by contacting Christopher Graham at christopher.graham@sdstate.edu.

Acknowledgments: The authors acknowledge all technicians and research interns that helped in conducting field experiments successfully.

Conflicts of Interest: The authors declare no conflict of interest.

References

1. Gu, B.; van Grinsven, H.J.; Lam, S.K.; Oenema, O.; Sutton, M.A.; Mosier, A.; Chen, D. A credit system to solve agricultural nitrogen pollution. *Innovation* **2021**, *2*, 100079. [CrossRef] [PubMed]
2. Harindintwali, J.D.; Zhou, J.; Muhoza, B.; Wang, F.; Herzberger, A.; Yu, X. Integrated eco-strategies towards sustainable carbon and nitrogen cycling in agriculture. *J. Environ. Manag.* **2021**, *293*, 112856. [CrossRef] [PubMed]
3. Wan, X.; Wu, W.; Shah, F. Nitrogen fertilizer management for mitigating ammonia emission and increasing nitrogen use efficiencies by 15N stable isotopes in winter wheat. *Sci. Tot. Environ.* **2021**, *790*, 147587. [CrossRef] [PubMed]
4. Sawyer, J.; Nafziger, E.; Randall, G.; Bundy, L.; Rehm, G.; Joern, B. Concepts and rationale for regional nitrogen rate guidelines for corn. PM 2015. *Iowa State Univ. -Univ. Ext. Ames Iowa* **2006**, *28*.
5. Yost, M.A.; Pound, C.A.; Creech, J.E.; Cardon, G.E.; Pace, M.G.; Kitchen, B.; Nelson, M.; Russell, K. Nitrogen requirements of first-year small grains after alfalfa. *Soil Sci. Soc. Am. J.* **2021**, *85*, 1698–1709. [CrossRef]

6. Haney, R.L.; Hons, F.M.; Sanderson, M.A.; Franzluebbers, A.J. A rapid procedure for estimating nitrogen mineralization in manured soil. *Biol. Fert. Soils* **2001**, *33*, 100–104. [CrossRef]
7. Haney, R.; Brinton, W.; Evans, E. Estimating soil carbon, nitrogen, and phosphorus mineralization from short-term CO_2 respiration. *Commun. Soil Sci. Plant Anal.* **2008**, *39*, 2706–2720. [CrossRef]
8. Haney, R.L.; Haney, E.B.; Hossner, L.H.; Arnold, J.G. Modifications to the new soil extractant H3A-1: A multinutrient extractant. *Commun. Soil Sci. Plant Anal.* **2010**, *41*, 1513–1523. [CrossRef]
9. Haney, R.L.; Franzluebbers, A.J.; Jin, V.L.; Johnson, M.V.; Haney, E.B.; White, M.J.; Harmel, R.L. Soil organic C:N vs. water-extractable organic C:N. *Open J. Soil Sci.* **2012**, *2*, 269–274. [CrossRef]
10. Franzluebbers, A.J. Short-term C mineralization (aka the flush of CO_2) as an indicator of soil biological health. *CAB Reviews: Pers Agric. Vet. Sci. Nutr. Nat. Res.* **2018**, *13*, 14. [CrossRef]
11. Harmel, R.D.; Haney, R.L. Initial field evaluation of the agro-economic effects of determining nitrogen fertilizer rates with a recently-developed soil test methodology. *Open J. Soil Sci.* **2013**, *3*, 91–99. [CrossRef]
12. Yost, M.A.; Veum, K.S.; Kitchen, N.R.; Sawyer, J.E.; Camberato, J.J.; Carter, P.R.; Ferguson, R.B.; Fernández, F.G.; Franzen, D.W.; Laboski, C.A.; et al. Evaluation of the Haney Soil Health Tool for corn nitrogen recommendations across eight Midwest states. *J. Soil Water Conserv.* **2018**, *73*, 587–592. [CrossRef]
13. Large, E.C. Growth stages in cereals. *Plant Patholog.* **1954**, *3*, 128–129. [CrossRef]
14. Graham, C.J.; Stockton, M. Winter wheat response to fertilizer type and timing in western South Dakota. *Agron. J.* **2019**, *11*, 1433–1440. [CrossRef]
15. Gerwing, J.; Gelderman, R. *Fertilizer Recommendations Guide—EC 750*; United States Department of Agriculture: Brookings, SD, USA. Available online: http://agbiopubs.sdstate.edu/articles/EC750.pdf.
16. Haney, R.L.; Haney, E.B.; Smith, D.R.; White, M.J. Estimating potential nitrogen mineralisation using the Solvita soil respiration system. *Open J. Soil Sci.* **2015**, *5*, 319–323. [CrossRef]
17. Franzluebbers, A.J. Should soil testing services measure soil biological activity? *Agric. Environ. Lett.* **2016**, *1*, 150009. [CrossRef]
18. Singh, S.; Jagadamma, S.; Yoder, D.; Yin, X.; Walker, F. Agroecosystem management responses to Haney soil health test in the southeastern United States. *Soil Sci. Soc. Am. J.* **2020**, *84*, 1705–1721. [CrossRef]
19. Pan, W.L.; Kidwell, K.K.; McCracken, V.A.; Bolton, R.P.; Allen, M. Economically optimal wheat yield, protein and nitrogen use component responses to varying N supply and genotype. *Front. Plant Sci.* **2020**, *10*, 1790. [CrossRef]
20. Schwalbert, R.A.; Amado, T.J.C.; Reimche, G.B.; Gebert, F. Fine-tuning of wheat (*Triticum aestivum*, L.) variable nitrogen rate by combining crop sensing and management zones approaches in southern Brazil. *Precis. Agric.* **2019**, *20*, 56–77. [CrossRef]
21. Cerrato, M.E.; Blackmer, A.M. Comparison of Models for Describing Corn Yield Response to Nitrogen Fertilizer. *Agron. J.* **1990**, *82*, 132–143. [CrossRef]
22. Fierer, N.; Schimel, J.P. A proposed mechanism for the pulse in carbon dioxide production commonly observed following the rapid rewetting of a dry soil. *Soil Sci. Soc. Am. J.* **2003**, *67*, 798–805. [CrossRef]
23. Maithani, K.; Arunachalam, A.; Tripathi, R.S.; Pandey, H.N. Nitrogen mineralization as influenced by climate, soil and vegetation in a subtropical humid forest in northeast India. *For. Ecol. Manag.* **1998**, *109*, 91–101. [CrossRef]
24. Song, W.; Chen, S.; Zhou, Y.; Lin, G. Rainfall amount and timing jointly regulate the responses of soil nitrogen transformation processes to rainfall increase in an arid desert ecosystem. *Geoderma* **2020**, *364*, 114197. [CrossRef]
25. Zhang, X.Y.; Li, Q.W.; Gao, J.Q.; Hu, Y.H.; Song, M.H.; Yue, Y. Effects of rainfall amount and frequency on soil nitrogen mineralization in Zoigê alpine wetland. *Eur. J. Soil Biol.* **2020**, *97*, 103170. [CrossRef]
26. Bean, G.M.; Kitchen, N.R.; Veum, K.S.; Camberato, J.J.; Ferguson, R.B.; Fernandez, F.G.; Franzen, D.W.; Laboski, C.A.; Nafziger, E.D.; Sawyer, J.E.; et al. Relating four-day soil respiration to corn nitrogen fertilizer needs across 49 US Midwest fields. *Soil Sci. Soc. Am. J.* **2020**, *84*, 1195–1208. [CrossRef]
27. Sanyal, D.; Wolthuizen, J.; Bly, A. Influence of Nitrogen Fertilization Rate on Soil Respiration: A Study Using a Rapid Soil Respiration Assay. *Nitrogen* **2021**, *2*, 218. [CrossRef]
28. Liang, G.; Cai, A.; Wu, H.; Wu, X.; Houssou, A.A.; Ren, C.; Wang, Z.; Gao, L.; Wang, B.; Li, S.; et al. Soil biochemical parameters in the rhizosphere contribute more to changes in soil respiration and its components than those in the bulk soil under nitrogen application in croplands. *Plant Soil* **2019**, *435*, 111–125. [CrossRef]
29. Bly, A.; Woodard, H.J. Nitrogen management: Foliar nitrogen application timing influence on grain yield and protein concentration of hard red winter and spring wheat. *Agron. J.* **2003**, *95*, 335–338. [CrossRef]
30. Wilson, T.L.; Guttieri, M.J.; Nelson, N.O.; Fritz, A.; Tilley, M. Nitrogen and sulfur effects on hard winter wheat quality and asparagine concentration. *J. Cereal Sci.* **2020**, *93*, 102969. [CrossRef]
31. Xue, C.; Matros, A.; Mock, H.P.; Mühling, K.H. Protein composition and baking quality of wheat flour as affected by split nitrogen application. *Front. Plant Sci.* **2019**, *10*, 642. [CrossRef]

Nitrogen

MDPI

Article

Effects of Drainage Water Management in a Corn–Soy Rotation on Soil N_2O and CH_4 Fluxes

Jacob G. Hagedorn [1,2,*], Eric A. Davidson [1], Thomas R. Fisher [3], Rebecca J. Fox [3,4], Qiurui Zhu [1], Anne B. Gustafson [3], Erika Koontz [3,5], Mark S. Castro [1] and James Lewis [6]

1 Appalachian Laboratory, University of Maryland Center for Environmental Science, Frostburg, MD 21532, USA; edavidson@umces.edu (E.A.D.); qzhu@umces.edu (Q.Z.); mcastro@umces.edu (M.S.C.)
2 Department of Environmental Studies, University of North Carolina, Asheville, NC 28804, USA
3 Horn Point Laboratory, University of Maryland Center for Environmental Science, Cambridge, MD 21613, USA; fisher@umces.edu (T.R.F.); rfox3@washcoll.edu (R.J.F.); abgust@umces.edu (A.B.G.); erika.Koontz@gmail.com (E.K.)
4 Department of Environmental Science and Studies, Washington College, Chestertown, MD 21620, USA
5 Smithsonian Environmental Research Center, Edgewater, MD 21037, USA
6 University of Maryland Extension Caroline County, Denton, MD 21629, USA; jlewis2@umd.edu
* Correspondence: jghagedo@unca.edu

Abstract: Drainage water management (DWM), also known as controlled drainage, is a best management practice (BMP) deployed on drainage ditches with demonstrated success at reducing dissolved nitrogen export from agricultural fields. By slowing discharge from agricultural ditches, subsequent anaerobic soil conditions provide an environment for nitrate to be reduced via denitrification. Despite this success, incomplete denitrification might increase nitrous oxide (N_2O) emissions and more reducing conditions might increase methanogenesis, resulting in increased methane (CH_4) emissions. These two gases, N_2O and CH_4, are potent greenhouse gases (GHG) and N_2O also depletes stratospheric ozone. This potential pollution swapping of nitrate reduction for GHG production could negatively impact the desirability of this BMP. We conducted three years of static chamber measurements of GHG emissions from the soil surface in farm plots with and without DWM in a corn–soybean rotation on the Delmarva Peninsula. We found that DWM raised the water table at the drainage ditch edge, but had no statistically significant effect on water-filled pore space in the field soil surface. Nor did we find a significant effect of DWM on GHG emissions. These findings are encouraging and suggest that, at least for this farm site, DWM can be used to remove nitrate without a significant tradeoff of increased GHG emissions.

Keywords: nitrous oxide; methane; greenhouse gases; denitrification; nitrogen; drainage water management; agricultural management; controlled drainage; best management practices

1. Introduction

Excess agricultural nitrogen (N) and phosphorus (P) addition to the environment has been a long-term problem in the United States, contributing to algal blooms and eutrophication [1]. Within the Chesapeake Bay watershed agricultural production on the Delmarva peninsula accounted for ~90% of the nitrogen inputs [2–4], much of which is derived from synthetic nitrogen fertilizer [5]. Despite the need for nitrogen to increase crop yields, more than half of the N applied can be lost to the environment in a variety of pathways such as leaching and gas emissions [6,7]. Surface ditches and tile drains installed to remove water from wet fields act as conduits for export of dissolved nutrients. Drainage water management (DWM) is a best management practice (BMP) designed to slow this movement and thereby reduce reactive N loss from agricultural fields that have drainage ditches [8,9].

Drainage water management (DWM) functions by placing drainage control structures (DCS) at the end of drainage ditches with the goal of slowing water discharge and raising water levels in ditches. By keeping water in the field for longer, the water table upstream of the ditch rises, creating more widespread anaerobic soil conditions [10]. When N, as nitrate, is exposed to more reducing conditions microbially driven denitrification is more likely [11,12]. Some studies have demonstrated that DWM has been effective at reducing nitrate loading into downstream waters [13–15], including preliminary results from the farm of present study [16]. However, incomplete denitrification can result in nitrous oxide (N_2O) production, a potent greenhouse gas (GHG) and a reactant in stratospheric ozone destruction [17,18]. In addition, nitrification has also been documented to produce N_2O as a byproduct [11]. Agricultural soil management, driven by anthropogenic nitrogen inputs, is responsible for 75% of the total N_2O emissions in the US and 45% of total N_2O emissions in North America [19,20]. In addition to N_2O, reducing conditions can also create higher methane (CH_4) emissions via methanogenesis [21]. Methane is the second most important GHG, with potency 25-fold greater than CO_2 in a 100 year time frame [22].

Broadly, soil N_2O fluxes are controlled primarily by inorganic-N content, carbon availability, soil moisture, microbial community abundances, soil temperature, soil pH, and soil texture [11,12]. Fertilizer N can be rapidly converted to nitrate, through nitrification, and nitrate can then become substrate for denitrification. Both nitrification and denitrification can contribute to large N_2O emission spikes [23], especially when high soil nitrate concentrations co-occur with high soil moisture conditions [24,25]. The combination of high N input with DWM that retains water in the field could result in lower hydrologic losses of N as nitrate through the ditch but with the tradeoff of higher losses of N_2O.

Previous work examining the impact of water table changes and DWM on GHG emissions in agriculture fields has yielded contradictory results. Some studies found that undrained fields had higher N_2O emissions than drained fields [26–29] while other studies found that there was no difference in N_2O emissions between drained and undrained fields [30–33]. For CH_4, one study [26] found higher emissions in undrained fields while others found no drainage effect on CH_4 emissions [30–32]. Some of these studies did not directly evaluate DWM manipulations, creating higher and lower water tables at specific times of the year, rather they explored drainage impact in general [26,27,32], or drainage impact with subsurface irrigation [29,31]. The variability in conclusions and paucity of direct assessments of DWM mean that further research is needed to understand the impacts of this particular BMP on the potential for increased GHGs. Here, we compare measured soil emissions of N_2O and CH_4 in corn and soybean fields with controlled drainage (DWM) and free drainage (non-DWM) at a farm in eastern Maryland, on the Delmarva Peninsula. We hypothesized that DWM would increase soil GHG emissions.

2. Materials and Methods

2.1. Study Site Location

The research site was located on a private farm within the Choptank watershed on the coastal plain of the Delmarva Peninsula on the Eastern Shore of Maryland (Figure 1, N 39.11973, W 75.8069). Historically a wet landscape, the land was made arable by installing drainage ditches. Harvested crops mostly contribute feed to nearby concentrated poultry production. Precipitation averages 110 cm per year. This study site is an operational farm with a collaborating farmer willing to accommodate our research needs while working within the timing of standard agricultural management. A DWM system was already in place and a direct evaluation of that system was possible in a relatively small area. The ditches split the site into four similarly sized fields ranging from 3 to 6 hectares each, with an average length of 220 m. The dominant soil type is a Fallsington series on 0–2 percent slopes classified as fine-loamy, mixed, active, mesic Typic Endoaquults.

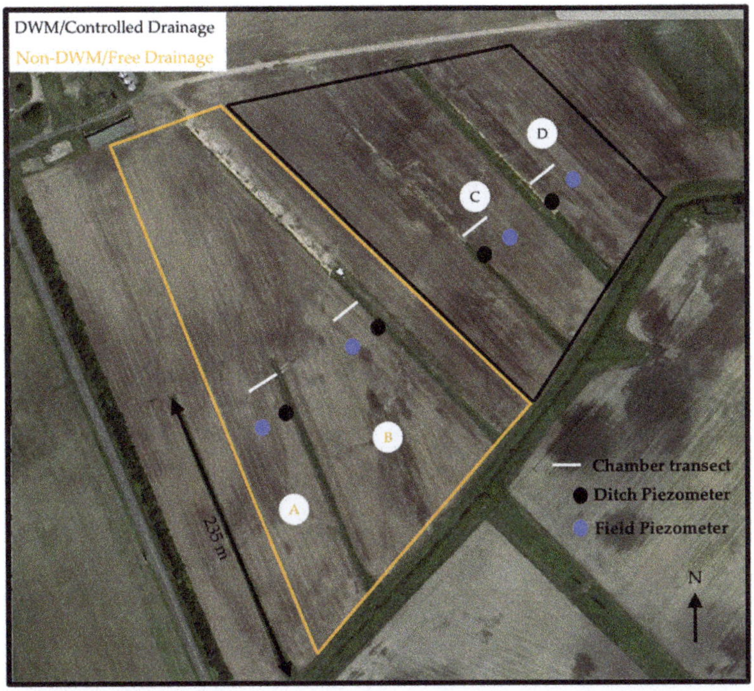

Figure 1. Aerial image of the research site showing the approximate piezometer locations, chamber transect locations, and treatment area borders. Fields A and B were free draining (non-DWM); fields C and D had controlled drainage (DWM).

2.2. Farm Management

The crop management plan was a no-till, corn–soybean rotation with a winter radish cover crop. The typical growing season ranged from April to October. Approximately 213 kg ha^{-1} of synthetic urea-ammonium nitrate (UAN) was applied during corn years and no N fertilizer was applied during soybean years. Approximately 22 kgN ha^{-1} was knifed-in during corn planting, and the other 191 kgN ha^{-1} was broadcast-applied in late May or early June. During corn planting, the fields received a sulfur addition of 15 kgS ha^{-1}. Liquid or dry potash was applied during corn and soybean years. Herbicide was applied prior to or soon after planting and sometimes in June to suppress undesired vegetative growth. Prior to harvest, daikon radish seed was applied by air to serve as the winter cover crop. Manure and lime were applied every 5 years, which did not occur during this project's duration. The site had pivot irrigation that was used to supplement natural precipitation events in the summer months. All the research fields had the same crop rotation schedule, N fertilization rate, and irrigation rate from 2017 to 2019.

2.3. DWM Management

The DWM strategy consisted of inserting and removing boards within the drainage control structures (DCS) to slow the ditch discharge while accommodating standard farm practices (Figure S1). For farm activities such as planting, N fertilization, and harvest, the DCS boards were removed and fields were drained to create a drier condition for equipment use. This management timeline was dependent on precipitation events, groundwater levels, and evapotranspiration; therefore the strategy was modified to adapt to changing environmental conditions. We refer to the time period when the boards were in place in the DWM DCS as "active" and the period when they were removed from the DWM DCS as "inactive".

2.4. Experimental Design

Figure 1 shows a site map depicting the static chamber transect locations and borders of study plots. Non-DWM, or free drainage, was implemented on two fields (A and B) and controlled drainage, or DWM, implemented in the other two fields (C and D). In each field the soil chamber sampling design consisted of a transect of eight chambers installed perpendicular to the ditch. Each transect was approximately 12 m in length and the soil chamber collars were equally spaced. The replicate plots were adjacent to each other in order to increase the horizontal fetch of micrometeorological measurements of a companion part of this study that will be reported elsewhere.

Soil and soil gas flux measurements were taken approximately monthly, depending on climatic conditions, but more temporally intensive sampling was done around planting, irrigation, N fertilization, and harvest to better characterize trace gas dynamics during those periods. Because preliminary studies showed some diel variation (Figure S2), the order of chamber sampling was randomized to ensure that the same field was not measured at the same time of day during every sampling date, thus minimizing potential sampling biases due to diel variation. Fluxes were sampled starting at approximately 8:00 EST and extended until approximately 17:00 EST.

Two slotted PVC (5 cm diameter) piezometers, used to measured groundwater levels, were installed approximately 1.5–2 m deep in the four research fields. One piezometer was installed within 1 m of the perimeter of the ditch while another was placed in the field 15 m perpendicular to the ditch piezometer (Figure 1). The piezometers were located within 15 m of the chamber transect. Solinst water level data loggers were placed in each piezometer which recorded pressure and temperature every 30 min. The pressure was then converted to depth of water below the surface by subtracting barometric pressure recorded separately by a Solinst barologger. Thirty-minute water depths were aggregated to daily mean for this analysis.

There were four primary controls on the statistical power of this experimental design: (1) plot replication, (2) number of chamber measurements in each plot, (3) number of days of sampling over three years and (4) number of gas concentration measurements used to calculate each flux. This experimental design was limited to two replicate plots per treatment because four was the maximum number of flux towers that could be established in the companion micrometeorological study. The within-plot statistical power of this study was provided by eight chambers per field, which was needed for measurements known to have high spatial variability, as is the case with N_2O and CH_4 [34]. This study had excellent statistical strength in the number of gas concentration measurements per flux calculation, as there was a single chamber headspace concentration recorded every second (1 Hz), which increased the flux estimation accuracy, lowered the flux detection limit, reduced the time needed for each flux measurement, and afforded improved statistical analysis [35].

2.5. Flux Measurements

A Picarro G2308 cavity ringdown spectrometer and a static soil chamber system were used to measure N_2O and CH_4 soil gas fluxes. The instrument specifications state that the raw precision (1σ) of the instrument was <25 ppb + 0.05% of the reading for N_2O and <10 ppb + 0.05% of the reading for CH_4. The chambers consisted of a PVC chamber top with a curled vent, 12 cm length and 1 mm ID, and a matching PVC chamber collar. The chamber was approximately 25 cm in diameter. The chamber collar height was 5 cm and was inserted 3–4 cm into the soil. The chamber top height was 2.5 cm and had a rubber gasket around the perimeter to ensure a tight fit onto the collar. The approximate total volume of the chamber, dependent on insertion depth into the soil, was 4 L. The flux measurement system used a recirculation technique through inlet and outlet tubing connected to the instrument, air pump, and the chamber top. Twenty-meter lengths of 6 mm OD polyethylene tubing connected the gas analyzer to the chamber top. Given the average flow rate of the pump (~230 mL min^{-1}), the tubing diameter, and length, it took approximately 45 s for gas to reach the analyzer from the chamber. Because of the high frequency (1 Hz) of reporting concentration by the instrument and its high sensitivity to change in concentration, the chamber top was left on each collar for only ~5 min, which was sufficient time to detect a non-zero slope of increasing or decreasing headspace concentrations, while minimizing a potential chamber artifact of altering the concentration gradient between soil and headspace gas. The slope was used to calculate a flux using Equation (1)

$$Flux\left(\mathrm{mol\ m^{-2}\ min^{-1}}\right) = \frac{\mathrm{dConc}}{\mathrm{dt}} * \frac{\mathrm{P} * \mathrm{V}}{\mathrm{A} * \mathrm{R} * \mathrm{T}} \quad (1)$$

where dConc/dt = slope (μL L^{-1} min^{-1}), P = air pressure (atm), R = gas law constant (L atm mol^{-1} K^{-1}), T = air temperature (K), V = chamber volume (L), and A = chamber area (m^2). Flux units of μmol m^{-2} min^{-1} were used for statistical analyses, but when flux interpolations and annual estimates were conducted, the flux units were converted to μg N_2O-N m^{-2} min^{-1}.

2.6. Soil Chemical and Physical Characteristics

Volumetric water content (VWC) of the soil was measured using a manual Decagon 5TM soil moisture probe. The probe was inserted vertically into the soil surface to a depth of 5 cm. At each chamber location (n = 32), three VWC measurements were taken outside the chamber area, and the mean calculated. The probe calibration was checked in the laboratory prior to field usage and the VWC was obtained using the Topp equation [36].

Soil samples used for gravimetric soil moisture, nitrate and ammonium concentration analyses were taken using a 2.2 cm diameter soil push probe during each gas flux measurement. At each chamber (n = 32) three, 5 cm deep soil plugs surrounding the chamber were collected and placed together into a labeled bag. The bag was placed in a refrigerated cooler, transported back to the laboratory and processed within 48 h. At the lab, the soil within each bag was homogenized and sieved using a 10 mesh (2 mm) pan. Six grams of the field-moist sieved soil were extracted in 30 mL of 2M potassium chloride (KCL), placed on an oscillating shaker table for at least 2 h, and vacuum filtered through a Whatman No. 42 filter paper. Extracts were then analyzed for nitrate, nitrite, and ammonium concentration using flow injection analysis Lachat autoanalyzer [37]. If samples were not processed in the analyzer within 12 h, they were stored at 4 °C until analysis within 30 days of soil collection. Soil push probe samples from five sampling dates selected over the three-year study were also analyzed for total soil carbon and total soil nitrogen using a Carlo Erba NC2500 elemental analyzer.

The surface bulk density of the soil was measured using a 5 cm length, 5 cm diameter PVC core. Air-dried soil was sieved with 10 mesh (2 mm) pan and weighed. The bulk density was calculated as the <2 mm air-dried mass divided by the volume. Two bulk density surface cores were taken in each field for two years of this study and averaged by field and year. Surface bulk density measurements were used to convert VWC to water-filled pore space (WFPS) by calculating the total porosity assuming a particle density of 2.65 g cm^{-3} [38]. Soil texture was assessed twice using hand augered soil samples at 10 cm intervals to an approximate depth of 1 m, twice, in each field. The soils were sieved using a 10 mesh (2 mm diameter) pan, air-dried for at least 72 h, and measured for soil texture using the hydrometer method [39].

2.7. Non-Parametric DWM Impact Model

The purpose of this model was to determine if the dependent variables (gas fluxes, piezometer depth, soil moisture) were impacted by DWM. Because of non-normality and heteroskedasticity of the dependent variable data, a non-parametric Wilcoxon test was used. The piezometer water depth, soil moisture and fluxes were assessed separately as dependent variables of interest. The independent variables were treatment (DWM/controlled drainage; versus non-DWM/free drainage) and board period (active—DCS boards inserted in the DWM plots; versus inactive—DCS boards removed for the farmer's management activities). A pairwise post hoc test was performed to estimate *p*-values assessing the difference between board periods under both treatment conditions.

2.8. Mixed-Effects Model

A generalized linear mixed-effects model (GLMER) in R program was used to assess the impact of treatment and other predictor variables on N_2O and CH_4 fluxes. Both N_2O and CH_4 fluxes were non-normal and right skewed, so a natural log transformation was applied. A small positive offset was added to make all values positive before log transformation [40]. A gamma log link distribution was applied in the GLMER model [41].

For N_2O, the independent variables of treatment, fertilization period, year, and board period were categorical variables assessed as fixed effects. Chamber location (1–8, with 1 being closest to the ditch and 8 being furthest) was a continuous variable serving as a measure of distance from ditch edge. After distance from the ditch was found to not be a significant fixed effect ($p > 0.05$), chamber number was then considered a random effect. Sampling dates were considered part of the fertilization period if they occurred within a window of 4 weeks after a fertilization event. Several studies have shown a peak of N_2O emissions within days of fertilization, followed by a gradual decline over 2–4 weeks [24,42,43]. Board period was the same categorical variable as described in Section 2.7. Interaction effects between all fixed effects were also evaluated. Final models were selected based on the lowest AIC value of multiple fixed effect combinations and interactions. An estimated marginal means post hoc test was conducted when *p*-values of main effects were less than 0.05. Model residuals were visually analyzed for normality and heteroskedasticity.

For CH_4, the independent variables of treatment, seasonal period, board period, and year were categorical variables assessed as fixed effects. Chamber location (1–8, with 1 being closest to the ditch and 8 being furthest) was initially tested as a continuous variable serving as a measure of distance from ditch edge but was not significant ($p > 0.05$), therefore distance was considered a random effect. Seasonal period was assigned a value based on inclusion within the data range of 1 April–30 June. These dates were selected because they corresponded to the beginning of the warmer spring season with wet soil conditions, prior to drier summer months. The final model was selected based on the lowest AIC value of multiple fixed effect combinations and interactions. An estimated marginal means post hoc test was conducted when *p*-values of main effects were less than 0.05. Model residuals were visually analyzed for normality and heteroskedasticity.

2.9. Regression Statistics

To examine how environmental variables controlled N_2O emissions, a stepwise linear regression was created using soil temperature, soil water-filled pore space (WFPS), and nitrate concentration as the predictor variables for N_2O emissions. The data were unaggregated, so each chamber flux measurement was an observation, for a total of 536 measurements with a complete suite of data for all dependent and independent variables. All data were tested for normality using the Shapiro–Wilk normality test and were log-transformed. For WFPS, a quadratic term was added to the model to account for an expected non-linear response [11]. The stepwise regression created multiple models with different combinations of predictor variables. The model with the lowest AIC was selected. Homoscedasticity and normality were assessed in the final model residuals.

2.10. Annual Flux Calculations

2.10.1. Seasonally-Based Flux Interpolation

To estimate an annual N_2O gas flux, each year was divided into 3–4 periods, detailed in Table S1, bounded by farm management events and observed seasonal impacts of patterns of measured fluxes. For periods in which no visually obvious N_2O spikes occurred, an arithmetic mean was taken of all the flux measurements within that time period and multiplied by the number of days within that period. For periods in which a spike of N_2O flux was observed followed by a sustained drop in flux values, usually within days or weeks after N fertilization in corn years or shortly after temperature and moisture increase in the soybean year, a decay curve and function were fitted to the data for the duration of those observations (Figure S3). Daily fluxes were then summed to obtain a total period value. The emission values in each period (both arithmetically derived and decay curve derived) were summed to obtain an annual flux measurement. There were 3–4 periods per year with 2 to 5 flux measurements per period. This procedure was repeated to calculate an annual flux for each chamber. The chamber annual fluxes were averaged by treatment and a 95% confidence interval (CI) was calculated, representing spatial flux variability.

For CH_4, each year was divided into 3 periods bounded by seasonal impacts on measured fluxes (Table S2). A mean flux was calculated for each period, multiplied by the number of days within that period, and period fluxes were summed to obtain an annual CH_4 flux. This procedure was repeated for each chamber and the chambers were averaged by treatment to obtain a 95% CI representing the spatial variation within the treatment.

2.10.2. Mixed-Effects Model Extrapolation

For an independent check on the interpolation method, annual CH_4 and N_2O fluxes were also calculated by extrapolating the mean flux values reported in the mixed-effects model results for each year. This more simplistic approach did not account specifically for observed spikes in gas fluxes but represented the spatial and temporal variation accounted for in the mixed-effects model. An error term for each annual estimate was calculated based on the standard errors of the estimates of the mixed-effects model output.

2.10.3. Regression Model N_2O Flux Interpolation

A regression-based model was used to obtain another independent annual N_2O flux estimate using environmental variables of soil moisture, soil nitrate, and soil temperature. Linear interpolations were made between each of the measured data points for these environmental drivers. The interpolated daily variable values were applied to a stepwise linear regression equation. Predicted hourly N_2O fluxes were summed to an annual flux. This was done for each chamber (n = 32) and an average and standard error were calculated by treatment.

3. Results

3.1. Environmental Variables

The following section describes the climatic and soil variables measured to complement the flux measurements (Figure 2).

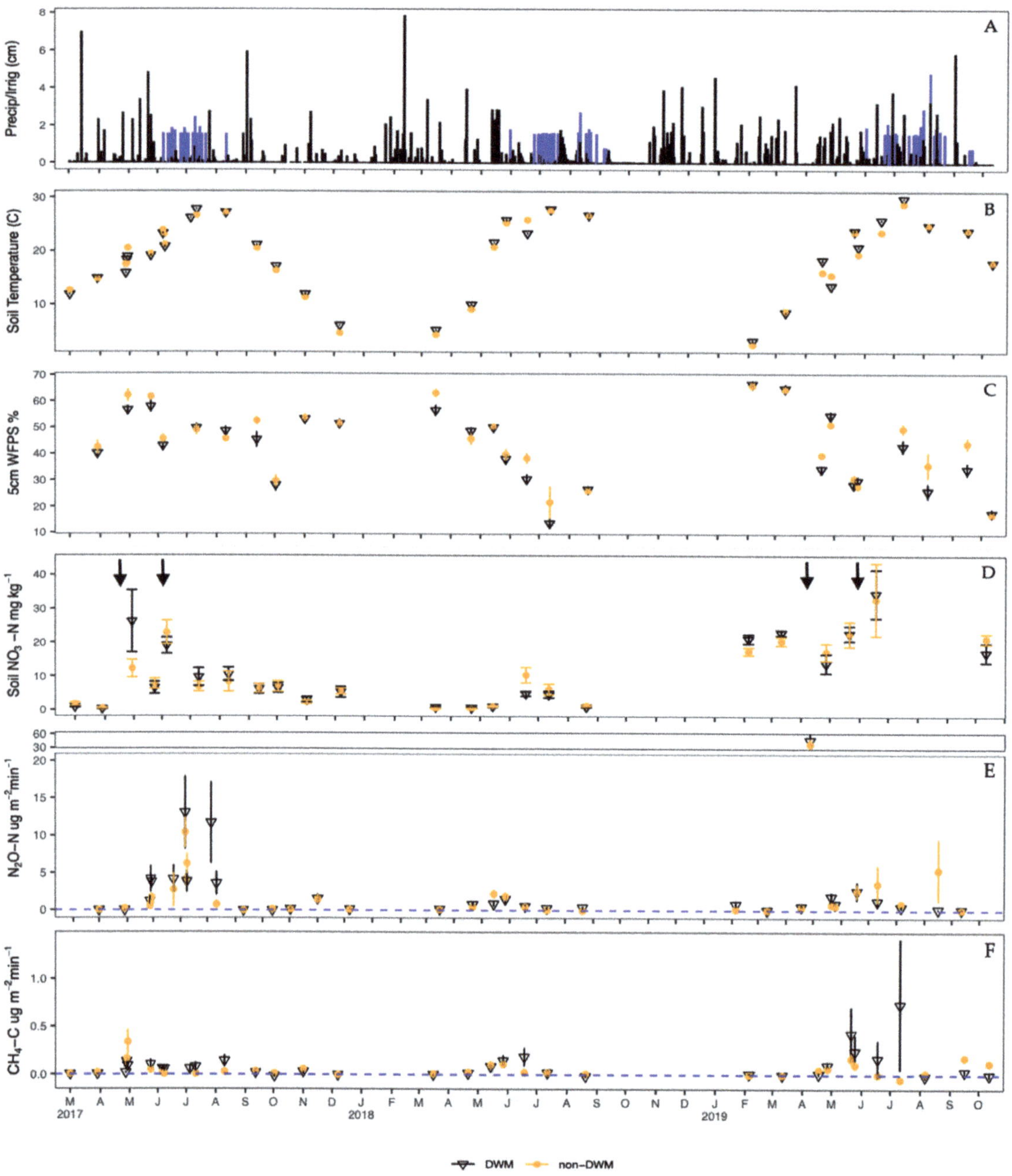

Figure 2. Climactic, soil, and flux variables measured over the course of this study. Arrows mark N fertilization events. Blue bars in panel (**A**) are irrigation events and black bars are precipitation events. The other measured variables are soil temperature (**B**), WFPS (**C**), soil nitrate (**D**), N_2O-N flux (**E**) and CH_4-C flux (**F**). Dashed blue lines in panels (**E**,**F**) mark the value of zero net flux. Mean + SE bars.

3.1.1. Precipitation and Air Temperature

Daily precipitation and irrigation amounts were recorded over the course of this study (Figure 2A). The highest daily rainfall was 7.84 cm in February 2018. Seasonally averaged over three study years (2017–2019), March–May had the highest average rainfall with 31.8 ± 1 cm, followed by December–February with 26.4 ± 3.6 cm, June–August with 19.1 ± 8.6 cm, and September–November with 9.0 ± 1.9 cm. The daily average air temperature over three years was 14.1 °C and ranged from −13.0 to 30.1 °C (data not shown).

3.1.2. Soil Temperature and Moisture

Soil temperature and moisture at 5 cm depth were measured at the same frequencies as chamber fluxes. Soil temperature ranged from 2.1 to 29.1 °C and followed expected seasonality trends (Figure 2B). The measured mean percent water-filled pore space (%WFPS) by treatment ranged from 13% to 67% over the study period (Figure 2C). The highest recorded soil moisture contents occurred in the winter (December–February) and spring (March–May), and the lowest soil moisture contents occurred during the summer (June–August) and fall (September–November). There was a particularly dry period in July 2018, when the lowest soil moisture content of this study was recorded. Statistical analysis of WFPS can be found in results Section 3.4.

3.1.3. Soil Extractable Nitrogen

The soil extractable nitrate in the top 5 cm, measured at each chamber sampling, ranged from 0.38 to 34.6 mg N kg^{-1} (Figure 2D). In April 2017 and 2019, 22 kg N ha^{-1} as UAN fertilizer were applied during planting and in June 2017 and 2019 200 kg N ha^{-1} as UAN were applied, as indicated by arrows in Figure 2. In general, there was a positive association between N fertilization and soil nitrogen concentration. Soil nitrate values increased soon after application of N fertilizer in April and June of 2017 and 2019. After this period, the soil nitrate concentrations decreased throughout the remainder of the year. In the 2018 soybean crop, the soil nitrate values rose in the spring but did not reach concentrations as high as the years with N fertilization. For soil extractable ammonium values, the concentrations remained below 10 mg N kg^{-1} during the study period, except one measurement in June 2018 (data not shown). Ammonium values were more variable than nitrate and higher during summer than during winter.

3.1.4. Soil Texture and Bulk Density

Soil texture varied spatially within and between the four fields. Augered soil cores in field C had higher clay and silt content compared to fields A, B, and D at depths ranging from 20 to 60 cm (Figure S4 and Table S3). The 5 cm surface bulk density values, %C, and %N also varied by field, with higher %C and %N values associated with higher clay content (Table S4).

3.2. Piezometers

The daily mean piezometer water depth was aggregated by treatment and piezometer location during the inactive and active DWM DCS board periods (Figure 3). In the ditch piezometers, in the DCS inactive (unshaded) periods, the DWM depth values were generally deeper than the non-DWM plot values, indicating spatial heterogeneity across the field. During the DCS active (shaded) periods, however, the lines on Figure 3 tend to converge, indicating that the DCS active installation of boards in the DWM plots increased the ditch piezometer water levels. In the field piezometers, the effect of board placement is less clear. The statistical evaluation of these piezometer data were combined analysis of WFPS data and are found in Section 3.4.

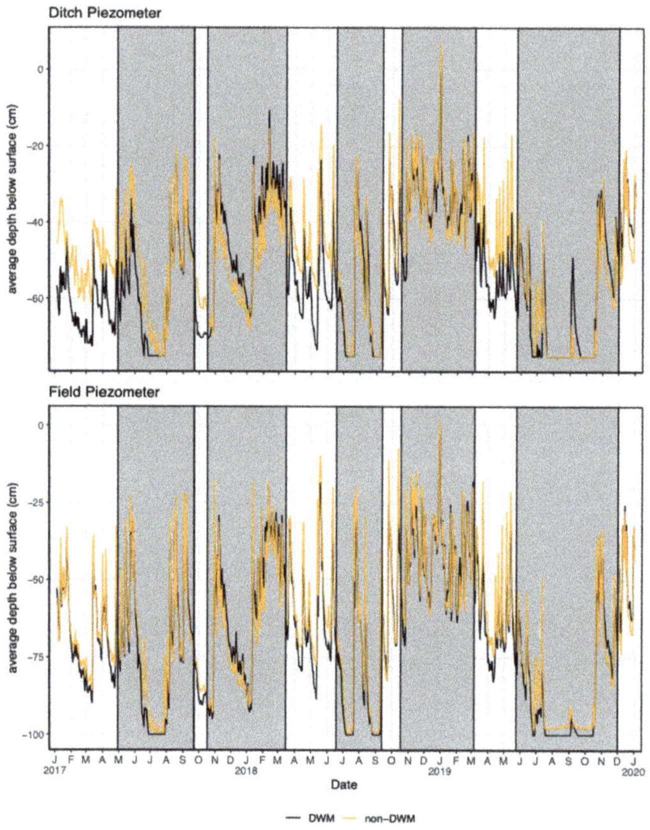

Figure 3. Mean piezometer water depth (cm) colored by treatment and piezometer location. The shaded time frame represents the active DWM DCS period and unshaded the inactive DWM DCS period.

3.3. Chamber Flux Summary

From 2017 to 2019, 959 chamber flux measurements were taken. Most fluxes were small and the data had right-skewed distribution. When calculating the fluxes using the change in slope over time, assessed by linear regression, 96% of the methane slopes were significantly different from zero ($p < 0.05$) and of those 21% were negative, indicating CH_4 consumption events. For N_2O, 85% of the N_2O slopes were significantly different from zero ($p < 0.05$) and 15% of those were negative. Because of the data skew, a log transformation was applied for parametric statistical analyses. An analysis of the minimum detectable flux (MDF) is provided in the supplementary materials (Table S5).

Figure 2E shows the N_2O-N flux data, averaged by treatment. The highest emissions occurred after N fertilization events (planting or broadcast) in the April–July period of 2017 and 2019, marked by arrows. In 2017, the larger N_2O peak occurred during the second fertilization and in 2019, after the first fertilization. The large response to the first fertilization in April 2019 N application may have resulted from co-occurrence of a precipitation event the day following application. Soil N_2O consumption events occurred at various times throughout this study at the individual chamber scale, but never as the average flux from all chambers at the field scale. This indicates small scale flux variability within a field, but not net negative fluxes at the field scale.

For CH_4, the highest CH_4 emissions occurred in the spring period (April–June; Figure 2F). Several individual chambers showed net consumption of atmospheric CH_4, but only rarely did this translate to net consumption as the mean at the field or treatment scale.

3.4. Non-Parametric DWM Model for Effects on Piezometers, WFPS, and Gas Fluxes

The non-parametric statistical analysis results for dependent variables of ditch piezometer, field piezometer, and surface WFPS showed differences in the interaction of treatment and board period (Figure 4). For the ditch piezometers, the mean water depth below ground was significantly deeper during the inactive DCS period than the active DCS period in the non-DWM sites. However, this trend was reversed in the DWM sites, with a higher mean water depth in the active DCS period compared to the inactive DCS period, indicating that the additional boards in the DWM sites raised the water table near the ditch, even during the dry summer. The field piezometer water levels showed the same trends as the ditch piezometers, except that the mean water level between active and inactive DCS periods was not significantly different in the DWM condition, indicating that the additional boards in the DWM sites had less of an effect on the water table in the mid-field, approximately 15 m from the ditch. The DWM only raised the water table in the mid-field piezometers to the same level as the non-DWM plots in the active DCS season, but did not exceed it, as was the case for the ditch piezometers (Figure 4). The WFPS showed no significant differences by treatment or board period, indicating the DWM did not affect near surface moisture content. The same non-parametric analysis for gas fluxes showed no significant differences by treatment or board period, indicating the DWM did not impact the soil gas fluxes in the fields (Figure 5).

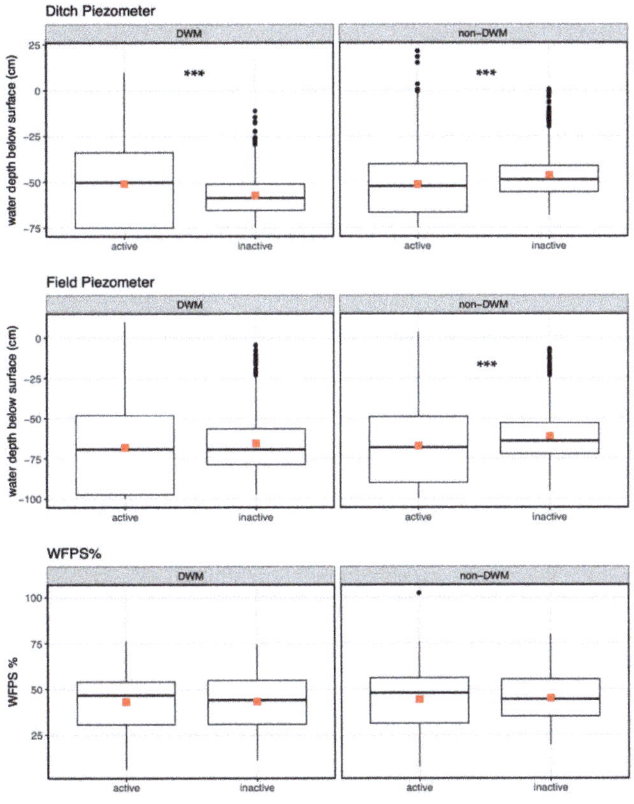

Figure 4. Box plot of the ditch piezometer, field piezometer and %WFPS in surface soil. Data were aggregated by treatment and board period. Asterisks indicate the significance of a Wilcoxon rank sum test for comparison between active DCS versus inactive DCS periods within each treatment (*** indicate $p < 0.001$; no asterisks indicate $p > 0.05$). Red squares represent means.

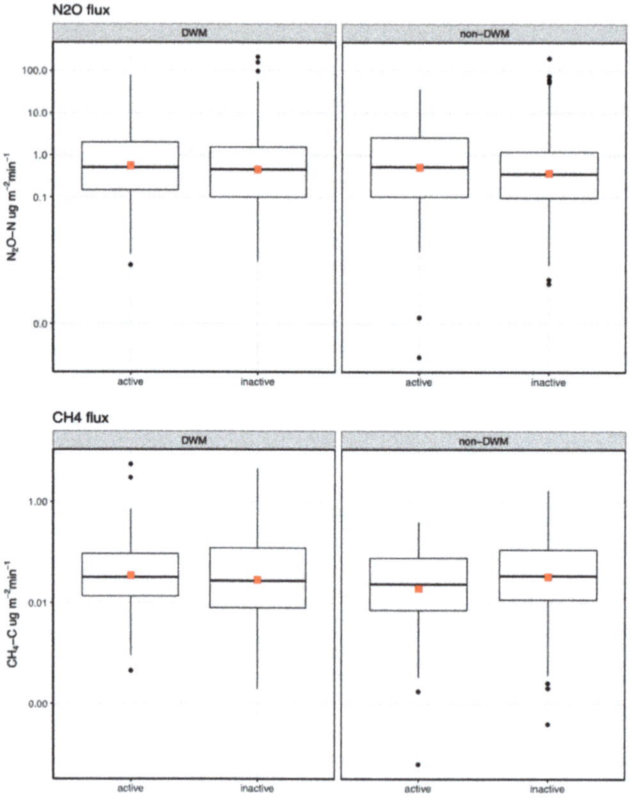

Figure 5. Box plot of gas flux data aggregated by board period and treatment. There were no significant differences ($p > 0.05$) between board period in either of the treatments according to the Wilcoxon rank sum test. Red squares represent means.

3.5. Mixed-Effects Model Results for N_2O and CH_4 Fluxes

The purpose of the mixed-effects model was to assess the impact of the DWM treatment and other dependent variables. For N_2O fluxes, the effects of the post-N fertilization period, year, chamber location, and interactions were assessed using two sub-models. In the first sub-model, the N fertilization effect was evaluated using only 2017 and 2019 data (corn years) because 2018 data (soybean year) did not have a N fertilization period. In the second sub-model the effects of treatment and year were assessed using all years' data, but N fertilization variable was not included. The mixed model results (Table 1) show that DWM did not significantly affect N_2O fluxes, which is consistent with the non-parametric model results (Figures 4 and 5). Year had a significant effect on N_2O fluxes, as the fertilized corn years of 2017 and 2019 had higher N_2O fluxes compared to the unfertilized soybean year of 2018. The post hoc test showed that the N_2O fluxes in 2017 and 2019 were not significantly different. Using the sub-model with only data from N fertilization years, the post-N fertilization period had significantly higher N_2O fluxes than the non-fertilization period (Table 2). No interaction effects between independent variables were significant.

Table 1. Mixed-effects model results for N_2O-N and CH_4-C mean fluxes (± SE) aggregated by year and treatment. For N_2O-N, years 2017 and 2019 were significantly different from 2018, but not significantly different from each other. The effects of year and DWM treatment were not significant for CH_4 fluxes.

	N_2O-N (µg N m^{-2} min^{-1})		CH_4-C (µg C m^2 min^{-1})	
Year	**DWM (A)**	**Non-DWM (A)**	**DWM (A)**	**Non-DWM (A)**
2017	0.94 ± 0.38 a	0.81 ± 0.38 a	0.03 ± 0.01 a	0.02 ± 0.01 a
2018	0.48 ± 0.40 b	0.39 ± 0.39 b	0.03 ± 0.01 a	0.02 ± 0.01 a
2019	1.20 ± 0.39 a	1.02 ± 0.39 a	0.04 ± 0.01 a	0.03 ± 0.01 a

Uppercase letters (A,B) indicate statistical difference ($p < 0.05$) by treatment, within each gas type. Lowercase letters (a,b) indicate statistical difference ($p < 0.05$) by year.

Table 2. Modeled mean ± SE of N_2O-N and CH_4-C fluxes aggregated by N fertilization period for years 2017 and 2019 for N_2O-N and by season for CH_4. The post-fertilization period had significantly higher N_2O emissions compared to the non-N fertilization period. The spring seasonal period had higher CH_4 emissions compared to the non-spring periods.

N_2O-N (µg N m^{-2} min^{-1})		CH_4-C (µg C m^{-2} min^{-1})	
N Fertilization Period	**Mean ± SE**	**Seasonal Period**	**Mean ± SE**
Post-fertilization	2.15 ± 0.38 a	Spring	0.04 ± 0.001 a
Non-fertilization	0.21 ± 0.38 b	Non-spring	0.02 ± 0.001 b

Lower case letters (a,b) next to mean ± SE indicate statistical difference ($p < 0.05$) by period within gas type.

For CH_4, DWM treatment, year and season were tested. CH_4 emissions did not have significant differences based on year or treatment (Table 1). The spring seasonal period had higher CH_4 emissions compared to the non-spring periods (Table 2).

3.6. *N_2O Multiple Regression*

There was a weak positive linear relationship ($r^2 = 0.12$, $p < 0.001$, Figure S5) between soil nitrate and N_2O fluxes. The largest N_2O fluxes occurred when soil nitrate concentrations were high, but some large N_2O fluxes also occurred at lower soil nitrate concentrations. For soil temperature, the higher N_2O emissions occurred above 12 °C, although it should be noted that N fertilization events also only occurred during periods above this temperature. For WFPS, the highest N_2O emission events occurred at intermediate WFPS values, but not all N_2O flux events between those WPFS values resulted in high emissions. The highest N_2O emissions typically occurred when the following combination of conditions existed: the soil temperature was above 12 °C, nitrate concentration was at least 10 mg N kg^{-1}, and WFPS was between 35% and 60% (Figure 6).

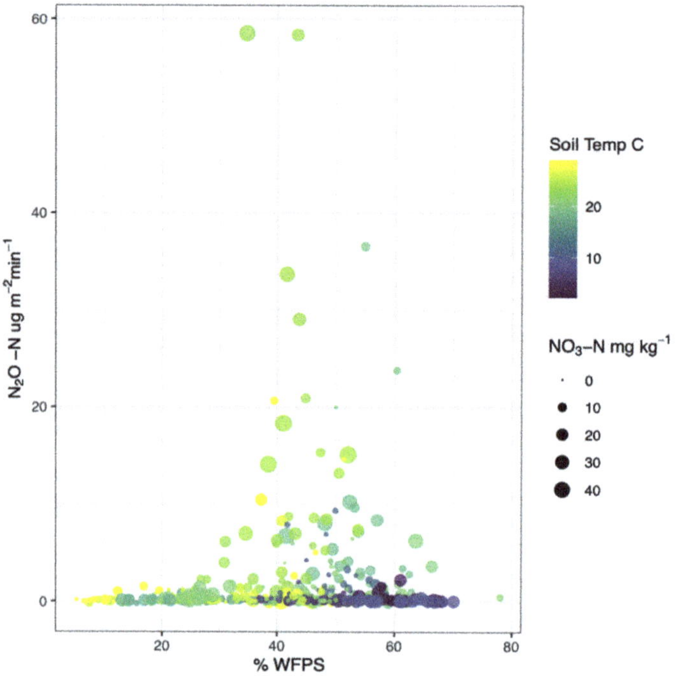

Figure 6. Bubble scatter plot showing the relationship between N_2O fluxes (y-axis), WFPS (x-axis), soil NO_3-N content (circle size), and soil temperature (color range).

The stepwise linear regression yielded the lowest AIC when the model included all terms, including a quadratic term for WFPS. The most important predictor variable was soil nitrate followed by soil temperature, and WFPS. The full model (Equation (2)) and each predictor variable were significant ($p < 0.001$), and the full model yielded an r^2 value of 0.23 (Table S6).

$$\ln\,(N_2O + 1) = (0.029 * NO_3N) + (0.021 * \text{soil temp}) + (0.056 * WFPS) - (0.001 * WFPS^2) - 1.37 \quad (2)$$

where N_2O is N_2O flux (μg N_2O-N m^{-2} min^{-1}), NO_3-N is concentration of nitrate (mg NO_3-N kg^{-1} dry soil), WFPS is water-filled pore space (%), soil temp is soil temperature (°C).

3.7. Annual Soil Emission Flux Estimates

Annual soil gas flux estimates were derived from the three methods described in Section 2.10 (Tables 3 and 4). All the methods indicated that the N_2O-N emissions in the corn years of 2017 and 2019 were elevated in both the DWM and non-DWM conditions compared to the soybean year of 2018. While the annual estimates show higher emissions in 2019 compared to 2017 which is a result of a higher peak N_2O emission and nitrate measurements after the first fertilization event in 2019 (Figure 2), the non-parametric and the mixed-effects model tests discussed above indicated that the difference between 2017 and 2019 fluxes was not significant. The seasonally-based interpolation method yielded higher annual flux estimates, but there was generally broad agreement between the three methods, given the uncertainty terms.

Table 3. Comparison between annual N_2O flux measurement methods. The SE associated with the seasonally-based interpolation and regression-based interpolation means represent spatial heterogeneity among chambers. The mixed-effects model mean extrapolation error represents the reported model mean error.

	N_2O-N kg ha^{-1} yr^{-1} (Mean ± SE)					
	Seasonally-Based Interpolation		Mx Effects Model Mean Extrapolated		Regression-Based Interpolation	
Year	DWM	Non-DWM	DWM	Non-DWM	DWM	Non-DWM
2017	6.4 ± 1.1	4.3 ± 0.8	4.9 ± 1.9	4.3 ± 1.9	3.9 ± 1.0	3.8 ± 0.7
2018	2.1 ± 0.3	1.2 ± 0.3	2.5 ± 2.1	2.0 ± 2.0	1.5 ± 0.2	1.8 ± 0.2
2019	7.7 ± 1.4	7.2 ± 3.5	6.3 ± 2.0	5.7 ± 2.0	5.7 ± 1.3	6.4 ± 2.3

Table 4. Summary of comparison between the two annual CH_4 flux calculation methods. The SE associated with the seasonally-based interpolation represents the spatial heterogeneity among chambers. The mixed-effects model mean extrapolation error was derived from the model output error term.

	CH_4-C kg ha^{-1} yr^{-1} (Mean ± SE)			
	Mx Effects Model Mean Extrapolated		Seasonally-Based Interpolation	
Year	DWM	Non-DWM	DWM	Non-DWM
2017	0.16 ± 0.05	0.11 ± 0.05	0.18 ± 0.10	0.19 ± 0.06
2018	0.16 ± 0.05	0.11 ± 0.05	0.17 ± 0.15	0.15 ± 0.06
2019	0.21 ± 0.05	0.16 ± 0.05	0.56 ± 0.73	0.42 ± 0.12

The seasonally-based interpolation method permitted partitioning fluxes into periods. Individual seasonal period estimates (mean ± 95% CI) for N_2O showed that brief (6–8 weeks) seasonal periods following the fertilization events, which were quantified by the decay function, contributed nearly half of the total annual N_2O flux (Table S1).

There were small net positive annual CH_4 fluxes in each year and treatment (Tables 4 and S2). Consistent with the mixed-effects model, there were no clear differences between years and treatments, except that the seasonally-based interpolation estimates were higher for 2019. This may be an artifact caused by unusually high fluxes measured in a few chambers during a period with sparse temporal coverage.

4. Discussion

4.1. Impact of Drainage Water Management

4.1.1. Groundwater Level

The DWM treatment raised the near-ditch water table when the DCS were active (Figure 4). This effect was also observed in the field piezometers but was weaker, indicating that DWM had an impact on the groundwater storage, with the strongest effect close to the ditch. In contrast, the DWM did not affect the surface soil moisture within the agricultural fields. The observed effectiveness of DWM to reduce nitrate load in previous studies may be due to denitrification occurring in the ditches or in the elevated groundwater leading to the ditches [13,16,44–48]. It is likely that the DWM reduced nitrate leaching into the ditches but did not significantly affect surface (5 cm) soil moisture and rates of denitrification within the surface soils of cropped fields between the ditches where the chamber flux measurements were conducted.

4.1.2. N_2O and CH_4 Fluxes

Soil GHG emissions were not significantly impacted by DWM or the interaction of board period and treatment (Figure 5 and Table 1). While DWM affected the depth of the water table, and preliminary results showed that hydrologic nitrate export was reduced [16], DWM did not impact the surface soil moisture or the GHG fluxes from the soil in the cropped areas (Figure 4 and Table 1). Hence, the results do not support the hypothesis that DWM would increase CH_4 or N_2O soil emissions. This null impact of DWM on GHG emissions confirms other literature findings, particularly those evaluating field scale DWM manipulations [30–33]. Other studies found some increase in CH_4 and N_2O emissions in undrained fields, but were not evaluating the effect of DWM manipulations per se, but rather the effect of tile drainage in general, by comparing drained fields to undrained fields. In those instances, the undrained fields had higher N_2O and CH_4 emissions, although the differences were seasonally variable and not consistent over multiple years [26,27].

4.2. Farm Management Impact on GHGs

For N_2O fluxes, in both DWM and non-DWM fields, the mixed model results (Tables 1 and 2) demonstrated that N fertilization during corn years had the greatest impact on N_2O emissions. Soon after each of the N fertilization events, we observed a pulsed increase in both soil nitrate concentrations and N_2O emissions, which then declined over a period of 1–2 months as soil nitrate declined during crop development (Figure 2D,E). This finding confirms existing literature that shows that N_2O fluxes are highest in the days and weeks following N fertilization [42,49–51].

For all N fertilization events in this study, fluxes were measured 1 day and again 3 or 4 days after. The largest fluxes were observed at 3–4 days after N fertilization. A study of corn/soybean in Brazil demonstrated a similar lag of a few days between fertilization and subsequent N_2O flux spikes, which they attributed to the time it took for nitrification to convert fertilizer-N to nitrate [42].

The multiple regression results align with the published literature demonstrating that key variables of soil nitrate, temperature, and soil moisture play a role in controlling N_2O emissions (Figure 6; Equation (2)) [42,52]. The variation in these parameters fit within the context found for N-fertilized agricultural fields [53].

There was not a specific farm management action that most controlled CH_4 fluxes in this study. Seasonality explained most of the CH_4 flux trends. The highest CH_4 emissions occurred during the period from April to July in all three years of this study. This period corresponded with a time when the temperature had already increased above winter lows but WFPS still remained high ($\geq$55%) due to spring rain and low rates of evapotranspiration before and for a few weeks following planting. The relatively warm and wet conditions were sufficiently conducive to methanogenesis within the soil to produce a period of somewhat elevated emissions relative to other seasons. No DWM effect occurred, and the spring seasonal effect occurred in both treatments.

4.3. Comparing Annual Soil Emission Estimates

For N_2O, all three interpolation methods produced lower annual estimates in the 2018 soybean year compared to the corn years of 2017 and 2019. The higher emissions in the corn years were driven by N fertilization. There was an appearance of a possible difference in treatment means (albeit with overlapping standard errors) in the seasonally-based interpolation in 2017, but this was not observed in 2018 or 2019, nor in the other two annual estimate techniques (Table 3). The non-parametric model and the mixed-effects model did not show significant differences in N_2O gas fluxes between treatments in any year, which suggests that the interpolated treatment means in 2017 should not be considered different. The regression-based interpolation produced similar annual estimates to the mixed-effects model mean extrapolation, and both were somewhat lower than the seasonally-based interpolation. A lower annual estimate for the regression model is likely because non-linear responses to soil nitrate and lack of continuous nitrate data may cause underestimation of the N_2O fluxes. Similarly, extrapolation of the means of the mixed-effects model gives more weight to mean responses rather than pulse (spike) events. Despite limitations for each of the three methods for estimating annual fluxes, they are in broad agreement and therefore represent the likely range of annual fluxes for each year and treatment. Because it captures the post-N fertilization N_2O peaks, we used the seasonally-based interpolation estimate for comparison to the literature in Section 4.4.

For CH_4, the seasonally-based annual interpolation had somewhat higher annual estimates compared to the mixed-effects model mean extrapolation, but all estimates indicated a very small net source, which is consistent with other studies of agricultural soils [54]. There were no differences in treatment in any year for either interpolation technique.

4.4. Comparison to Regional and Global Nitrogen Budget Measures

4.4.1. N_2O Emissions Factor

The nitrous oxide emission factor (EF) was calculated by dividing the seasonally-based interpolated N_2O annual flux, averaged over treatment and year, for the corn years reported in Table 3, by the average N fertilization rate of 213 kg N ha^{-1}. This resulted in an EF of 3.0%. This calculation method does not account for N_2O emissions from a background, unfertilized agriculture site, likely leading to an EF overestimate. To partially account for this, the soybean N_2O emissions in 2018 were used as an unfertilized control, and subtracted from the averaged N_2O emissions in 2017 and 2019. This may overcompensate for the expected flux from an unfertilized soil control because the previous year's fertilization could have had a residual impact on soil N availability in 2018 and because biological N fixation likely added soil N during the soybean year. However, subtracting the entire annual flux from the soybean year provides a lower bracket estimate for the EF during the corn years. This correction yields an EF of 2.2%. An N_2O EF in fertilized agriculture is influenced by a range of environmental variables such as soil type, N input type, and precipitation. The current IPCC guidelines use a default EF range of 1.3–1.9% for synthetic fertilizer agriculture in climates with greater than 1000 mm precipitation [55,56]. The EF estimate from this research is slightly higher than the upper limit of the uncertainty range of IPCC recommendation.

4.4.2. Nitrogen Balance

We calculated a nitrogen balance and nitrogen use efficiency (NUE) based upon average data from the two corn years. The average total amount of nitrogen applied during fertilized corn years was 213 kg N ha^{-1}. The average corn yield was 13,800 kg ha^{-1} (~220 bushels ac^{-1}). The harvested corn had an average (across treatments) nitrogen content of 0.84 ± 0.18% (n = 8), resulting in removal of 91–141 kg N ha^{-1}, the range representing the variation in the measured nitrogen content. The nitrogen balance (also known as N surplus), defined as the amount of N applied minus the amount of N removed via harvest, was estimated to be 97 ± 20 kg N ha^{-1}. Using these estimates of N inputs and outputs, and assuming that annual N deposition rate at this site of 6 kg N ha^{-1} [57] had a negligible impact, and that N fixation was only applicable during soybean crop years, the average nitrogen use efficiency (NUE) for the corn years, defined as the N removed in the harvest divided by the N additions, was 54 ± 11%. This value is somewhat low for the U.S. but falls within the range of values calculated nationally and globally in irrigated maize systems [6,58–61]. Ours is the first study known to us on N budget estimates from corn grown in the U.S. Mid-Atlantic region. Compared to the dataset compiled by [62] from the U.S. cornbelt, which supported a non-linear correlation between annual N_2O emissions and N balance, the annual N_2O emission estimates from the present study fall within the range reported by [62], but are higher than predicted based on their reported relationship with N balance (Figure S6). This result is consistent with the lower than average NUE reported here.

5. Conclusions

The DWM condition did not increase soil N_2O and CH_4 emissions. While there was evidence that DWM impacted the groundwater table near the drainage ditches, this impact was not evident in the surface soil moisture nor the surface soil GHG fluxes themselves. This finding spanned multiple years of corn and soybean crop rotation.

Three independent methods for deriving annual flux estimates revealed consistent estimates within their uncertainty ranges. Comparing the annual N_2O flux estimates and related measures of N cycling to other studies on corn fields in the U.S., this farm had above average N surplus, below average NUE, and above average annual N_2O fluxes, although these values fell within reported regional and global ranges.

Follow-up studies will address the possibility of N_2O accumulation within the groundwater near the ditch areas and degassing of ditch water and will also report on the effect of DWM on export of dissolved N and P. For the cropland soil surface studied here, DWM did not have an unintended consequence of increased greenhouse gas emissions. This study provides evidence that DWM may be an effective BMP to reduce hydrologic nitrate export from crop fields without concerns of significant pollution swapping of increased greenhouse gas emissions from the soil surface.

Supplementary Materials: The following supporting information can be downloaded at: https://www.mdpi.com/article/10.3390/nitrogen3010010/s1, Figure S1: Detail of DCS and DWM technique; Figure S2: Summary of diel variation of N_2O fluxes; Figure S3: Visual of decay curve fitting for annual N_2O estimate; Figure S4: Clay size fraction by depth and field; Figure S5: Summary of individual regression results for input variables into multiple linear regression; Figure S6: Plot of N balance versus N_2O emissions; Table S1: N_2O seasonally-based annual estimate details; Table S2: CH_4 seasonally-based annual estimate details; Table S3: USDA soil textural class description; Table S4: Soil characterization results; Table S5: Summary of minimal detectable flux calculations; Table S6: Summary results of multiple linear regression model. References [63–66] are cited in the Supplementary Materials.

Author Contributions: Conceptualization, E.A.D., T.R.F., R.J.F., M.S.C., J.L. and J.G.H.; methodology, J.G.H., E.A.D., T.R.F., R.J.F., Q.Z., A.B.G., E.K., M.S.C. and J.L.; formal analysis, J.G.H. and E.A.D.; investigation, J.G.H., T.R.F., R.J.F., Q.Z., A.B.G., E.K. and M.S.C.; data curation, J.G.H.; writing—original draft preparation, J.G.H. and E.A.D.; writing—review and editing, J.G.H., E.A.D., T.R.F., R.J.F., Q.Z., A.B.G., E.K., M.S.C. and J.L.; visualization, J.G.H. and E.A.D.; supervision, E.A.D., T.R.F., R.J.F., M.S.C. and J.L.; project administration, E.A.D., T.R.F., R.J.F., M.S.C. and J.L.; funding acquisition, E.A.D., T.R.F., R.J.F., M.S.C., J.L. and J.G.H. All authors have read and agreed to the published version of the manuscript.

Funding: This research was funded by USDA NIFA, grant number 2016-67019-25263, Maryland SeaGrant R/E-22F, Maryland Water Resources Research Center Z9212103, and the UMCES Appalachian Laboratory.

Institutional Review Board Statement: Not applicable.

Informed Consent Statement: Not applicable.

Data Availability Statement: Data available in a publicly accessible repository. The data presented in this study are openly available in [FigShare] at [10.6084/m9.figurshare.19349828].

Acknowledgments: We would like to acknowledge training and technical support from Katie Kline at UMCES Appalachian Laboratory. We acknowledge data collection work performed by Michelle Lepori-Bui through UMCES Horn Point Laboratory. We acknowledge Nick Nickerson for the loaning of the Eosense automated chamber gas flux equipment for a period of this research.

Conflicts of Interest: The authors declare no conflict of interest. The funders had no role in the design of the study; in the collection, analyses, or interpretation of data; in the writing of the manuscript, or in the decision to publish the results.

Sample Availability: Samples are not available.

References

1. Gilbert, P.M. From hogs to HABs: Impacts of industrial farming in the US on nitrogen and phosphorus and greenhouse gas pollution. *Biogeochemistry* **2020**, *150*, 139–180. [CrossRef] [PubMed]
2. Wieczorek, M.E.; LaMotte, A.E. *Attributes for NHDPlus Catchments (Version 1.1) for the Conterminous United States: Nutrient Application (Phosphorus and Nitrogen) for Fertilizer and Manure Applied to Crops (Cropsplit), 2002*; Data Series DS-490-08; U.S. Geological Survey Digital: Reston, VA, USA, 2010.
3. Ator, S.W.; Denver, J.M. *Understanding the Nutrients in the Chesapeake Bay Watershed and Implications for Management and Restoration: The Eastern Shore*; Circular Series 1406; U.S. Geological Survey Digital: Reston, VA, USA, 2015; p. 84.

4. Sabo, R.D.; Clark, C.M.; Bash, J.; Sobota, D.; Cooter, E.; Dobrowolski, J.P.; Houlton, B.Z.; Rea, A.; Schwede, D.; Morford, S.L.; et al. Decadal shift in nitrogen inputs and fluxes across the contiguous United States: 2002–2012. *J. Geophys. Res. Biogeosciences* **2019**, *124*, 3104–3124. [CrossRef]
5. Smil, V. Nitrogen in crop production: An account of global flows. *Glob. Biogeochem. Cycles* **1999**, *13*, 647–662. [CrossRef]
6. Lassaletta, L.; Billen, G.; Grizzetti, B.; Anglade, J.; Garnier, J. 50 year trends in nitrogen use efficiency of world cropping systems: The relationship between yield and nitrogen input to cropland. *Environ. Res. Lett.* **2014**, *9*, 105011. [CrossRef]
7. Zhang, X.; Zou, T.; Lassaletta, L.; Mueller, N.D.; Tubiello, F.N.; Lisk, M.D.; Lu, C.; Conant, R.T.; Dorich, C.D.; Gerber, J.; et al. Quantification of global and national nitrogen budgets for crop production. *Nat. Food* **2021**, *2*, 529–540. [CrossRef]
8. Skaggs, R.W.; Breve, M.A.; Gilliam, J.W. Hydrologic and water quality impacts of agricultural drainage. *Crit. Rev. Environ. Sci. Technol.* **1994**, *24*, 1–32. [CrossRef]
9. Evans, R.O.; Skaggs, R.W.; Gilliam, J.W. Controlled versus conventional drainage effects on water quality. *J. Irrig. Drain. Eng.* **1995**, *121*, 271–276. [CrossRef]
10. Skaggs, R.W.; Fausey, N.R.; Evans, R.O. Drainage water management. *J. Soil Water Conserv.* **2012**, *67*, 167A–172A. [CrossRef]
11. Firestone, M.K.; Davidson, E.A. Microbiological basis of NO and N_2O production and consumption in soil. *Exch. Trace Gases Terr. Ecosyst. Atmos.* **1989**, *47*, 7–21.
12. Nömmik, H. Investigations on denitrification in soil. *Acta Agric. Scand.* **1956**, *6*, 195–228. [CrossRef]
13. Carstensen, M.V.; Hashemi, F.; Hoffmann, C.C.; Zak, D.; Audet, J.; Kronvang, B. Efficiency of mitigation measures targeting nutrient losses from agricultural drainage systems: A review. *Ambio* **2020**, *49*, 1820–1837. [CrossRef]
14. Drury, C.F.; Tan, C.S.; Reynolds, W.D.; Welacky, T.W.; Oloya, T.O.; Gaynor, J.D. Managing tile drainage, subirrigation, and nitrogen fertilization to enhance crop yields and reduce nitrate loss. *J. Environ. Qual.* **2009**, *38*, 1193–1204. [CrossRef]
15. Sunohara, M.D.; Craiovan, E.; Topp, E.; Gottschall, N.; Drury, C.F.; Lapen, D.R. Comprehensive nitrogen budgets for controlled tile drainage fields in eastern Ontario, Canada. *J. Environ. Qual.* **2014**, *43*, 617–630. [CrossRef]
16. Hagedorn, J.; Davidson, E.; Fox, R.; Koontz, E.; Fisher, T.; Castro, M.; Zhu, Q.; Gustafson, A. Pollution Swapping of N_2O and CH_4 Emissions with Dissolved Nitrogen and Phosphorus Export in Drainage Water Managed Agricultural Field. In Proceedings of the AGU 2019 Fall Meeting, San Francisco, CA, USA, 9–13 December 2019. [CrossRef]
17. Kanter, D.; Wagner-Riddle, C.; Groffman, P.M.; Davidson, E.A.; Galloway, J.N.; Gourevitch, J.D.; van Grinsven, H.J.M.; Houlton, B.Z.; Keeler, B.L.; Ogle, S.M.; et al. Improving the social cost of nitrous oxide. *Nat. Clim. Change* **2021**, *11*, 1008–1010. [CrossRef]
18. Prather, M.J.; Hsu, J.; DeLuca, N.M.; Jackman, C.H.; Oman, L.; Douglass, A.R.; Fleming, E.L.; Strahan, S.E.; Steenrod, S.D.; Søvde, O.A.; et al. Measuring and modeling the lifetime of nitrous oxide including its variability. *J. Geophys. Res. Atmos.* **2015**, *120*, 5693–5705. [CrossRef]
19. US-EPA. In *Inventory of US Greenhouse Gas Emissions and Sinks, 1990–2019; US Environmental Protection Agency, 2021.* Available online: https://www.epa.gov/ghgemissions/inventory-us-greenhouse-gas-emissions-and-sinks (accessed on 13 November 2021).
20. Xu, R.; Tian, H.; Pan, N.; Thompson, R.L.; Canadell, J.G.; Davidson, E.A.; Nevison, C.; Winiwarter, W.; Shi, H.; Pan, S.; et al. Magnitude and uncertainty of nitrous oxide emissions from North America based on bottom-up and top-down approaches: Informing future research and national inventories. *Geophys. Res. Lett.* **2021**, *48*, e2021GL095264. [CrossRef]
21. Knowles, R. Methane: Processes of production and consumption. *Agric. Ecosyst. Eff. Trace Gases Glob. Clim. Change* **1993**, *55*, 145–156.
22. Pachauri, R.K.; Meyers, L.A. *Climate Change 2014: Synthesis Report. Contribution of Working Groups I, II and III to the Fifth Assessment Report of the Intergovernmental Panel on Climate Change*; IPCC: Geneva, Switzerland, 2014; p. 151.
23. Bateman, E.J.; Baggs, E.M. Contributions of nitrification and denitrification to N2O emissions from soils at different water-filled pore space. *Biol. Fertil. Soils* **2005**, *41*, 379–388. [CrossRef]
24. Fisher, T.R.; Fox, R.J.; Gustafson, A.B.; Lewis, J.; Millar, N.; Winstene, J.R. Fluxes of nitrous oxide and nitrate from agricultural fields on the Delmarva Peninsula: N biogeochemistry and economics of field management. *Agric. Ecosyst. Environ.* **2018**, *254*, 162–178. [CrossRef]
25. Dobbie, K.E.; McTaggart, I.P.; Smith, K.A. Nitrous oxide emissions from intensive agricultural systems: Variations between crops and seasons, key driving variables, and mean emission factors. *J. Geophys. Res. Atmos.* **1999**, *104*, 26891–26899. [CrossRef]
26. Kumar, S.; Nakajima, T.; Kadono, A.; Lal, R.; Fausey, N. Long-term tillage and drainage influences on greenhouse gas fluxes from a poorly drained soil of central Ohio. *J. Soil Water Conserv.* **2014**, *69*, 553–563. [CrossRef]
27. Fernández, F.G.; Venterea, R.T.; Fabrizzi, K.P. Corn nitrogen management influences nitrous oxide emissions in drained and undrained soils. *J. Environ. Qual.* **2016**, *45*, 1847–1855. [CrossRef] [PubMed]
28. Kliewer, B.A.; Gilliam, J.W. Water table management effects on denitrification and nitrous oxide evolution. *Soil Sci. Soc. Am. J.* **1995**, *59*, 1694–1701. [CrossRef]
29. Elmi, A.; Burton, D.; Gordon, R.; Madramootoo, C. Impacts of water table management on N_2O and N_2 from a sandy loam soil in southwestern Quebec, Canada. *Nutr. Cycl. Agroecosystems* **2005**, *72*, 229–240. [CrossRef]
30. Nangia, V.; Sunohara, M.; Topp, E.; Gregorich, E.; Drury, C.; Gottschall, N.; Lapen, D. Measuring and modeling the effects of drainage water management on soil greenhouse gas fluxes from corn and soybean fields. *J. Environ. Manag.* **2013**, *129*, 652–664. [CrossRef] [PubMed]
31. Crézé, C.M.; Madramootoo, C.A. Water table management and fertilizer application impacts on CO_2, N_2O and CH_4 fluxes in a corn agro-ecosystem. *Sci. Rep.* **2019**, *9*, 1–13.

32. Datta, A.; Smith, P.; Lal, R. Effects of long-term tillage and drainage treatments on greenhouse gas fluxes from a corn field during the fallow period. *Agric. Ecosyst. Environ.* **2013**, *171*, 112–123. [CrossRef]
33. Van Zandvoort, A.; Lapen, D.; Clark, I.; Flemming, C.; Craiovan, E.; Sunohara, M.; Boutz, R.; Gottschall, N. Soil CO_2, CH_4, and N_2O fluxes over and between tile drains on corn, soybean, and forage fields under tile drainage management. *Nutr. Cycl. Agroecosystems* **2017**, *109*, 115–132. [CrossRef]
34. Kravchenko, A.N.; Robertson, G.P. Statistical challenges in analyses of chamber-based soil CO_2 and N_2O emissions data. *Soil Sci. Soc. Am. J.* **2015**, *79*, 200–211. [CrossRef]
35. Levy, P.E.; Gray, A.; Leeson, S.R.; Gaiawyn, J.; Kelly, M.P.C.; Cooper, M.D.A.; Dinsmore, K.J.; Jones, S.K.; Sheppard, L.J. Quantification of uncertainty in trace gas fluxes measured by the static chamber method. *Eur. J. Soil Sci.* **2011**, *62*, 811–821. [CrossRef]
36. Topp, G.C.; Davis, J.L.; Annan, A.P. Electromagnetic determination of soil water content: Measurements in coaxial transmission lines. *Water Resour. Res.* **1980**, *16*, 574–582. [CrossRef]
37. Maynard, D.G.; Kalra, Y.P.; Crumbaugh, J.A. Nitrate and exchangeable ammonium nitrogen. In *Soil Sampling and Methods of Analysis*; CRC Press: Boca Raton, FL, USA, 1993; Volume 1.
38. Robertson, G.P.; Coleman, D.C.; Sollins, P.; Bledsoe, C.S. *Standard Soil Methods for Long-Term Ecological Research*; Oxford University Press: Oxford, UK, 1999; Volume 2.
39. Gavlak, R.; Horneck, D.; Miller, R.O. *Soil, Plant and Water Reference Methods for the Western Region*; WREP-125; Western Region Extension Publication: Logan, UT, USA, 2005.
40. Van der Weerden, T.J.; Noble, A.N.; Luo, J.; de Klein, C.A.M.; Saggar, S.; Giltrap, D.; Gibbs, J.; Rys, G. Meta-analysis of New Zealand's nitrous oxide emission factors for ruminant excreta supports disaggregation based on excreta form, livestock type and slope class. *Sci. Total Environ.* **2020**, *732*, 139235. [CrossRef]
41. Bolker, B.M.; Brooks, M.E.; Clark, C.J.; Geange, S.W.; Poulsen, J.R.; Stevens, M.H.H.; White, J.-S.S. Generalized linear mixed models: A practical guide for ecology and evolution. *Trends Ecol. Evol.* **2009**, *24*, 127–135. [CrossRef]
42. Jankowski, K.; Neill, C.; Davidson, E.A.; Macedo, M.N.; Costa, C.; Galford, G.L.; Santos, L.M.; Lefebvre, P.; Nunes, D.; Cerri, C.E.P.; et al. Deep soils modify environmental consequences of increased nitrogen fertilizer use in intensifying Amazon agriculture. *Sci. Rep.* **2018**, *8*, 1–11. [CrossRef]
43. Matson, P.A.; Billow, C.; Hall, S.; Zachariassen, J. Fertilization practices and soil variations control nitrogen oxide emissions from tropical sugar cane. *J. Geophys. Res. Atmos.* **1996**, *101*, 18533–18545. [CrossRef]
44. Fisher, T.R.; Jordan, T.E.; Staver, K.W.; Gustafson, A.B.; Koskelo, A.I.; Fox, R.J.; Sutton, A.J.; Kana, T.; Beckert, K.A.; Stone, J.P.; et al. The Choptank Basin in transition: Intensifying agriculture, slow urbanization, and estuarine eutrophication. In *Coastal Lagoons: Systems of Natural and Anthropogenic Change*; Kennish, M.J., Paerl, H.W., Eds.; CRC Press: Boca Raton, FL, USA, 2010; pp. 135–165.
45. Williams, M.R.; King, K.W.; Fausey, N.R. Drainage water management effects on tile discharge and water quality. *Agric. Water Manag.* **2015**, *148*, 43–51. [CrossRef]
46. Lavaire, T.; Gentry, L.E.; David, M.B.; Cooke, R.A. Fate of water and nitrate using drainage water management on tile systems in east-central Illinois. *Agric. Water Manag.* **2017**, *191*, 218–228. [CrossRef]
47. Smith, E.L.; Kellman, L.M. Nitrate loading and isotopic signatures in subsurface agricultural drainage systems. *J. Environ. Qual.* **2011**, *40*, 1257–1265. [CrossRef]
48. Lalonde, V.; Madramootoo, C.A.; Trenholm, L.; Broughton, R.S. Effects of controlled drainage on nitrate concentrations in subsurface drain discharge. *Agric. Water Manag.* **1996**, *29*, 187–199. [CrossRef]
49. Parkin, T.B.; Kaspar, T.C. Nitrous oxide emissions from corn–soybean systems in the Midwest. *J. Environ. Qual.* **2006**, *35*, 1496–1506. [CrossRef]
50. Baggs, E.M.; Stevenson, M.; Pihlatie, M.; Regar, A.; Cook, H.; Cadisch, G. Nitrous oxide emissions following application of residues and fertiliser under zero and conventional tillage. *Plant Soil* **2003**, *254*, 361–370. [CrossRef]
51. Bremner, J.M.; Breitenbeck, G.A.; Blackmer, A.M. Effect of nitrapyrin on emission of nitrous oxide from soil fertilized with anhydrous ammonia. *Geophys. Res. Lett.* **1981**, *8*, 353–356. [CrossRef]
52. Ullah, S.; Moore, T.R. Biogeochemical controls on methane, nitrous oxide, and carbon dioxide fluxes from deciduous forest soils in eastern Canada. *J. Geophys. Res. Biogeosciences* **2011**, *116*. [CrossRef]
53. Venterea, R.T.; Coulter, J.A. Split application of urea does not decrease and may increase nitrous oxide emissions in rainfed corn. *Agron. J.* **2015**, *107*, 337–348. [CrossRef]
54. Cowan, N.; Maire, J.; Krol, D.; Cloy, J.M.; Hargreaves, P.; Murphy, R.; Carswell, A.; Jones, S.K.; Hinton, N.; Anderson, M.; et al. Agricultural soils: A sink or source of methane across the British Isles? *Eur. J. Soil Sci.* **2021**, *72*, 1842–1862. [CrossRef]
55. Intergovernmental Panel on Climate Change. 2019 Refinement to the 2006 IPCC Guidelines for National Greenhouse Gas Inventories, Volume 4—Agriculture, Forestry and Other Land Use, Chapter 3. *Consistent Representation of Lands.* 2019. Available online: https://www.ipcc-nggip.iges.or.jp/public/2019rf/vol4.html (accessed on 13 November 2021).
56. Hergoualc'H, K.; Mueller, N.; Bernoux, M.; Kasimir, Ä.; van der Weerden, T.J.; Ogle, S.M. Improved accuracy and reduced uncertainty in greenhouse gas inventories by refining the IPCC emission factor for direct N_2O emissions from nitrogen inputs to managed soils. *Glob. Change Biol.* **2021**, *27*, 6536–6550. [CrossRef]
57. Fisher, T.R.; Fox, R.J.; Gustafson, A.B.; Koontz, E.; Lepori-Bui, M.; Lewis, J. Localized water quality improvement in the choptank estuary, a tributary of Chesapeake Bay. *Estuaries Coasts* **2021**, *44*, 1274–1293. [CrossRef]

58. Conant, R.T.; Berdanier, A.B.; Grace, P.R. Patterns and trends in nitrogen use and nitrogen recovery efficiency in world agriculture. *Glob. Biogeochem. Cycles* **2013**, *27*, 558–566. [CrossRef]
59. Zhang, X.; Davidson, E.A.; Mauzerall, D.L.; Searchinger, T.D.; Dumas, P.; Shen, Y. Managing nitrogen for sustainable development. *Nature* **2015**, *528*, 51–59. [CrossRef]
60. Basso, B.; Shuai, G.; Zhang, J.; Robertson, G.P. Yield stability analysis reveals sources of large-scale nitrogen loss from the US Midwest. *Sci. Rep.* **2019**, *9*, 1–9. [CrossRef]
61. Roy, E.D.; Wagner, C.R.H.; Niles, M.T. Hot spots of opportunity for improved cropland nitrogen management across the United States. *Environ. Res. Lett.* **2021**, *16*, 035004. [CrossRef]
62. Eagle, A.J.; McLellan, E.L.; Brawner, E.M.; Chantigny, M.H.; Davidson, E.A.; Dickey, J.B.; Linquist, B.A.; Maaz, T.M.; Pelster, D.E.; Pittelkow, C.M.; et al. Quantifying on-farm nitrous oxide emission reductions in food supply chains. *Earth's Future* **2020**, *8*, e2020EF001504. [CrossRef]
63. Verchot, L.V.; Davidson, E.; Cattanio, J.H.; Ackerman, I.L.; Erickson, H.E.; Keller, M. Land use change and biogeochemical controls of nitrogen oxide emissions from soils in eastern Amazonia. *Glob. Biogeochem. Cycles* **1999**, *13*, 31–46. [CrossRef]
64. Courtois, E.A.; Stahl, C.; Burban, B.; Van den Berge, J.; Berveiller, D.; Bréchet, L.; Soong, J.L.; Arriga, N.; Peñuelas, J.; Janssens, I.A. Automatic high-frequency measurements of full soil greenhouse gas fluxes in a tropical forest. *Biogeosciences* **2019**, *16*, 785–796. [CrossRef]
65. Christiansen, J.R.; Outhwaite, J.; Smukler, S.M. Comparison of CO_2, CH_4 and N_2O soil-atmosphere exchange measured in static chambers with cavity ring-down spectroscopy and gas chromatography. *Agric. For. Meteorol.* **2015**, *211*, 48–57. [CrossRef]
66. Nickerson, N. *Evaluating Gas Emission Measurements using Minimum Detectable Flux (MDF)*; Eosense Inc.: Dartmouth, Canada, 2016.

Nitrogen

Review

Beyond Soil Inoculation: Cyanobacteria as a Fertilizer Replacement

Michael S. Massey [1] and Jessica G. Davis [2,*]

[1] Environment Canterbury, Christchurch 8140, New Zealand; michael.massey@ecan.govt.nz
[2] Agricultural Experiment Station, Colorado State University, Fort Collins, CO 80523, USA
* Correspondence: jessica.davis@colostate.edu

Abstract: Nitrogen-fixing bacteria such as cyanobacteria have the capability to fix atmospheric nitrogen at ambient temperature and pressure, and intensive cultivation of cyanobacteria for fertilizer could lead to its use as an "environmentally friendly" replacement or supplement for nitrogen (N) fertilizer derived from the Haber–Bosch process. Prior research has focused on the use of N-fixing bacteria as a soil inoculum, and while this can improve crop yields, yield improvements are generally attributed to plant-growth-promoting substances produced by the bacteria, rather than to biological N fixation. The intensive cultivation of cyanobacteria in raceways or bioreactors can result in a fertilizer that provides N and organic carbon, as well as potentially similar growth-promoting substances observed in prior research work. On-farm or local production of cyanobacterial fertilizer could also circumvent infrastructure limitations, economic and geopolitical issues, and challenges in distribution and transport related to Haber–Bosch-derived N fertilizers. The use of cyanobacterial N fertilizer could have many agronomic and environmental advantages over N fertilizer derived from the Haber–Bosch process, but study of cyanobacteria as a replacement for other N fertilizers remains very limited. Scientific and practical challenges remain for this promising but as-yet unproven approach to maintaining or improving soil N fertility.

Keywords: biological N fixation; Haber–Bosch process; cyanobacteria

1. Going beyond Soil Inoculation

Nitrogen (N) is the major nutrient needed by crops in the largest quantities. Sources of N fertilizer have historically included mined Chilean nitrate ($NaNO_3$), guano from birds and bats, and farmyard manure. After the Haber–Bosch process was developed in the early 1900s to convert N_2 from the atmosphere into NH_3 (gas), NH_3-based fertilizers have dominated the market, including urea, ammonium nitrate, and mixed fertilizers, such as mono- or di-ammonium phosphates. The global human population has increased from fewer than 2 billion people in the early 1900s to more than 8 billion people today, magnifying the demand for N fertilizer to grow crops and raise livestock for human consumption.

Cyanobacteria have long been recognized to occur naturally in flooded rice paddies and to benefit rice through biological N fixation [1]. The inoculation of soil with N-fixing bacteria has been considered as a method of increasing N availability for dryland crops, not only those grown in paddies. In particular, heterotrophic N-fixing bacteria such as *Azospirillum* sp. have been used as an inoculum [2,3] and consume organic matter in the soil for their growth, as well as to power the fixation of "reactive" nitrogen (e.g., ammonium or nitrate) from atmospheric N_2 with the nitrogenase enzyme. Inoculation of soil with heterotrophic N-fixing bacteria can increase yields [2,4], but yield increases were attributed to promotion of root growth, rather than to biological N fixation [4]. More recent research supports the finding that inoculation with *Azospirillum* sp. can improve plant growth and increase resilience, even if its use as a N fertilizer is limited [5]. Cyanobacteria are found in soil crusts and have been studied for their potential to improve the physical properties of soil, in addition to potential chemical changes [6–8]. Others investigated the inoculation of soil with cyanobacteria as a fertilizer [9–11], but the inoculum could not compete with native microorganisms [9]. In a pot study, it was reported that cyanobacterial inoculation enhanced soil inorganic N by less than 3 mg kg^{-1}, a very small amount compared to typical crop N uptake rates of 50–100 mg kg^{-1} or more [12]. In other research, inoculation with cyanobacteria improved rice yields, but issues with contamination, growing season limitations, and slow growth rate reportedly made the practice unpopular with farmers [1]. Soil can also be inoculated with other microorganisms (e.g., mycorrhizal fungi such as Glomeromycetes, Ascomycetes, or Basidiomycetes) which can enhance plant N uptake, but which do not themselves appreciably increase the quantity of soil N available for plant uptake.

Symbiotic N fixation (e.g., with leguminous plants and N-fixing *Rhizobium* sp.) can, in contrast, provide for a plant's entire N requirement, and can substantially enhance soil N fertility. However, symbiotic N fixation comes at a cost in both time and water use and might, therefore, be impractical in situations where water is scarce or growing seasons are limited. The focus of this work is to look beyond the inoculation of soil, and beyond symbiotic N fixation, to the use of intensively grown autotrophic N-fixing bacteria (i.e., cyanobacteria) as a replacement for other, more common N fertilizers such as urea. Objectives of this review are to explain some of the disadvantages of N fertilizers produced through the Haber–Bosch process, to elucidate the potential benefits of cyanobacterial fertilizer, and to explore the challenges for research needed to bring this opportunity to fruition.

2. Disadvantages of Fertilizers Produced by the Haber–Bosch Process

High-temperature (circa 500 °C), high-pressure (circa 300 atm) fixation of ammonia (NH_3) from atmospheric N_2 via the Haber–Bosch process provides the crucial feedstock for production of other common N fertilizers such as urea. Although almost 40% of the human population depends upon food grown using fertilizer produced through the Haber–Bosch process [13], on the other hand, the amount of anthropogenic N in the Earth system is roughly equal to the amount of natural N—that is, humans have doubled the amount of reactive N in the Earth system since the advent of the Haber–Bosch process [14–16]. This doubling of N pools and fluxes has come with environmental consequences [17], not the least of which is the amount of energy consumed by the process, in addition to energy used in distribution and transport. Depending on the country, fertilizer production, distribution, and application can account for a large proportion of energy use in modern diets. For example, Sandström et al. [18] found that fertilizer production and use accounts for about 14% of the total carbon footprint of the diets of people in the European Union. Process energy for urea, i.e., the energy used in the urea production process, amounted to circa 5.3 GJ/metric ton in 2019; carbon dioxide emissions from N fixation amount to circa 2.0 t CO_2 per ton of fixed NH_3 [19,20]; yet process energy reportedly only accounted for 10% of the total energy used for N fertilizer in agriculture [19]. Between manufacturing and distribution, energy usage and CO_2 emissions are major drawbacks to the fertilizers produced by the Haber–Bosch process. Once fertilizer is applied it can drive further CO_2 emissions from increased mineralization of soil organic matter [21], as well as emission of nitrogenous greenhouse gases such as nitrous oxide [22].

There are also economic and geopolitical implications, as well as infrastructure limitations to the manufacture and distribution of fertilizers produced by the Haber–Bosch process. Methane (natural gas) is the most common source of hydrogen (H) for the production of NH_3 from N_2 gas, and natural gas can also be used as an energy source for N fixation in the Haber–Bosch process. When natural gas prices rise, such as when supplies are restricted due to geopolitical instability, fertilizer prices also tend to rise, with wide-ranging impacts. Nitrogen fixation requires a reliable supply of H, as well as power, so cannot occur in places where adequate supplies are not available. Therefore, the product must be distributed after manufacture, and the point of use might be very remote from the point of manufacture. Distribution requires infrastructure which might not be available at all in some areas, or if infrastructure is available, the costs of transport could be considerably higher than in industrialized nations with extensive transportation networks. Infrastructure limitations contribute to high fertilizer prices, limited fertilizer availability, and limited fertilizer use in some areas of the world. Indeed, these factors contribute to observed imbalances in agricultural development, with some areas facing persistent soil nutrient deficits, while other regions have substantial soil nutrient surpluses [23] which, in turn, result in negative impacts to water quality and human and ecological health [17].

Recent research efforts have focused on improvements to the Haber–Bosch process, such as the development of new catalysts for lower-temperature N fixation [24,25]. Additionally, progress has been made on moving away from fossil fuel use for N fixation, often also with a focus on the use of new catalysts or renewable energy sources [20,26], or electrochemical reduction of N_2 to NH_3 [27]. While these approaches are promising, they only address issues with the production of common fertilizers such as urea—while not addressing difficulties in distribution, or issues that occur after the fertilizer is applied to the soil. In contrast, distributed production of biologically based N fertilizer with some organic matter content has the potential to at least partially address some of the drawbacks of fertilizers made using NH_3 from the Haber–Bosch process.

3. Potential Benefits of Cyanobacterial Fertilizer

Cyanobacteria are a diverse phylum of Gram-negative bacteria that exhibit different morphological characteristics (single-celled, filamentous, etc.) and inhabit many environments (terrestrial, freshwater aquatic, marine, etc.), though they might be dormant in certain conditions or at certain times. Cyanobacteria can photosynthesize, and some can use the energy gained from photosynthesis to fix inert N_2 gas from the atmosphere into biologically reactive forms of N such as NH_3. Unlike the Haber–Bosch process, cyanobacterial N fixation occurs at ambient temperature and pressure. In cyanobacteria, photosynthesis and N fixation are separated either in space (for example, through specialized cells where N fixation takes place) or in time (daytime photosynthesis and nighttime N fixation) within individual bacterium. By intensively growing and harnessing cyanobacteria as a fertilizer, we can effectively use solar power through photosynthesis to fix N into plant-available forms and decrease or eliminate the use of fossil fuels for N fertilizer production. Cyanobacteria production methods can be applied across a variety of scales, from a large-scale industrial facility to small-scale, on-farm production. Therefore, a distributed model where farmers or farming communities grow their own fertilizer should be explored (Figure 1). If cyanobacterial fertilizer was grown on-farm or within the local community, the use of fossil fuels for distribution and transport, and the associated costs, could be greatly reduced.

Photosynthesis also effectively sequesters CO_2 [28], and additional C from application of live cyanobacteria to land often results in increases in microbial biomass, while also improving soil health parameters such as aggregate stability and water retention [29]. Along with decreased CO_2 emissions, improvements in CO_2 sequestration in soil might therefore result from application of cyanobacteria to soil as fertilizer.

On-farm cyanobacterial fertilizer production offers flexibility in both its production and use. For example, it can be produced as certified organic fertilizer [30] or using conventional sources of nutrients [31]. In addition, it can be utilized directly and applied as a liquid as described above, or it can be dried down and applied as a solid fertilizer [32,33]. When aeration and mixing is stopped, cyanobacterial biomass sinks in the production ponds, the supernatant can be pumped off, and the process is repeated until there is little remaining liquid. Then fans can be utilized to air-dry the cyanobacterial biomass for use as a solid fertilizer.

The N fixation process that takes place in cyanobacterial cells is facilitated by the nitrogenase enzyme and results in the production of amino acids (e.g., glutamine, glutamic acid). Since plants generally take up nitrate and ammonium, the amino acids must be broken down through the mineralization process to be converted into these inorganic forms. A 140 d laboratory incubation was used to compare the N mineralization of solid cyanobacterial biofertilizer with compost [34]. The N availability (N in the nitrate and ammonium forms) from cyanobacterial biofertilizer was 31.6% at the end of the incubation, while compost demonstrated only 15.5% of the applied N to be plant-available in the same period of time. The incubation study was followed by a greenhouse study which showed that cyanobacterial biofertilizer resulted in a significantly higher yield of kale (*Brassica oleracea*) as compared to compost application [34].

Figure 1. Producing cyanobacterial fertilizer on-farm. (**a**) Hoop house in background where cyanobacteria are grown. (**b**) Cyanobacterial production in a raceway inside of a hoop house. (**c**) Paddle wheel used to mix and aerate cyanobacterial culture. Photos by J.G. Davis.

In addition to the reduction in fossil fuel use for N fertilizer production through on-site N fixation powered by photosynthesis, application of cyanobacterial fertilizer through an irrigation system (fertigation) with no dehydration required has also been shown to reduce greenhouse gas (N_2O and CO_2) and NH_3 emissions [35,36]. These field studies evaluated solid (blood meal, feather meal) and liquid organic fertilizers (cyanobacterial fertilizer, fish emulsion) applied to vegetable crops; solid fertilizers were applied in a pre-plant subsurface band, and liquid fertilizers were applied in small doses throughout the growing season through a surface drip irrigation system. The reduced emissions from cyanobacterial fertilizer were likely due to its application as a liquid through fertigation.

Furthermore, cyanobacteria have been shown to solubilize inorganic phosphates such as hydroxyapatite in agar [37], in submerged conditions such as in rice paddies [38,39], and in incubations of moist soils at field capacity [40,41]. *Anabaena, Nostoc,* and *Westiellopsis* sp. have demonstrated the ability to solubilize tricalcium phosphate and hydroxyapatite [29], and several different mechanisms have been proposed [28]. Afkairin et al. [41] demonstrated the effectiveness of *Anabaena* sp. in increasing bioavailability of phosphorus (P) from organic P fertilizers, such as bone meal and rock phosphate.

Cyanobacteria also release extracellular phytohormones including gibberellins, cytokinin, auxin, and abscisic acid [28]. Alvarez et al. [29] thoroughly summarized the scientific literature on the production of phytohormones by cyanobacterial species and plant growth and performance responses to those cyanobacteria. The most commonly reported effects on plants include improving germination, root and shoot length and/or weight, leaf number, or flower parameters. It is difficult to separate the influence of nutrients in cyanobacterial fertilizer from the impact of its accompanying phytohormones on plant growth or physiology. Wenz et al. [42] evaluated whether management of the cyanobacterial production process influenced phytohormone concentrations; they reported that culture inoculation density was a significant variable in predicting the concentrations of indole-3-acetic acid, indole acetamide, and salicylic acid concentrations. As we learn more, it may become possible to manage the cyanobacterial fertilizer production process to optimize phytohormone levels for different crops.

Cyanobacterial fertilizer has also been reported to result in nutritional benefits. For example, Sukor et al. [43] compared cyanobacterial fertilizer with a no-fertilizer control and three other organic fertilizers applied to lettuce. They reported that cyanobacterial fertilizer increased β-carotene concentration in lettuce leaves in one of the two study years. However, fish emulsion increased β-carotene concentration in both years. The application of indole-3-acetic acid (IAA, a plant growth hormone) was highest in the fish emulsion treatment, and IAA application in the fertilizers was positively correlated with the β-carotene concentration in the lettuce [43].

Zinc (Zn) and iron (Fe) are important minerals in human health, and their concentrations in plants have shown variable responses to cyanobacterial fertilizer application. Both Zn and Fe are usually included in the nutrient solutions commonly used to grow cyanobacteria, and so might end up in the fertilizer that is eventually applied. In a greenhouse study evaluating both liquid and solid cyanobacterial fertilizers, available Zn and Fe concentrations in soil were significantly increased by both fertilizer types, and leaf Zn and Fe were increased in response to solid cyanobacterial fertilizer in kale, pepper, and maize, while the liquid fertilizer increased leaf Zn and Fe concentrations in pepper and maize only [32]. In field studies, the results have been less conclusive. Yoder and Davis [44] reported no significant effect of cyanobacterial fertilizer on Zn and Fe concentrations in kale. Sterle et al. [45] evaluated cyanobacterial fertilizer on peach trees and found that the leaf chlorophyll concentration (measured nondestructively with a soil plant analysis development meter) was higher in trees receiving cyanobacterial fertilizer. Chlorophyll was also positively correlated with leaf Fe concentration [45]. In field studies of lettuce and sweet corn, an interaction was observed between inorganic N sources (NO_3^- and NH_4^+) in organic fertilizers and plant Fe concentrations; this interaction complicated the interpretation of the results [46]. Nonetheless, in some cases, cyanobacterial fertilizer applications have resulted in higher Zn and/or Fe concentrations.

Plant water use efficiency (WUE) can also be influenced by fertilizer type. Field WUE (fWUE) is defined as crop yield divided by water used, while instantaneous WUE (iWUE) is defined as the ratio of net photosynthetic rate to transpiration rate. Sukor et al. [47] compared cyanobacterial fertilizer to other organic fertilizers (fish emulsion, blood meal, and feather meal) applied to sweet corn and found that cyanobacterial fertilizer had the highest yield and the highest WUE. In an effort to understand the mechanism behind the impact of fertilizer type on the WUE, these investigators correlated the amounts of plant growth hormones applied with the fertilizers and found that the amount of salicylic acid applied was positively correlated with both fWUE and iWUE. The cyanobacterial fertilizer had the highest amount of salicylic acid applied, an order of magnitude higher than the other fertilizers evaluated [47]. Cyanobacterial fertilizer might therefore improve WUE, potentially through the application of salicylic acid in addition to N.

Several studies have recorded shifts in the soil microbial population following cyanobacterial applications [29]. For example, Rogers and Burns [48] reported increases in some bacterial groups, actinomycetes, and fungi in a greenhouse study after inoculating a soil with *Nostoc muscorum* [48]. Afkairin [49] evaluated *Anabaena cylindrica* applications in two organic cropping systems (an annual vegetable rotation and an untilled peach orchard) grown on different soil types. *A. cylindrica* cyanobacterial fertilizer significantly impacted the biomass and structure of soil microbial communities in both cropping systems. The bacteria and actinomycete populations increased in the vegetable system, and the soil fungi and arbuscular mycorrhizal fungi increased in the peach orchard. Afkairin [49] also demonstrated that cyanobacterial fertilizers can affect soil microbial communities, even in short periods of time (<1 year).

In addition to shifts in microbial communities, cyanobacteria can also have antagonistic effects against plant diseases through their production of bioactive compounds with antibacterial, antifungal, and antialgal potential [28]. Singh et al. [28] and Alvarez et al. [29] summarized the extant literature regarding cyanobacterial production of these compounds, but it is clear that more work remains to be carried out to be able to harness cyanobacteria as natural antidisease agents. In addition to biotic stresses, cyanobacteria have also been reported to aid plants facing abiotic stresses, such as salinity, drought, and heavy metal pollution [50]. Most research on both biotic and abiotic stresses has been carried out in laboratory environments to date, with little or no information available from field-scale work.

4. Challenges for Future Research

There are a number of trade-offs to be considered regarding cyanobacterial biofertilizer production in the future. For example, some scientists have chosen to work on low-tech systems to simplify cyanobacteria production for on-farm systems (Figure 2). However, such low-tech systems, which are usually open systems, are more vulnerable to contamination of the cyanobacterial culture with other microorganisms. In addition, low-tech systems generally have slower growth rates than more advanced systems and would therefore require a greater land base than a high-tech system. This is a serious limitation because land is often the primary resource of smallholder farmers, especially in developing countries. On the other hand, low-tech on-farm systems can often make use of locally available materials to minimize dependence on imported materials and keep expenses low [31,51,52]. In general, more advanced cyanobacteria production systems are more productive but also more expensive. In addition, high-tech systems usually require a higher level of management and training and might have increased costs for operations and maintenance.

Figure 2. Prototype cyanobacteria raceways in production in Ethiopia, created using local materials and knowledge. The cyanobacteria were cultured from local soils. Photo by J.G. Davis. S. Yigrem pictured.

Chemical challenges also exist to maintain a favorable growth environment for cyanobacterial fertilizer cultures. Even simply sourcing nutrients for cyanobacterial fertilizer cultivation might be a challenge in some parts of the world. In addition, the chemistry of the nutrients themselves can limit availability for cyanobacterial fertilizer; for example, P solubility can be low at the neutral to slightly alkaline pH optimal for cyanobacteria growth. Stability of the pH can itself become an issue if high rates of photosynthesis remove pH-buffering alkalinity from the water in which the cyanobacteria grow. After the removal of residual alkalinity, the pH can increase to extremely alkaline levels, resulting in the loss of the cyanobacterial culture. These challenges will need to be met or managed if widespread production of cyanobacterial fertilizer is to be realized.

The production system and level of technology utilized to grow cyanobacteria depend on the value of the product and the level of control needed over the production process [53]. Open systems that use sunlight have disadvantages such as contamination by other microorganisms or algae, or predation by grazers (including microbiological predators). Additionally, open systems are more vulnerable to diurnal fluctuation of environmental parameters. In contrast, closed systems that utilize sunlight are less susceptible to contamination and can achieve higher cell densities while also taking advantage of solar radiation; however, the cost of these systems is much higher due to the need for transparent materials, gas management (i.e., oxygen and CO_2), and temperature maintenance [53]. Finally, closed systems that utilize artificial light as a source of energy can also be used to cultivate cyanobacteria. These systems provide real-time control and optimization of parameters, but the cost increases significantly. On the other hand, large amounts of high-quality cyanobacterial biomass can be produced, and since escape to the environment is improbable, cultivation of genetically modified cyanobacteria is possible [53]. When cyanobacterial cultures are of high value and purity is essential, then these high-tech systems may be worth the additional cost; however, when cyanobacteria are being grown to use as fertilizer, the value is not perceived to be as high, and purity is not as necessary. Therefore, systems that use natural light are preferable in most cases for growing cyanobacteria as N fertilizer.

While an on-farm production model for cyanobacterial fertilizer has a number of advantages, another option could be a distributed network of "fertilizer farms" where cyanobacterial biofertilizer is grown for sale to local farmers. This approach would allow a greater technological investment and a higher level of management, while minimizing fossil fuels used in transportation. Additionally, since cyanobacterial N fertilizer production is not dependent on supplies of natural gas, as NH_3 production with the Haber–Bosch process can be, cyanobacteria can potentially be grown without the infrastructure limitations associated with conventional N fertilizer production.

There are a variety of options for sourcing cyanobacterial cultures. In order to prevent environmental contamination in case of a breach in a raceway or pond, many scientists have chosen to culture local cyanobacteria, such as *Anabaena cylindrica* [30] or *Anabaena oscillariodes* [31], from local soil or water sources [30,31,51]. Others have chosen marine cyanobacteria because marine cyanobacteria have fewer potential contaminants. Although some cyanobacterial species can be grown in seawater, the application of these cyanobacteria as fertilizer for terrestrial plants could result in toxic effects on plants due to excessive salinity. The utility of cyanobacteria for various purposes can be enhanced through genetic manipulation or modification which can potentially increase growth, photosynthetic efficiency, biomass yield, or other important parameters [28]; the potential for environmental contamination by genetically modified organisms (GMO) can be minimized through the use of closed bioreactors. However, many farmers avoid growing or utilizing GMOs, especially farmers in countries with GMO regulations or those following organic standards that prohibit the use of GMOs. From a biosecurity perspective, the use of locally cultured cyanobacteria is probably the safest option, but locally cultured cyanobacteria might not be the most productive in terms of N fixation or production of plant growth promoting substances.

Cyanotoxins are another potential risk of cyanobacterial fertilizer use. Reports of cyanobacterial blooms and accompanying cyanotoxins have grown in frequency in recent years [54,55]. Cyanotoxins pose threats to both human and animal health, including livestock and wildlife, usually through exposure to toxin-containing water. The most common cyanotoxins include microcystins, cylindrospermopsin, and anatoxins [55]. Concentrations of all cyanotoxin groups generally increase with increasing concentrations of N and P in the water body [56]. It is critically important that cyanobacteria used as fertilizer do not contain toxins, since there is a chance of cyanotoxins being introduced into the food chain through this pathway. One way to reduce the risk of cyanotoxins is to produce cyanobacterial biofertilizers in water with low levels of N and P; conveniently, biological N fixation of N_2 from the atmosphere is also maximized when N levels in the water are low. On the other hand, efforts have been made to recycle nutrients from industrial wastewaters to grow cyanobacteria for use as a fertilizer or soil amendment [57]; the use of nutrient-rich wastewaters for cyanobacterial production might enhance the risk from cyanotoxins, due to the increased potential for cyanotoxin production. To ensure that no toxins are present, it is good practice to analyze cyanobacterial fertilizers for cyanotoxins prior to their use [42]. Wenz et al. [42] used mass spectroscopy to analyze their *Anabaena* sp. cyanobacterial biofertilizer for anatoxin-a (LOD > 0.06 $\mu g\ L^{-1}$), microcystins (LOD > 0.05 $\mu g\ L^{-1}$), and cylindrospermopsin (LOD > 0.1 $\mu g\ L^{-1}$); none of these cyanotoxins were detected.

Economic feasibility is essential for the adoption and long-term success of cyanobacterial fertilizers. Start-up costs include capital costs such as land, water, construction of raceways and aerators or mixers, and power to operate the aerators and mixers. Investments required in capital and operational costs are very high and necessitate efforts to reduce production costs to achieve profitability [53]. An economic evaluation of the production of microalgal biomass as an energy source (not on cyanobacterial fertilizer) reported that of the scenarios they evaluated, the only potentially profitable scenario was the extraction and commercialization of concentrated proteins [58]. Unfortunately, no economic evaluation of cyanobacterial fertilizer compared with urea is available in the scientific literature.

For operation of the production system, cyanobacterial cultures, nutrients to feed the cultures (e.g., P and Fe), maintenance costs, and trained personnel to manage the system are all important. It is essential to minimize the opportunity cost of the land set aside for cyanobacterial biofertilizer production by siting the production units either on nonarable land or on land with low agricultural productivity. When these costs have been defined for a specific locale, then the time required to break even can also be calculated so that an investor can plan accordingly. Another vital aspect to long-term economic success will be a market analysis and distribution plan, so that there are enough farmers in close proximity to the production units to ensure profitability.

As cyanobacterial fertilizer production systems become optimized, quantification of the carbon footprint and economic costs and benefits of this fertilizer system will be critical to full-scale adoption. To become a successful replacement for urea and other Haber–Bosch-based fertilizers, cyanobacterial fertilizers will need to outcompete urea both in the marketplace and in favorable environmental outcomes. Whether this eventuality comes to pass remains to be seen.

5. Conclusions

In light of the many disadvantages of N fertilizers produced through the Haber–Bosch process, the potential benefits of cyanobacterial fertilizer use are worthy of further investigation and investment. Better plant growth, increased crop yields, more efficient nutrient and water use, decreased geopolitical risks, reduced emissions from fertilizer distribution and transport, lower greenhouse gas emissions following field application, and more sustainable food production are some of the potential benefits that might be harvested from the intensive cultivation of cyanobacteria as N fertilizer. Remaining challenges include the optimization of the cyanobacterial fertilizer production system for different environments and precise economic analyses to optimize its competitiveness around the world.

Author Contributions: Conceptualization, M.S.M. and J.G.D.; writing—original draft preparation, J.G.D. and M.S.M.; writing—review and editing, M.S.M. and J.G.D. All authors have read and agreed to the published version of the manuscript.

Funding: This research received no external funding.

Data Availability Statement: No new data were collected or created.

Conflicts of Interest: The authors declare no conflict of interest.

References

1. Mishra, U.; Pabbi, S. Cyanobacteria: A potential biofertilizer for rice. *Resonance* **2004**, *9*, 6–10. [CrossRef]
2. Okon, Y.; Labandera-Gonzalez, C.A. Agronomic applications of Azospirillum: An evaluation of 20 years of worldwide field inoculation. *Soil Biol. Biochem.* **1994**, *26*, 1591–1601. [CrossRef]
3. Dobbelaere, S.; Croonenborghs, A.; Thys, A.; Ptacek, D.; Vanderleyden, J.; Dutto, P.; Labandera-Gonzalez, C.; Caballero-Mellado, J.; Aguirre, J.F.; Kapulnik, Y.; et al. Responses of agronomically important crops to inoculation with *Azospirillum*. *Aust. J. Plant Physiol.* **2001**, *28*, 871–879. [CrossRef]
4. Smith, R.L.; Schank, S.C.; Milam, J.R.; Baltensperger, A.A. Responses of Sorghum and Pennisetum species to the N_2-fixing bacterium *Azospirillum brasilense*. *Appl. Environ. Microbiol.* **1984**, *47*, 1331–1336. [CrossRef]
5. Fukami, J.; Cerezini, P.; Hungria, M. *Azospirillum*: Benefits that go far beyond biological nitrogen fixation. *AMB Expr.* **2018**, *8*, 73. [CrossRef]
6. Obana, S.; Miyamoto, K.; Morita, S.; Ohmori, M.; Inubushi, K. Effect of *Nostoc* sp. on soil characvteristics, plant growth and nutrient uptake. *J. Appl. Phycol.* **2007**, *19*, 641–646. [CrossRef]
7. Chamizo, S.; Mugnai, G.; Rossi, F.; Certini, G.; De Philippis, R. Cyanobacteria inoculation improves soil stability and fertility on different textured soils: Gaining insights for applicability in soil restoration. *Front. Environ. Sci.* **2018**, *6*, 49. [CrossRef]
8. Roncero-Ramos, B.; Román, J.R.; Gómez-Serrano, C.; Cantón, Y.; Acién, F.G. Production of a biocrust-cyanobacteria strain (*Nostoc commune*) for large-scale restoration of dryland soils. *J. Appl. Phycol.* **2019**, *31*, 2217–2230. [CrossRef]
9. Silva, P.G.; Silva, H.J. A retained viability index (RVI10) for evaluation of a cyanobacterial powder fertilizer. *World J. Microbiol. Biotechnol.* **2003**, *19*, 891–895. [CrossRef]
10. Silva, P.G.; Silva, H.J. Effect of mineral nutrients on cell growth and self-flocculation of *Topothrix tenuis* for the production of a biofertilizer. *Bioresour. Technol.* **2007**, *98*, 607–611. [CrossRef]
11. Silva, P.G.; Ferrari, S.G.; Silva, H.J. Preservation methods of *Tolypothrix tenuis* for use as a cyanobacterial fertilizer. *J. Appl. Phycol.* **2007**, *19*, 239–246. [CrossRef]
12. Maqubela, M.P.; Mnkeni, P.N.S.; Issa, O.M.; Pardo, M.T.; D'Acqui, L.P. *Nostoc* cyanobacterial inoculation in South African agricultural soils enhances soil stucture, soil fertility, and maize growth. *Plant Soil* **2009**, *315*, 79–92. [CrossRef]
13. Smil, V. Detonator of the population explosion. *Nature* **1999**, *400*, 415. [CrossRef]
14. Gruber, N.; Galloway, J.N. An Earth-system perspective of the global nitrogen cycle. *Nature* **2008**, *451*, 293–296. [CrossRef]
15. Fowler, D.; Coyle, M.; Skiba, U.; Sutton, M.A.; Cape, J.N.; Reis, S.; Sheppard, L.J.; Jenkins, A.; Grizzetti, B.; Galloway, J.N.; et al. The global nitrogen cycle in the twenty-first century. *Phil. Trans. R. Soc. B* **2013**, *368*, 20130164. [CrossRef]
16. Erisman, J.W.; Sutton, M.A.; Galloway, J.; Klimont, Z.; Winiwarter, W. How a century of ammonia synthesis changed the world. *Nat. Geosci.* **2008**, *1*, 636–639. [CrossRef]
17. Erisman, J.W.; Galloway, J.N.; Seitzinger, S.; Bleeker, A.; Dise, N.B.; Petrescu, A.M.R.; Leach, A.M.; de Vries, W. Consequences of human modification of the global nitrogen cycle. *Phil. Trans. R. Soc. B* **2013**, *368*, 20130116. [CrossRef]
18. Sandström, V.; Valin, H.; Krisztin, T.; Havlík, P.; Herrero, M.; Kastner, T. The role of trade in the greenhouse gas footprints of EU diets. *Glob. Food Secur.* **2018**, *19*, 48–55. [CrossRef]
19. Daramola, D.A.; Hatzell, M.C. Energy demand of nitrogen and phosphorus based fertilizers and approaches to circularity. *ACS Energy Lett.* **2023**, *8*, 1493–1501. [CrossRef]
20. Chen, J.G.; Crooks, R.M.; Seefeldt, L.C.; Bren, K.L.; Bullock, R.M.; Darensbourg, M.Y.; Holland, P.L.; Hoffman, B.; Janik, M.J.; Jones, A.K.; et al. Beyond fossil fuel-driven nitrogen transformations. *Science* **2018**, *360*, eaar6611. [CrossRef]
21. Sainju, U.M.; Jabro, J.D.; Stevens, W.B. Soil carbon dioxide emission and carbon content as affected by irrigation, tillage, cropping system, and nitrogen fertilization. *J. Environ. Qual.* **2008**, *37*, 98–106. [CrossRef] [PubMed]
22. Reay, D.S.; Davidson, E.A.; Smith, K.A.; Smith, P.; Melillo, J.M.; Dentener, F.; Crutzen, P.J. Global agriculture and nitrous oxide emissions. *Nat. Clim. Chang.* **2012**, *2*, 410–416. [CrossRef]
23. Vitousek, P.M.; Naylor, R.; Crews, T.; David, M.B.; Drinkwater, L.E.; Holland, E.; Johnes, P.J.; Katzenberger, J.; Martinelli, L.A.; Matson, P.A.; et al. Nutrient imbalances in agricultural development. *Science* **2009**, *324*, 1519–1520. [CrossRef] [PubMed]
24. Guo, W.; Zhang, K.; Liang, Z.; Zou, R.; Xu, Q. Electrochemical nitrogen fixation and utilization: Theories, advanced catalyst materials and system design. *Chem. Soc. Rev.* **2019**, *48*, 5658–5716. [CrossRef] [PubMed]
25. Guo, X.; Gu, J.; Lin, S.; Zhang, S.; Chen, Z.; Huang, S. Tackling the activity and selectivity challenges of electrocatalysts toward the nitrogen reduction reaction via atomically dispersed biatom catalysts. *J. Am. Chem. Soc.* **2020**, *142*, 5709–5721. [CrossRef]
26. Fu, X.; Pedersen, J.B.; Zhou, Y.; Saccoccio, M.; Li, S.; Sazinas, R.; Li, K.; Andersen, S.Z.; Xu, A.; Deissler, N.H.; et al. Continuous-flow electrosynthesis of ammonia by nitrogen reduction and hydrogen oxidation. *Science* **2023**, *379*, 707–712. [CrossRef]
27. Du, H.-L.; Chatti, M.; Hodgetts, R.Y.; Cherepanov, P.V.; Nguyen, C.K.; Matuszek, K.; McFarlane, D.R.; Simonov, A.N. Electroreduction of nitrogen with almost 100% current-to-ammonia efficiency. *Nature* **2022**, *609*, 722–727. [CrossRef]
28. Singh, J.S.; Kumar, A.; Rai, A.N.; Singh, D.P. Cyanobacteria: A precious bio-resource in agriculture, ecosystem, and environmental sustainability. *Front. Microbiol.* **2016**, *7*, 529. [CrossRef]
29. Alvarez, A.L.; Weyers, S.L.; Goemann, H.M.; Peyton, B.M.; Gardner, R.D. Microalgae, soil and plants: A critical review of microalgae as renewable resources for agriculture. *Algal Res.* **2021**, *54*, 102200. [CrossRef]
30. Barminski, R.; Storteboom, H.; Davis, J.G. Development and evaluation of an organically-certifiable growth medium for cyanobacteria. *J. Appl. Phycol.* **2016**, *28*, 2623–2630. [CrossRef]

31. Wolde, G.; Asmamaw, M.; Sido, M.Y.; Yigrem, S.; Wolde-meskel, E.; Chala, A.; Storteboom, H.; Davis, J.G. Optimizing a cyanobacterial biofertilizer manufacturing system for village-level production in Ethiopia. *J. Appl. Phycol.* **2020**, *32*, 3983–3994. [CrossRef]
32. Asmamaw, M.; Wolde, G.; Yohannes, M.; Yigrem, S.; Woldemeskel, E.; Chala, A.; Davis, J.G. Comparison of cyanobacterial bio-fertilizer with urea on three crops and two soils of Ethiopia. *Afr. J. Agric. Res.* **2019**, *14*, 588–596.
33. Wickham, A.; Davis, J.G. Optimizing organic carrot yield and quality using fish emulsions, cyanobacterial fertilizer, and seaweed extracts. *Agronomy* **2023**, *13*, 1329. [CrossRef]
34. Jama, A.; Widiastuti, D.P.; Gafur, S.; Davis, J.G. Azolla biofertilizer is an effective replacement for urea fertilizer in vegetable crops. *Sustainability* **2023**, *15*, 6045. [CrossRef]
35. Toonsiri, P.; Del Grosso, S.J.; Sukor, A.; Davis, J.G. Greenhouse gas emissions from solid and liquid organic fertilizers applied to lettuce. *J. Environ. Qual.* **2016**, *45*, 1812–1821. [CrossRef]
36. Erwiha, G.M.; Ham, J.; Sukor, A.; Wickham, A.; Davis, J.G. Organic fertilizer source and application method impact ammonia volatilization. *Commun. Soil Sci. Plant Anal.* **2020**, *51*, 1469–1482. [CrossRef]
37. Cameron, H.J.; Julian, G.R. Utilization of hydroxyapatite by Cyanobacteria as their sole source of phosphate and calcium. *Plant Soil* **1988**, *109*, 123–124. [CrossRef]
38. Mandal, B.; Das, S.C.; Mandal, L.N. Effect of growth and subsequent decomposition of cyanobacteria on the transformation of phosphorus in submerged soils. *Plant Soil* **1992**, *143*, 289–297. [CrossRef]
39. Vaishampayan, A.; Sinha, R.P.; Häder, D.P.; Dey, T.; Gupta, A.K.; Bhan, U.; Rao, A.L. Cyanobacterial biofertilizers in rice agriculture. *Bot. Rev.* **2001**, *67*, 453–516. [CrossRef]
40. Yandigeri, M.S.; Yadav, A.K.; Srinivasan, R.; Kashyap, S.; Pabbi, S. Studies on mineral phosphate solubilization by cyanobacteria *Westiellopsis* and *Anabaena*. *Microbiology* **2011**, *80*, 558–565. [CrossRef]
41. Afkairin, A.; Ippolito, J.A.; Stromberger, M.; Davis, J.G. Solubilization of organic phosphorus sources by cyanobacteria and a commercially available bacterial consortium. *Appl. Soil Ecol.* **2021**, *162*, 103900. [CrossRef]
42. Wenz, J.; Davis, J.G.; Storteboom, H. Influence of light on endogenous phytohormone concentrations of a nitrogen-fixing *Anabaena* sp. cyanobacterium culture in open raceways for use as fertilizer for horticultural crops. *J. Appl. Phycol.* **2019**, *31*, 3371–3384. [CrossRef]
43. Sukor, A.; Amer, F.S.M.; Vanamala, J.; Davis, J.G. Phytohormones in organic fertilizers influence β-carotene concentration and marketable yield of lettuce (*Lactuca sativa*). *Acta Hortic.* **2022**, *1348*, 15–22. [CrossRef]
44. Yoder, N.; Davis, J.G. Organic fertilizer comparison on growth and nutrient content of three kale cultivars. *HortTechnology* **2020**, *30*, 176–184. [CrossRef]
45. Sterle, D.G.; Stonaker, F.; Ela, S.; Davis, J.G. Cyanobacterial biofertilizer as a supplemental fertilizer for peaches: Yield, trunk growth, leaf nutrients and chlorosis. *J. Am. Pomol. Soc.* **2021**, *75*, 165–175.
46. Sukor, A. Organic Nitrogen Fertilizers Influence Nutritional Value, Water Use Efficiency, and Nitrogen Dynamics of Drip Irrigated Lettuce and Sweet Corn. Ph.D. Dissertation, Colorado State University, Fort Collins, CO, USA, 2016.
47. Sukor, A.; Qian, Y.; Davis, J.G. Organic nitrogen fertilizer selection influences water use efficiency in drip-irrigated sweet corn. *Agriculture* **2023**, *13*, 923. [CrossRef]
48. Rogers, S.L.; Burns, R.G. Changes in aggregate stability, nutrient status, indigenous microbial populations, and seedling emergence, following inoculation of soil with *Nostoc muscorum*. *Biol. Fert. Soils* **1994**, *18*, 209–215. [CrossRef]
49. Afkairin, A. Cyanobacteria Biofertilizer Solubilizes Soil Phosphorus and Alters Soil Microbial Communities. Ph.D. Dissertation, Colorado State University, Fort Collins, CO, USA, 2022.
50. Poveda, J. Cyanobacteria in plant health: Biological strategy against abiotic and biotic stresses. *Crop Prot.* **2021**, *141*, 105450. [CrossRef]
51. Tsedeke, A.; Degefu, T.; Wolde-meskel, E.; Davis, J. The effect of pond depth and lining plastic color on nitrogen fixing capacity of the cyanobacteria, *Anabaena* sp. E3. *Afr. J. Biotechnol.* **2016**, *15*, 1442–1451. [CrossRef]
52. Hurisso, T.T.; Davis, J.G.; Chala, A.; Getachew, A.; Wolde-meskel, E. Impacts of grinding and acidification of animal bones with coffee wastewater on plant dry matter yield and recovery of phosphorus. *Commun. Soil Sci. Plant Anal.* **2021**, *52*, 1076–1088. [CrossRef]
53. Pathak, J.; Rajneesh; Maurya, P.K.; Singh, S.P.; Häder, D.-P.; Sinha, R.P. Cyanobacterial farming for environment friendly sustainable agriculture practices: Innovations and perspectives. *Front. Environ. Sci.* **2018**, *6*, 7. [CrossRef]
54. Buratti, F.M.; Manganelli, M.; Vichi, S.; Stefanelli, M.; Scardala, S.; Testai, E.; Funari, E. Cyanotoxins: Producing organisms, occurrence, toxicity, mechanism of action and human health toxicological risk evaluation. *Arch. Toxicol.* **2017**, *91*, 1049–1130. [CrossRef]
55. Svirčev, Z.; Lalić, D.; Savić, G.B.; Tokodi, N.; Backović, D.D.; Chen, L.; Meiluoto, J.; Codd, G.A. Global geographical and historical overview of cyanotoxin distribution and cyanobacterial poisonings. *Arch. Toxicol.* **2019**, *93*, 2429–2481. [CrossRef]
56. Dolman, A.M.; Rücker, J.; Pick, F.R.; Fastner, J.; Rohrlack, T.; Mischke, U.; Wiedner, C. Cyanobacteria and cyanotoxins: The influence of nitrogen versus phosphorus. *PLoS ONE* **2012**, *7*, e38757. [CrossRef]
57. Do Nascimento, M.; Battaglia, M.E.; Rizza, L.S.; Ambrosio, R.; Di Palma, A.A.; Curatti, L. Prospects of using biomass of N_2-fixing cyanobacteria as an organic fertilizer and soil conditioner. *Algal Res.* **2019**, *43*, 101652. [CrossRef]
58. Bravo-Fritz, C.P.; Sáez-Navarrete, C.A.; Herrera-Zeppelin, L.A.; Varas-Concha, F. Multi-scenario energy-economic evaluation for a biorefinery based on microalgae biomass with application of anaerobic digestion. *Algal Res.* **2016**, *16*, 292–307. [CrossRef]

Nitrogen

MDPI

Review

Efficiency and Management of Nitrogen Fertilization in Sugar Beet as Spring Crop: A Review

Ivana Varga [1,*], Jurica Jović [2], Mirta Rastija [1], Antonela Markulj Kulundžić [3], Vladimir Zebec [2], Zdenko Lončarić [2], Dario Iljkić [1] and Manda Antunović [1]

[1] Department of Plant Production and Biotechnology, Faculty of Agrobiotechnical Sciences Osijek, Josip Juraj Strossmayer University of Osijek, Vladimira Preloga 1, 31000 Osijek, Croatia; mirta.rastija@fazos.hr (M.R.); dario.iljkic@fazos.hr (D.I.); manda.antunovic@fazos.hr (M.A.)

[2] Department of Agroecology and Environment Protection, Faculty of Agrobiotechnical Sciences Osijek, Josip Juraj Strossmayer University of Osijek, Vladimira Preloga 1, 31000 Osijek, Croatia; jurica.jovic@fazos.hr (J.J.); vladimir.zebec@fazos.hr (V.Z.); zloncaric@fazos.hr (Z.L.)

[3] Department of Industrial Plants, Agricultural Institute Osijek, Južno predgrađe 17, 31000 Osijek, Croatia; antonela.markulj@poljinos.hr

* Correspondence: ivana.varga@fazos.hr; Tel.: +385-31-55-49-11; Fax: +385-31-554-853

Abstract: Sugar beet fertilization is a very complex agrotechnical measure for farmers. The main reason is that technological quality is equally important as sugar beet yield, but the increment of the root yield does not follow the root quality. Technological quality implies the concentration of sucrose in the root and the possibility of its extraction in the production of white table sugar. The great variability of agroecological factors that directly affect root yield and quality are possible good agrotechnics, primarily by minimizing fertilization. It should be considered that for sugar beet, the status of a single plant available nutrient in the soil is more important than the total amounts of nutrients in the soil. Soil analysis will show us the amount of free nutrients, the degree of soil acidity and the status of individual elements in the soil so that farmers can make a compensation plan. An estimate of the mineralizing ability of the soil, the N min, is very important in determining the amount of mineral nitrogen that the plant can absorb for high root yield and good technological quality. The amount of N needed by the sugar beet crop to be grown is an important factor, and it will always will be in the focus for the producers, especially from the aspect of trying to reduce the N input in agricultural production to preserve soils and their biodiversity but also to establish high yields and quality.

Keywords: sugar beet root yield; root quality; amount of nitrogen; vegetative growth; soil analyses

1. Introduction

Sugar beet (*Beta vulgaris* L. subsp. *vulgaris* var. *altissima* Döll.) and sugar cane (*Saccharum officinarum* L.) are the plants that are mostly grown in the world as raw materials for processing into sugar. Sugar cane cultivation for sugar production began in India in 2000 BC, while sugar beet is a relatively young plant and has been grown since the late 18th century. Sugar beet is biologically adapted to the temperate climate zone and is grown in the area from 30° to 60° north latitude and from 25° to 35° south latitude. Nowadays, sugar beet accounts for 20% of world sugar production, while the remaining 80% of sugar is from sugar cane [1]. The most suitable geographical conditions for beet growing are found where the soil is deep, friable, well drained and contains a certain proportion of lime. Sugar beet is usually cultivated as a rotation crop. Even though there is a need for agrotechnical measures, it would be interesting to try the adoption of conservation tillage, which provides considerable environmental benefits while improving soil fertility [2–6]. Sugar beet yields under reduced tillage are similar to those of beets grown in plowed soil when the N application increases [7].

The sugar content in sugar beet root is usually 13–20% [8,9]. The total biomass of the sugar beet plant is divided into root and leaf dry matter. According to Hoffmann and Kenter [10], sugar beet root yield has increased in the past decades by 1.5% per year, but the sugar content has increased at the expense of the plant canopy (or leaf dry matter). More than 98% of total root sugar is sucrose, fructose and glucose in very small amounts [11].

Sugar beet fertilization significantly impacts yield formation and sugar beet root quality. Nitrogen is probably the most studied nutrient for sugar beet because N is the most limiting nutrient directly related to sugar beet yield and quality. Many researchers reported that N fertilization greatly influences excessive N fertilization, resulting in more lush development of the leaves and crowns, root maturity is slowed down, and the sugar content in the root is lower. In addition, root quality may be reduced due to the increased content of melas-forming elements [12–15].

As one of the main building blocks, nitrogen has a special significance in plants because it is an integral part of proteins from which protoplasm, cells and plant tissues are formed [16,17]. In addition, N is the main yielding element in plant production. The range of N concentration in the leaf blade of sugar beet is from 2.2 to 3.5% and in the leaf stalk from 1.0 to 1.5% [18]. Sugar beet needs a large amount of N as a nutrient. The amount of N applied to the soil should be performed according to the chemical and microbiological properties of the soil. Furthermore, weather conditions should be considered to avoid unused amounts of N by plants that will, due to extreme instability of this nutrient, move into the deeper layers of the soil and bring about groundwater eutrophication [19,20]. Moreover, excessive amounts of N fertilizers negatively affect the content of melas-forming substances and primarily alpha-amino [21–23].

2. Determination of N Rate: Importance of EUF and N Min Methods

Chemical analytical methods and techniques are used to make crop fertilization recommendations based on nutrient status in the soil [24]. The choice of analytical methods should consider that any analytical method is good if there is a well-designed system of interpretation of analytical values. The interpretation of the results uses tables of limit values, which give more indicative results because nutrient bioavailability is not a static value. It depends on a number of factors: biological, physical, hydrological, agrotechnical [25]. Classical methods of soil analysis are basically one-off extractions that estimate only the available nutrients in the soil and not the reserves of nutrients. Determining the available amount of nutrients in the soil is a common basis for recommending fertilization and application of fertilizers. Therefore, analytical and prognostic methods are required to provide insight into the amount of available nutrients status in the soil, the rate of change in the amount of identified nutrients during the growing season (e.g., for absorption, rinsing or fixation) and the amount of nutrients to be added for optimal plant nutrition. In some cases, the calculation also considers the need to increase the level of some nutrients in the soil to the target value.

The electro-ultrafiltration (EUF) method is the extraction of both the inorganic (EUF-NO_3) and organic (EUF-N_{org}) EUF-N fractions, but it can also extract synthetic amino acids dissolved in water [26,27]. The plant nutrients determined are separated from solutions in the form of ions, cations (Ca_2^+, K^+, Mg_2^+, NH_4+, Mn_2^+, Fe_3^+) and anions (NO_3^-, SO_4^{2-}, BO_3^{3-}). By changing the temperature and voltage of the current in soil–water suspensions from the soil, weaker or more tightly bound nutrients are separated. The extraction medium is distilled water, and during an extraction under the influence of electric fields (direct current) from the suspension, differently charged ions and particles are detached from the anodes and cathodes [28]. This process first separates easily soluble nutrients and then bound nutrient fraction, which is gradually mobilized during the growing season from the soil. For the EUF method, the extracted inorganic N (main nitrate) and organic N are the basis for the N fertilizer recommendation [29], since sugar beet uses both N fractions (EUF-NO_3 and EUF-N_{org}). The EUF method for soil analysis was introduced in Austria and Germany in 1974 and 1982 [30]. Since its introduction, more than 1.2 million soil samples have been analyzed by EUF [31]. Except for sugar beet, the method was used for other field crops, such as cereals, potatoes and vineyards.

For EUF calculation of the N fertilizer requirements, it is suggested that the sum of the EUF-extractable N amounts be used. For sugar beet grown as a spring crop in the climatic conditions of western and central Europe, the N supply during the summer months (May–September) is very important because it determines the yield and root quality. In Austria, Yugoslavia and Denmark, it was determined that over three years, the EUF-N value of 1 mg 100 g^{-1} soil in the period between June and September was equivalent to 40 kg ha^{-1} N [32]. Additionally, in addition to the EUF, mechanical analysis is often used, and in the spring planting, N status is often checked by the N min method.

With the N min method, the concentration and total amount of mineral N forms in the soil—nitrate (N-NO_3), ammonia (N-NH_4^+) and soil moisture—were determined [33,34]. The N min method determines the amount of mineral forms of N (available N) for spring crops before sowing and before sowing and harvesting for winter cereals in the zone that is reached by the root system. Therefore, instead of an individual determination, the N min method represents a synthesis of several factors of N availability (previous crop residual N, harvest residues, soil and climatic conditions [35]). The N min calculator is intended for N fertilization and N starter fertilization of winter cereals, corn and sugar beet. Computer model fertilization recommendation based on the N min method considers all necessary parameters (the content of mineralized N in the soil, soil moisture, sampling time and soil texture) for estimating N doses. The disadvantages of the N min method are the impracticality of the short time between sampling and performing the analysis until the application of the recommendation for fertilization, and with the EUF method, a great variability of results, especially on clay soils [36]. For sugar beet as a spring crop, the soil samples for the N min method need to be sampled in the third week of February and in the first decade of March. A total of 20–25 individual samples were taken from an area of no more than 10–20 ha from the same pre-crop, the same type of soil and uniform terrain configuration and composited into a 0.5–1 kg sample. Individual samples were taken from a depth of 0–30 cm, 30–60 cm, 60–90 cm.

3. Nitrogen Fertilization Influence on Vegetative Growth

Malnou et al. [16] point out that it is desirable for sugar beet to reach 85% of the canopy, which covers the soil, as soon as possible to maximize the use of solar radiation, while Jaradat and Rinke [37] point out that newer genotypes can achieve up to 90% soil coverage. According to a two-year study in Germany (Göttingen), Hoffman and Kluge-Severin [38] state that under temperate climatic conditions, sugar beet as a spring crop absorbed only 24% of the incoming radiation in May and up to 83% in June.

The most intense leaf formation in European production area is from the beginning of June, when sugar beet leaves close the rows and last until mid-July [39]. After that, the plant forms leaves of smaller dimensions, and the leaves gradually die off by the end of the vegetation. The growth of sugar beet roots through vegetative stages has a linear trend, and the largest root growth occurs from mid-July to mid-August.

The leaf rosette of sugar beet reaches its maximum at the end of July and the beginning of August, while toward the end of the vegetation, it gradually decreases [6,40]. According to Kristek and Liović [41] and Jelić et al. [42], in agroecological conditions of southeast Europe, the largest daily increase in sugar beet leaves is from mid-June to mid-July.

Manderscheid et al. [43] state that a higher amount of N (126 and 156 kg ha^{-1} N) compared to a smaller one (63 and 78 kg ha^{-1} N) resulted in an increase in fresh leaf mass and an increase in individual leaves through the vegetation (11th, 16th, 26th and 31st leaf), while statistically significant differences were not found between different N fertilizations and the number of sugar beet leaves. Lüdecke [44] states that the development of sugar beet in conditions of reduced N supply at the beginning of vegetation is the same because beets have enough N available. The author emphasizes that sugar beet matures earlier in conditions of lower N supply, and phenotypically, the leaves are lighter in color, while the root lags behind in growth.

Newer genotypes have erect leaves that allow growth in a smaller vegetation area [45]. Vukadinović et al. [46] point out that hybrids with more upright leaves tolerate shading better, i.e., they can be grown in higher density. In addition to the above, in terms of photosynthesis, Müller-Linow et al. [47] emphasize the importance of the vertical distribution of leaves, i.e., the angle at which the leaves are located on the rosette of sugar beet.

In spring sowing, the dying of sugar beet leaves is more intense after July. Kenter et al. [48] point out that air temperature and radiation have a very significant impact ($p < 0.001$) on the growth of sugar beet leaves in the first 65 days after sowing (early June) by growing sugar beet at 27 locations in Germany.

In the phase of intensive leaf growth (mid-June to the end of July), the total leaf area of one plant is 2000 to 6000 cm^2 [12,40,49], while the optimal leaf area index (LAI) of sugar beets in this period is from 3 to 4 m^2 m^{-2} [12,50–53]. With the optimal leaf area index of sugar beet (3 to 4 m^2 m^{-2}), the outer leaves use almost all of the sun's energy. If the LAI is higher than optimal, the photosynthesis of the lower leaves is reduced, the crop is not drought resistant and has a greater need for nutrients. Manderscheid et al. [43] emphasize the importance of N fertilization and state that depending on N fertilization, a different number of dry sugar beet leaves was found in August and September. Furthermore, the authors point out that sugar beet had an average of 1.58 dry leaves in late June and early July, 5.58 in August and 9.55 dry leaves in the second half of September.

Based on the analysis of eleven hybrids of the Institute of Field and Vegetable Crops in Novi Sad, Putnik-Delić [54] emphasizes the importance of the genotype itself on the formation of the leaf rosette. The author points out that in young beets in the 6–12 leaves phase, the leaf area per plant differs depending on the genotype (414.24 cm^2 to 1099.89 cm^2). According to Drachovská and Šandera [49], the differences in leaf area between genotypes in the phase of intensive root growth (11 July–11 August) were not so pronounced, which was confirmed by the example of leaf area of yield and sugar genotype (average 2741.5 cm^2 and 2892.3 cm^2). In the analysis of LAI, Tsialtas and Maslaris [52] again emphasize the importance of genotype. According to their research in Greece on 12 sugar beet genotypes, it was found that in the phase of the largest leaf area, in early July, LAI ranged from 1.78 m^2 m^{-2} (hybrid Ramona) to 5.03 m^2 m^{-2} (hybrid Dorothea).

Kosterj and Repka [55] obtained an LAI value from 0.38 m^2 m^{-2} to 5.36 m^2 m^{-2} during vegetative growth in the first year of vegetation. In 11 sampling dates (from June to October), Tsialtas and Maslaris [56] found very significant changes in LAI (hybrid Rizor). The authors also state that toward the end of the vegetation, the leaves become smaller in size (leaf area, length decreases, width and radius), and the leaf blade becomes more rounded.

Kristek and Halter [39] conclude that with a smaller number of plants per unit area, sugar beet cannot make full use of vegetation space, and due to a higher nutrient supply, the N uptake was higher, the root sugar content was reduced, and sugar yield was lower.

As with most field crops, N is an important nutrient in sugar beet. Sugar beet requires higher amounts of N to achieve maximum yields, so fertilization is one of the most important agronomic measures in sugar beet production. Excessive as well as scarce fertilization, especially N, can lead to loss of yield or reduced quality of sugar beet roots [57]. In conditions of scarce N supply, the sugar beet develops a smaller assimilation area and a smaller root with more sucrose, but the sugar yield per unit area is lower due to lower root yield [35].

Considering the needs of sugar beet for N, Pospišil [16] points out that sugar beet needs the most N in the phase of intensive leaf growth, which, in our agroecological conditions, most often occurs from early June, i.e., from the closing of rows, until mid-July. On the other hand, there is the least need for N in beets at the time of intensive accumulation of sugar in the root, so N's availability in the soil should be lower to prevent the formation of undesirable N compounds that prevent sugar separation.

Malnou et al. [16] state that sugar beet needs 0.04 g cm^{-2} N leaf to develop 85% of the leaf rosette, which corresponds to 120 kg ha^{-1} N, so it is very important to supply the plant with enough N so that the leaf rosette can better harness solar energy.

Sugar beet also absorbs the ammonium form of N (NH_4^+) well, which, due to physiological processes within the root cells, can negatively affect the sugar content in the root, breaking it down or preventing the synthesis of sucrose. Therefore, in the spring, before sowing and in the feeding of sugar beet, fertilizers that have an ammonium form of N, such as urea ($CO(NH_2)_2$, 46% N) and UAN (urea + NH_4NO_3 + water, 30% N), should be avoided. This is confirmed by Brentrup et al. [58] who stated that the yield of sugar beet when applying 115 kg ha^{-1} N in the form of calcium ammonium nitrate (CAN) was 47.7 t ha^{-1}, and the yield of sugar was 8.49 t ha^{-1}, while with the same amount of N from urea, root yield was lower at 44.2 t ha^{-1} and sugar yield at 7.31 t ha^{-1}.

After two years of research in Croatia, Kristek et al. [57] state that basic fertilization with 100 kg ha of urea increases root yield from an average of 57.6 t ha^{-1} (control) to 61.4 t ha^{-1}, root sugar content from an average of 15.7% (control) to 15.8% and sugar yield from an average of 7.7 t ha^{-1} (control) to 8.3 t ha^{-1}. Furthermore, the same authors state that top dressing leaves in phase 2–4 with 50 kg ha^{-1} CAN without basic fertilization in the autumn gave, on average, the best results: root yield 68.1 t ha^{-1}, sugar content in root 15.9% and sugar yield 9.31 t ha^{-1}.

Draycott [59] states that sugar beet root at harvest contains phosphorus in the amount of 0.8 kg t^{-1} of fresh matter, potassium 1.7 kg t^{-1} fresh matter, while the whole plant in the vegetation can take a total of 1.9 kg t^{-1} fresh matter phosphorus and potassium in the amount of 7.9 kg t^{-1} fresh matter.

4. Sugar Beet N Fertilization Management

In most soils in Croatia, sugar beet fertilization is 140–160 kg ha^{-1} N, 80–130 kg ha^{-1} P_2O_5 and 150–250 kg ha^{-1} K_2O [17]. In sugar beet production, phosphorus and potassium are usually applied in the fall, while one-half to one-third of the total required N was added in the fall and the rest in the spring.

Märländer et al. [50] state that in Germany, in the last 20 years, the use of phosphorus fertilizers in sugar beet cultivation has been reduced from 65 to 20 kg ha^{-1} and potassium fertilizers from 200 to 60 kg ha^{-1}. The authors point out that in order to achieve satisfactory yields, N was applied in the amount of 120 kg ha^{-1} N, and such practice is in use in most countries of the European Union. The main reason for that is that phosphorus fertilizer resources worldwide are very small in relation to nitrogen and potassium fertilizers [51–54]. Fertilizers should be used only if there is a proven lack in the soil to avoid overapplication and environmental pollution [55–58]. Soil analysis must be performed before use of fertilizers, and the dosage can be partly reduced by adding manure [59–69].

Draycott [70] points out that in many European countries, it is the practice to add all the N needed to achieve maximum sugar beet yields in the spring in the amount of 30 to 40 kg ha^{-1}, which should be added before sowing, while the rest is added in top dressing in phase 2–4 leaves. In our conditions, 60–80 kg of ha^{-1} N is most often added before sowing [14,16,50].

The greatest need of sugar beet for nutrients is in the phase of intensive leaf growth (from early June to mid-July). Sugar yield is linearly related to dry matter yield and total N uptake. However, Last et al. [71] point out that increasing N above 200 kg ha^{-1} of N does not increase sugar yield, and the maximum sugar yield over six years of research was obtained by applying 125 kg ha^{-1} N or less.

Starke and Hoffmann [72] state that the differences in the share of root dry matter depending on N fertilization were not very pronounced, and according to the results of their research, the average share of root dry matter was 22.5%, with fertilization of 150 kg ha^{-1} N 23.6%, while at an even higher amount of N than 300 kg ha^{-1}, dry matter was reduced to the level of the control treatment (22.5%). According to another study, Starke and Hoffmann [73] point out the positive effect of N on the dry matter yield of sugar beet leaves. Namely, the authors conclude that with higher fertilization, with as much as 300 kg ha^{-1} N, the yield of leaf dry matter increased to over 31 t ha^{-1}, while without N fertilization, it was lower and amounted to 26 t ha^{-1}.

Märländer et al. [50] also emphasize the importance of soil mineralization. Namely, although the average needs of sugar beet for N are about 200 to 250 kg ha^{-1}, the authors state that approximately 100 to 150 kg ha^{-1} of N was mineralized in the soil through vegetation, which reduces the need for larger amounts of N fertilization. About 75% of the total area of sugar beet cultivation in Germany adds less than 120 kg ha^{-1} N by fertilization, while in the rest of the area, even less was applied, i.e., about 80 kg ha^{-1} N.

Jaćimović et al. [14], using a combination of different doses of NPK fertilizers in a two-year study, determined the highest yield of roots (98.86 t ha^{-1}) and sugar (8.91 t ha^{-1}) in the fertilization treatment of 100 kg ha^{-1} N, 150 kg ha^{-1} P_2O_5 and 150 kg ha^{-1} K_2O.

The connection between N fertilization and N content in the plant is emphasized by Monreal et al. [74]. The authors found that N fertilization with 480 kg ha^{-1} N resulted in 569 kg ha^{-1} N in sugar beet plants, while unfertilized beets had 420 kg ha^{-1} N.

Vielemeyer et al. [75] state that increased N fertilization (0, 120 and 240 kg ha^{-1} N) decreased sugar content from 17.1% in the control treatment to 15.7%. However, the yield of crystallized sugar remained unchanged (average 5.7 t ha^{-1}). Furthermore, the authors state that the increase in N fertilization to 240 kg ha^{-1} N led to an increase in leaf yield as well as crowns and root yield (82.8 t ha^{-1} and 42.9 t ha^{-1}, respectively) compared to the control treatment of 42.7 t ha^{-1} and 38.2 t ha^{-1}, respectively).

5. Nitrogen Use Efficiency (NUE) in Sugar Beet Production

Increasing nitrogen use efficiency can be achieved by using the right combination of nutrients, fertilizing at the right time and avoiding nutrient loss [76]. Nitrogen is a very mobile element and can be lost in several ways. Losses are often in the air volatile but also as leaching in the deeper layers of the soil by rainfall and groundwater. Both cases cause economic losses but also environmental problems. Nitrogen losses are affected by nitrogen form (nitrate, ammonium or urea), as well as soil properties (pH, texture, temperature, moisture, cation exchange ability, organic matter) and fertilizer management (time and dosage) [77,78].

Another important approach to sugar beet production is organic farming. Organic farming is one of the most important agricultural practices focused on food safety and biodiversity. This means producing food from a healthy environment of land, plants and animals. Organic fertilizers, genetically modified organisms, pesticides and other synthetic chemicals are not used in organic production. Organic agricultural production reduces human impact on the environment, improves soil quality in the long run and contributes to increasing biodiversity [79].

Therefore, a major challenge for breeders is developing sugar beet cultivars with higher NUE without creating nitrogen deficits and/or reducing yields to reduce environmental pollution and production costs. To create N-efficient cultivars, a selection can be carried out directly via N stress conditions or indirectly under optimal production conditions [80–84]. Direct selection in the target environment has the advantage to detect the interactions between the genotype and production system. Precision farming with the use of geographical information system (GIS) will also, in the future, have a great impact on improving NUE in sugar beet production [85–87].

6. Influence of Soil Type and N Fertilization on Sugar Beet Yield and Quality

The sugar beet production should always note that the land and weather conditions determine the framework and implement agrotechnical measures for the production level. Sugar beet can use large quantities of N, and the optimal amount moves within narrow limits. With an increased supply of N for the same root yield, beets can adopt up to 70% more N from the required amounts, negatively impacting root quality. Due to these facts, N fertilization should be done only to replenish stocks in the ground. As N in the soil occurs in the organic and mineral form, determining the stock of plants' available N represents a problem, and because of the great variability in the soil, the problem is the time of soil sampling for analysis. Sugar beet depends on several factors, such as soil and weather, plant distribution in the field and number per unit area, a form of added N [88], as well as the fact that beets absorb uneven N throughout different growth phases in vegetation.

Today, we apply N based on the established stocks of N min in the soil before tillage and pre-sowing fertilization when fertilizing sugar beet. Early determination of N fertilization assumes the impossibility of accurately predicting N's mobilization (subsequent delivery) from organic matter soil, since the period elapsed between the optimal fertilization period and harvest period is extremely long. This is important because the yield and quality of beets also depend on the availability of N during the second half of the vegetation. In practice, N fertilization is very often the cause of bad production results. In this research, the aim was to determine the effect of early N addition in the form of urea and increased fertilization through yield fertilization and root quality. Precipitation and air temperature over the season significantly affect N fertilization and achieved production results. Kristek et al. [68] point out that soil type and especially N application time significantly affect yield elements and sugar beet root quality. Sugar beet prefers the nitrate form of N(NO_3^-), so top dressing in production is usually performed in the phase 3–4 of leaves with mineral fertilizer calcium ammonium nitrate (CAN), which contains 27% N.

Last and Draycott [88] found in a three-year study that fertilization (0, 125, 250 kg ha^{-1} N) in carbonate loamy soils did not affect the increase in dry matter yield of roots. Icreasing the dry matter yield of roots requires a lower dose of fertilizer on clayey loamy soil than on sandy loamy soil. They further found that in the area with an average annual rainfall (around 600 mm for sugar beet), the loss of N by leaching was negligible but that the losses were significant during very humid springs, averaging 40 kg ha^{-1} N. Pospišil et al. [89] state that fertilization of 120 kg ha^{-1} N in soils of poor fertility (2 mg N-min/100 g soil, at a depth of 0–60 cm) and a more abundant rainfall during spring and summer significantly increase the leaf area of sugar beet. Marinković et al. [90] state that autumn fertilization of chernozem (Novi Sad, Serbia) with different doses of NPK (50, 100 and 150 kg ha^{-1} N, P, K) increased the average yield of root (58.5 t ha^{-1}) in comparison with unfertilized plots (35.6 t ha^{-1}). They further state that increasing the dose of N in the fertilizer of 100 kg ha^{-1} N significantly increases the sugar yield (8.69 t ha^{-1}), while further increase in the amount of N of 150 kg ha^{-1} N reduces the sugar yield by 1.69 t ha^{-1}. Hoffmann and Märländer [91] analyzed 57 genotypes from 22 locations in Germany (mainly luvisol and chernozem) with fertilization up to 160 kg N ha^{-1}, stating that the concentration of total soluble N in beets ranges from 25 to 65 mmol kg^{-1} and amino N from 6 to 24 mmol kg^{-1}. Malnou et al. [92], during a three-year experiment in Great Britain (Broom's Barn) on sandy soil with favorable water–air relations, state that autumn fertilization of N (0 to 220 kg ha^{-1} N) with 160 kg ha^{-1} N significantly increases average sugar beet root yield (71.5 t ha^{-1}) and sucrose content (17.4%), compared to the root yield and sucrosecontent without N fertilization, (55.7 t ha^{-1}, 17.7% respectively). In the Republic of Croatia, according to Kristek et al. [68], after a large amount of precipitation in the winter period (in 2009/10), in soil with good physical properties in the spring, the amount of N min in the layer 0–60 cm was significantly smaller than in the previous year, when the amount of precipitation was almost half less and relative to soil with a differentiated profile. In fertile soil, urea application in the autumn with tillage was neither justified nor necessary, since the maximum root yield (74.9 t ha^{-1}) and sugar yield (10.1 t ha^{-1}) were achieved in the treatment without fertilization urea and top dressing (N-0 + 27 kg ha^{-1} N). Marinković and Crnobarac [93] determined the highest yield of roots (62.7 t ha^{-1}) and sugar by fertilizing chernozem (Serbia, Rimski Šančevi) with 150, 100 and 50 kg N, P and K ha^{-1}, respectively 41 t ha^{-1}, with half of N (75 kg ha^{-1} N) applied in the autumn and the other half pre-sowing.

7. Cercospora Leaf Spot and N Fertilizers

In sugar beet production, the most important disease is the Cercospora leaf spot caused by the fungus *Cercospora beticola* Sacc., which occurs regularly in the production [94–97]. Several diseases are not so economically important, and they are caused by the fungus *Ramularia beticola* Futr (Ramularia leaf spot) [98,99], *Alternaria tenuis* Nees (Alternaria leaf spot) [100–102], *Phoma betae* Frank (zoned spot) [103,104], *Uromyces betae* Perset Lev. (sugar beet rust) and *Pseudomonas syringae* pathotype aptata Brown and Jamieson, Stevens (bacterial spot) [105].

Additionally, numerous authors' results state that fertilization generally and high amounts of N fertilizers favor the development of the disease of the crops, affecting the plant's disease risk either directly or indirectly through changes in the community structure or nitrogen [106–109].

The CLS is the most detrimental disease of sugar beet in temperate climates worldwide [110,111], which affects the above-ground parts of plants, including seeds. Symptoms are observed on older leaves in the form of gray spots inside and with a brown edge. The infected leaves become necrotic and eventually fall to the ground but remain attached to the root head. According to Kristek et al. [45], the damage caused by *Cercospora beticola* Sacc. reduces the photosynthesizing surface of the leaves. The yield of sugar beet root may be reduced up to 60% and sucrose content by 3–7%. These decrements in sugar beet root quality ultimately lead to a 42% reduction in technological sugar yield [112–114] up to 50% [115].

Ontario [116] suggests that the N application rate could reduce the need for fungicide applications to control Alternaria leaf blight and Cercospora leaf spot diseases in the field-grown carrots because lower N treatments resulted in fewer live leaves per plant at harvest. Makheti Mutebi and Atieno Ondede [117] found that for mulberry (*Morus alba* L.), plant fertilization amount of 200 kg ha^{-1} N is an effective approach for suppressing Cercospora leaf spot of mulberry and can be recommended to the farmers.

According to Kristek et al. [3], by applying non-symbiotic and associative N bacteria in sugar beet cultivation, there is a possible reduction in mineral N fertilizers.

8. Leaf N Content

According to Bergmann [118], the dry matter of a fully developed mature sugar beet leaf, 50 to 60 days after germination (June or July), optimally contains 4.0–6.0% N, 0.35–0.60% phosphorus, 3, 5–6.0% potassium and 0.30–0.70% magnesium. Similar values in the content of macro-elements in the dry matter of sugar beet leaves were shown by Stanaćev [40], according to which the N content is usually from 1.80 to 2.46%, phosphorus from 0.57 to 0.71%, potassium from 3.0 to 4.46%, sodium from 1.15 to 2.40% and magnesium from 0.37 to 0.75%.

The N concentration in the sugar beet plant depends on the amount of N in the soil. Draycott and Chirstenson [13] state that in the phase of fully expanded leaves, the N concentration in a sugar beet plant can range from 1.0 to 3.5%, in the root from 0.5 to 0.8%, in the leaf blade from 2.2 to 3.5% and in the stalk from 1.0 to 1.5%, but that in soils where beets are grown for a long time, there is N in the soil either in excessive or insufficient amounts and, therefore, N concentrations may be outside the specified limits.

Reviewing the results of several studies, Draycott and Chirstenson [13] state that the dry matter of sugar beet leaves at harvest contains about 0.3% phosphorus, 3% potassium, 2.5% sodium and 0.4% magnesium, while the dry matter of the root contains about 0.1 phosphorus, 0.8% potassium and 0.1% sodium. Furthermore, according to the experience in France and Great Britain, the authors state that at the time of the largest leaf area (July and August), the above-ground mass of sugar beet has the highest amount of phosphorus, from 20 to 25 kg ha^{-1} P_2O_5, while in the root, the amount of phosphorus increases during vegetation and at the time of extraction can be up to 40 kg ha^{-1} P_2O_5.

In recent breeding programs, the amino N concentration is the only selection criterion with regard to N components in the beet [119]. To achieve high-yield and high-quality sugar beet roots, sugar beet has high nutrient requirements, both macro and micronutrients. Fertilizing the soil with N is usually performed in the spring before sowing. Nitrates are present in the environment, i.e., in the air, food (mostly in fruits and vegetables) and water. Additionally, nitrates are used as fertilizers in agriculture. Nitrate or nitrite is always investigated for protecting the environment in water, agriculture and food. Nitrogen carries the offering but is also a nutrient that affects root quality. Both leaf blades and the sugar beet petioles leaf are important in determining the plants' nutrient status [59].

Leaf blades and petioles of healthy sugar beet plants contain a large amount of N in the nitrate form. The lowest concentration of any nutrient can change a little with sugar beet variety, climate or soil conditions, moisture stress, disease, etc. Bilir and Saltalı [120] state that different N fertilization can greatly influence nitrate concentration in the sugar beet leaves, which needs to be analyzed if the leaves are used as animal fed.

Grzebisz et al. [121], according to the average of a two-year study in Poland, found that sugar beet leaf in the growth phase of six pairs of leaves (BBCH16) contains, on average, the most potassium (42.99 g^{-1} kg^{-1} dry matter), followed by N (33.59 g^{-1} kg^{-1} dry matter), magnesium (5.64 g^{-1} kg^{-1} dry matter) and the least phosphorus (3.79 g^{-1} kg^{-1} dry matter). Later, in the fully developed leaf rosette (BBCH43), the leaf dry matter contains the most N (42.53 g^{-1} kg^{-1} dry matter). At the same time, the content of other macro-elements, potassium, magnesium and phosphorus, was reduced (36.34 g^{-1} kg^{-1}, 5.22 g^{-1} kg^{-1} and 3.09 g^{-1} kg^{-1} dry matter, respectively).

Sodium does not play a major role in the nutrition of sugar beet, but sodium affects the quality of sugar beet roots. If the root contains higher amounts of sodium (>0.65 mmol 100 g^{-1} beets), the yield of sugar from the root is reduced. Since sugar beet is a halophytic species, sodium can partially replace potassium in the nutrition of sugar beet [70,120]. For example, Hampe and Marschner [122] state that the cultivation of sugar beet when Na + ions were applied to the substrate instead of K + ions increased in dry matter, leaf area and the number of shoots, while net photosynthesis remained unchanged. Sugar beet requires nutrients throughout the growing season, and if there are not enough of them, the plant will lag behind in growth. Samples of plant material through the period of growth in weekly or monthly intervals can give a lot of information about the needs of plants for nutrients.

As a spring crop, sugar beet generally absorbs the most nutrients in July, when water needs are the highest. This is confirmed by Malnou et al. [92] who, on the example of N fertilization (0, 40, 80, 120 and 160 kg ha^{-1} N), state that the sugar beet crop in the phase of the highest leaf area index takes up from 102 to 147 kg ha^{-1} N in mid-July. Stanaćev [40] points out that the absorption of N from the soil slows down in sugar beet in the second half of July.

Draycott [70] considers that the amount of nutrients required for sugar beet is highly variable. Namely, the author points out that sugar beet absorbs the least Mg (23 kg ha^{-1}), then P (5–100 kg ha^{-1}) and Na (60–100 kg ha^{-1}) and macronutrients N (100–220 kg ha^{-1}), and the highest needs are for K (168–660 kg ha^{-1}).

9. Sugar Beet Root Quality

Sugar beet fertilization is specific in relation to other field crops. As crop which have high yield, technological quality is also expected to be high. Proper fertilization management includes several essential questions, such as which fertilizer to apply, what amount of nutrients, in what ratio and when to apply them. With increasing rates of N, there is an increased content of α-amino N and another non-sugar matter at the root, which is unfavorable, reflecting sugar crystallization during processing.

Last et al. [88], during a six-year field study (1973–78), examined the impact of ammonium nitrate fertilization (0, 41, 82, 124, 166 and 207 kg N ha^{-1}) and irrigation on the yield and quality of sugar beet. The authors stated that in the arable soil layer, the optimal concentration of N min was about 40 mg kg^{-1} N soil during May. Later in the growing season, N greatly increased dry matter, and root yield was also reflected in sugar yield.

Sugar beet root yield linearly increase in relation to accumulation of dry matter. However, from the data of several experiments (Table 1), increasing in N resulted in higher root yield in most cases, but on the contrary, higher N rate even decreased sucrose content in the root.

Table 1. Different amounts of N and their influence on final sugar beet yield.

Country	Nitrogen Amount	Root Yield (t ha^{-1})	Sucrose Content (%)	Reference
Serbia	50	57.7	13.8	Marinković and Crnobarac [123] [1]
	100	55.9	13.4	
	150	59.9	13.0	
Serbia	90	70.8	17.0	Milić et al. [124] [2]
	120	82.9	16.6	
	150	93.7	14.9	
	180	94.7	15.5	
United Kingdom	0	67.7	16.8	Malnou et al. [16] [3]
	40	72.2	16.8	
	80	76.8	16.7	
	120	78.2	16.7	
	160	78.6	16.5	
Nebraska (SAD)	0	56.2	17.2	Hergert and Nielsen [125] [4]
	39	61.3	16.8	
	79	64.4	16.6	
	118	64.8	16.1	
	157	63.3	16.2	
	196	64.3	16.1	
	235	61.8	14.5	
Poland	75	57.4	16.6	Barłóg et al. [126] [5]
	125	54.7	16.8	
Republic of Croatia	46	61.4	15.8	Kristek et al. [68] [6]
	60	62.8	15.8	
	73	61.2	15.6	
	87	65.2	15.6	
	100	63.3	15.4	
Egypt	0	29.5	16.3	Abdelaal and Sahar [127] [7]
	35	40.8	16.6	
	70	61.4	16.9	
	105	63.4	17.1	
Czech Republic	0	45.3	14.3	Pulkrábek et al. [128] [8]
	70	50.7	14.6	
	130	53.3	14.2	
Egypt	165	59.9	20.4	Leilah and Khan [129] [9]
	220	67.2	18.9	
	275	74.3	18.0	
Poland	0	65.2	18.1	Pogłodziński et al. [130] [10]
	40	72.3	17.9	
	80	75.4	18.0	
	120	76.4	17.8	
	160	73.3	17.6	
	200	72.5	17.5	

[1] average of 1993–1998; [2] year 2004; [3] average of 2000–2001 and two sites (Broom's Barn and Sutton Bonington); [4] average of 2006–2008; [5] average of 1997–1999; [6] average of 2009–2010; [7] average of 2012/2013–2013/2014; [8] year 2020; [9] average of 2014/2015 and 2015/2016; [10] average of 20015–2016 and two sites (Nowa Wieś Królewska and Neryngowo).

Kristek et al. [68] state that fertilization with urea in the year with insufficient rainfall (around 540 mm from April to October) was also not justified, as the highest root yield (61.53 t ha^{-1}) and sugar (9.02 t ha^{-1}) were achieved in the variant with top dressing

N-0 + 13.5 (50 kg ha^{-1} CAN), while in a wet year (around 900 mm from April to October), in that soil, the highest root yield (75.13 t ha^{-1}) and sugar (9.54 t ha^{-1}) were achieved with fertilization urea (46% N) and without the addition of N in the vegetation. Based on the data from the literature, the regression analyses show that increment of N fertilizer amount can slightly increase the root yield, but for the sucrose content, the linear model has a negative trendline (Figure 1). There is evidence that sugar nitrogen fertilization has a great influence in increasing the biomass of the roots, but the increased amount of nitrogen has negative influence on sugar beet root quality and increases the amount of harmful nitrogen in the root and decreases the sucrose content at harvest. Even though the regression coefficient is not significant, the trendline shows negative influence on root quality, and there is still need for further experiments of nitrogen fertilization in sugar beet production.

Sugar beet fertilization is very specific because it needs to achieve high root yield, high sugar content, and small amounts of soluble non-sugars in the root. The deficit, as well as excessive amounts of nutrients, affect the final yield but also the quality of the roots; therefore, proper fertilization is crucial for a successful sugar beet production. The main reason for less sucrose content was an increased amount of brei impurities [116,117,131–138], especially the amount of amino N. In the case of N, it is impossible to build up long-term N reserves in the soil, so the excess or lack of available N during the year is likely. Sugar beet does have a mechanism to regulate N uptake, but the root can absorb as much N as is provided.

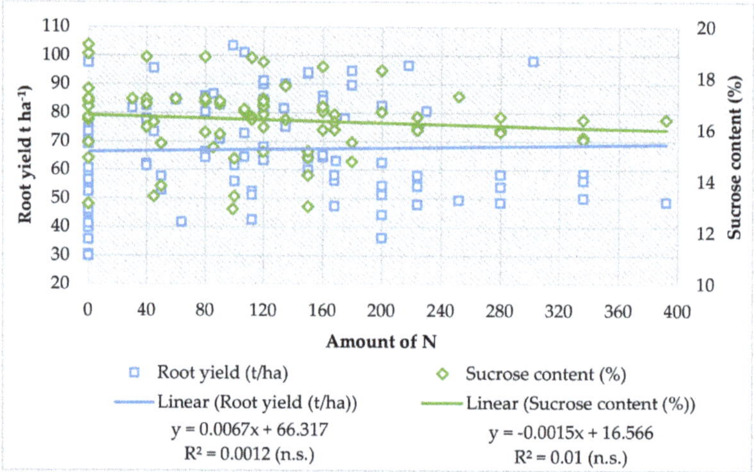

Figure 1. Linear regression models of sugar beet root yield and sucrose content (data obtained from: Malnou et al. [16], Marinković et al. [90,93], Milić et al. [124], Pogłodziński et al. [130], Varga et al. [134], Barłóg et al. [85], Lentz and Lehrsch [137] Carter and Traveller [23], Tarkalson et al. [22], Lauer [138]).

10. Conclusions

This study comprehensively describes N's influence on sugar beet vegetative growth and root yield and quality. The main elements of nutrition N, phosphorus and potassium, have the greatest importance in plant nutrition and the greatest influence on the creation of yield and increased technological value of the sugar beet root. As a large consumer of plant assimilates, sugar beet will react strongly to enhanced nutrition with N, phosphorus and potassium used in the form of minerals nutrients. Even though the N needs of sugar beet have been the main focus for decades, studies with N fertilization of sugar beet are still very important due to the development of new genotypes with different accumulation capacities. In sugar beet production, the most important factors are proper and comprehensive plant nutrition and adoption of certain plant assimilates and their role in synthesizing organic matter in plant tissues and the accumulation of sugar in the roots. There is no unique plant nutrition system because the needs of sugar beet for certain elements depend on weather conditions, soil, cultivation methods (natural water regime, irrigation) but also on the genotypes. Increasing N increases the root yield of sugar beet but also the content of harmful N, which reduces the technological root value. Thus, soil analyses and fertilization recommendations represent one of the most important factors in successful production, especially in relation to sugar beet root quality.

Author Contributions: Conceptualization, I.V. and V.Z.; investigation, I.V., J.J., M.R. and M.A.; writing—original draft preparation, I.V., V.Z., J.J. and A.M.K.; writing—review and editing, I.V., A.M.K., Z.L. and D.I. All authors have read and agreed to the published version of the manuscript.

Funding: This research was part of project KK.01.1.1.04.0052: "Innovative production of organic fertilizers and substrates (INOPROFS)", co-financed by the European Union from the European Regional Development Fund within the Operational programme Competitiveness and Cohesion 2014–2020 of the Republic of Croatia.

Institutional Review Board Statement: Not applicable.

Conflicts of Interest: The authors declare no conflict of interest.

References

1. Bruhns, J.; Baron, O.; Maier, K. *Sugar Economy Europe 2004*; No. Ed. 50; Verlag Dr. Albert Bartens KG: Berlin, Germany, 2003.
2. Jug, D.; Jug, I.; Brozović, B.; Vukadinović, V.; Stipešević, B.; Đurđević, B. The role of conservation agriculture in mitigation and adaptation to climate change. *Poljoprivreda* **2018**, *24*, 35–44. [CrossRef]
3. Kristek, S.; Brkić, S.; Jović, J.; Stanković, A.; Ćupurdija, B.; Brica, M.; Karalić, K. The application of nitrogen-fixing bacteria in order to reduce the mineral nitrogen fertilizers in sugar beet. *Poljoprivreda* **2020**, *26*, 65–71. [CrossRef]
4. Stošić, M.; Brozović, B.; Vinković, T.; Ravnjak, B.; Kluz, M.; Zebec, V. Soil resistance and bulk density under different tillage system. *Poljoprivreda* **2020**, *26*, 17–24. [CrossRef]
5. Rašovský, M.; Pačuta, V.; Černý, I.; Ernst, D.; Michalska-Klimczak, B.; Wyszyňski, Z. Monitoring of Influence of Biopreparates, Weather Conditions and Variety on Production Parameters of Sugar Beet. *Listy Cukrov. A Řepařské* **2021**, *137*, 154.
6. Varga, I.; Lončarić, Z.; Pospišil, M.; Rastija, M.; Antunović, M. Dynamics of sugar beet root, crown and leaves mass with regard to plant densities and spring nitrogen fertilization. *Poljoprivreda* **2020**, *26*, 32–39. [CrossRef]
7. Jurišić, M.; Radočaj, D.; Plaščak, I.; Rapčan, I. A Comparison of Precise Fertilization Prescription Rates to a Conventional Approach Based on the Open Source GIS Software. *Poljoprivreda* **2021**, *27*, 52–59. [CrossRef]
8. Antunović, M.; Varga, I.; Stipešević, B.; Ranogajec, L. Analýza chorvatského cukrovarnického sektoru a produkce cukrové řepy. *Listy Cukrov. A Řepařské* **2021**, *137*, 383–386.
9. Ernst, D.; Černý, I.; Pačuta, V.; Zapletalová, A.; Rašovský, M.; Skopal, J.; Vician, T.; Šulík, R.; Gažo, J. Yield and Sugar Content of Sugar Beet Depending on Different Soil Tillage Technologies. *Listy Cukrov. A Řepařské* **2021**, *137*, 319–324.
10. Turesson, H.; Andersson, M.; Marttila, S.; Thulin, I.; Hofvander, P. Starch biosynthetic genes and enzymes are expressed and active in the absence of starch accumulation in sugar beet tap-root. *BMC Plant Biol.* **2014**, *14*, 104. [CrossRef]
11. Hoffmann, C.M.; Kenter, C. Yield potential of Sugar beet—Have we hit the ceiling? *Front. Plant Sci.* **2018**, *9*, 289. [CrossRef]
12. Franzen, D.W. Delineating nitrogen management zones in a sugarbeet rotation using remote sensing—A review. *J. Sugar Beet Res.* **2004**, *41*, 47–60. [CrossRef]
13. Draycott, A.P.; Christenson, D.R. *Nutrients for Sugar Beet Production: Soil-Plant Relationships*; CABI: Wallingford, UK, 2003.
14. Jaćimović, G.; Marinković, B.; Crnobarac, J.; Bogdanović, D.; Kovačev, L.; Danojević, D. Influence of fertilization and nitrate-nitrogen position in soil profile on the sugar beet root yield and quality. *J. Agric. Sci.* **2008**, *53*, 83–90.
15. Kristek, S.; Kristek, A.; Evačić, M. Influence of nitrogen fertilization on sugar beet root yield and quality. *Cereal Res. Commun.* **2008**, *36*, 371–374.
16. Malnou, C.S.; Jaggard, K.W.; Sparkes, D.L. A canopy approach to nitrogen fertilizer recommendations for the sugar beet crop. *Eur. J. Agron.* **2006**, *25*, 254–263. [CrossRef]
17. Pospišil, M. *Ratarstvo II. Dio—Industrijsko Bilje*; Zrinski: Čakovec, Croatia, 2013.
18. Lundegårdh, H. *Plant Physiology*; Oliver and Boyd: Edinburgh, UK; London, UK; T. & A. Constable Ltd.: Edinburgh, UK, 1966.
19. Hoffmann, C.M.; Kluge-Severin, S. Growth analysis of autumn and spring sown sugar beet. *Eur. J. Agron.* **2011**, *34*, 1–9. [CrossRef]
20. Pevalek-Kozlina, B. (Ed.) *Fiziologija Bilja*; Profil International: Zagreb, Croatia, 2003.
21. Klotz, K.L.; Finger, F.L. Impact of temperature, length of storage and postharvest disease on sucrose catabolism in sugarbeet. *Postharvest Biol. Technol.* **2004**, *34*, 1–9. [CrossRef]
22. Tarkalson, D.D.; Bjorneberg, D.L.; Camp, S.; Dean, G.; Elison, D.; Foote, P. Improving nitrogen management in Pacific Northwest sugarbeet production. *J. Sugar Beet Res.* **2016**, *53*, 14–36. [CrossRef]

23. Carter, J.N.; Traveller, D.J. Effect of time and amount of nitrogen uptake on sugarbeet growth and yield. *Agron. J.* **1981**, *73*, 665–671. [CrossRef]
24. Mary, B.; Recous, S. Measurement of nitrogen mineralization and immobilization fluxes in soil as a means of predicting net mineralization. *Eur. J. Agron.* **1994**, *3*, 291–300. [CrossRef]
25. Mengel, K. Turnover of organic nitrogen in soils and its availability to crops. *Plant Soil* **1996**, *181*, 83–93. [CrossRef]
26. Nemeth, K. Recent advances in EUF research (1980–1983). *Plant Soil* **1985**, *83*, 1–19. [CrossRef]
27. Nemeth, K.; Bartels, H.; Vogel, M.; Mengel, K. Organic nitrogen compounds extracted from arable and forest soils by electro-ultrafiltration and recovery rates of amino acids. *Biol. Fertil. Soils* **1988**, *5*, 271–275. [CrossRef]
28. Natesan, S.; Ranganathan, V.; Nemeth, K.; Krishnan, V. EUF-analysis of tea soils of Southern India and tea productivity. *Plant Soil* **1985**, *83*, 191–198. [CrossRef]
29. Akinrinde, E.A.; Obigbesan, G.O.; Gaiser, T. Electro-ultrafiltration (EUF) technique in relation to conventional methods of soil testing for the determination of available P, Ca, Mg and NO_3-N in some tropical soils. *J. Agron.* **2006**, *5*, 375–381. [CrossRef]
30. Wiedeman, H. Soil analysis and N-fertilizer recommendations for growing sugar beet in southern Germany and Austria. In Proceedings of the International Institute for Beet Research 57th Winter Congress, Bruxelles, Belgium, 16–17 February 1994.
31. Horn, D.; Fürstenfeld, F. Nitrogen fertilizer recommendation for sugar beet according to the EUF soil testing system. In *Plant Nutrition*; Springer: Dordrecht, The Netherlands, 2001.
32. Wiklicky, L. Application of the EUF procedure in sugar beet cultivation. *Plant Soil* **1982**, *64*, 115–127. [CrossRef]
33. Scharpf, H.C.; Wehrmann, J. Bedeutung des Mineralstickstoffvorrates des Bodens zu Vegetationsbeginn fur die Bemessung der N Dungung zu Winterweizen. Landwirtschaftliche Forschung. *Sonderheft* **1976**, *52*, 109–126.
34. Wehrmann, J.V.; Scharpf, H.C. Der Mineralstickstoffgehalt des Bodens als Maßstab für den Stickstoffdüngerbedarf (N min-Methode). *Plant Soil* **1979**, *52*, 109–126. [CrossRef]
35. Vukadinović, V.; Lončarić, Z. *Ishrana Bilja. Poljoprivredni Fakultet u Osijeku*; Josip Juraj Strossmayer University of Osijek: Osijek, Croatia, 1997.
36. Bertić, B.; Vukadinović, V. *Filozofija Gnojidbe Sve Što Treba Znati o Gnojidbi*; Studio HS Internet d.o.o. Osijek: Osijek, Croatia, 2013.
37. Jaradat, A.A.; Rinke, J. Modeling sugar content of farmer-managed sugar beets (*Beta vulgaris* L.). *Commun. Biometry Crop Sci.* **2012**, *7*, 23–34.
38. Hoffmann, C.M.; Kluge-Severin, S. Light absorption and radiation use efficiency of autumn and spring sown sugar beets. *Field Crops Res.* **2010**, *119*, 238–244. [CrossRef]
39. Kristek, A.; Halter, J. Djelovanje vegetacijskog prostora na porast lišća šećerne repe i prinos korijena. *Agron. Glas.* **1988**, *2–3*, 79–94.
40. Stanaćev, S. *Šećerna Repa*; Nolit: Beograd, Serbia, 1979.
41. Kristek, A.; Liović, I. Ritam rasta šećerne repe u uvjetima 1987. godine. *Poljopr. Aktualnosti* **1988**, *30*, 173–185.
42. Jelić, S.; Antunović, M.; Bukvić, G.; Varga, I.; Iljkić, D. Impact of plant density on growth, yield and quality of sugar beet. *Listy Cukrov. A Řepařské* **2019**, *135*, 107.
43. Manderscheid, R.; Pacholski, A.; Weigel, H.J. Effect of free air carbon dioxide enrichment combined with two nitrogen levels on growth, yield and yield quality of sugar beet: Evidence for a sink limitation of beet growth under elevated CO_2. *Eur. J. Agron.* **2010**, *32*, 228–239. [CrossRef]
44. Lüdecke, H. *Šećerna Repa*; Poljoprivredni Nakladni Zavod Zagreb: Zagreb, Croatia, 1956.
45. Kristek, A.; Kristek, S.; Varga, I.; Drmić, Z. Rezultati u proizvodnji šećerne repe u zavisnosti od izbora hibrida i broja tretiranja fungicida. *Poljoprivreda* **2015**, *21*, 15–22. [CrossRef]
46. Vukadinović, V.; Jug, I.; Đurđević, B. Ekofiziologija Bilja. In *Poljoprivredni Fakultet u Osijeku*; Sveučilište Josipa Jurja Strossmayera u Osijeku Zebra: Vinkovci, Croatia, 2014.
47. Müller-Linow, M.; Pinto-Espinosa, F.; Scharr, H.; Rascher, U. The leaf angle distribution of natural plant populations: Assessing the canopy with a novel software tool. *Plant Methods* **2015**, *11*, 11. [CrossRef] [PubMed]
48. Kenter, C.; Hoffman, C.M. Seasonal patterns of sucrose concentration in relation to other quality parameters of sugar beet (*Beta vulgaris* L.). *J. Sci. Food Agric.* **2006**, *86*, 62–70. [CrossRef]
49. Drachovská, M.; Sandera, K. *Fysiologie Cukrovky*; Nakladatelství Československé Akademie věd: Prague, Czech Republic, 1959.
50. Märländer, B.; Hoffmann, C.M.; Koch, H.J.; Ladewig, E.; Merkes, R.; Petersen, J.; Stockfisch, N. Environmental Situation and Yield Performance of the Sugar Beet Crop in Germany: Heading for Sustainable Development. *J. Agron. Crop Sci.* **2003**, *189*, 201–226. [CrossRef]
51. Tóth, M.; Pokrivčák, J.; Smutka, L.; Dvořák, M.; Pulkrábek, J. Economic aspects of sugar beet production and biodiversity: Effects of ban on neonicotinoids use. *Listy Cukrov. Řepařské* **2022**, *138*, 116–120.
52. Pospišil, A.; Pospišil, M. The Effect of Organic Fertilizers on the Spelt Yield and the Yield of its Components. *Poljoprivreda* **2021**, *27*, 37–43. [CrossRef]
53. Tan, Z.X.; Lal, R.; Wiebe, K.D. Global soil nutrient depletion and yield reduction. *J. Sustain. Agric.* **2005**, *26*, 123–146. [CrossRef]
54. Pan, B.; Lam, S.K.; Mosier, A.; Luo, Y.; Chen, D. Ammonia volatilization from synthetic fertilizers and its mitigation strategies: A global synthesis. Agriculture. *Ecosyst. Environ.* **2016**, *232*, 283–289. [CrossRef]
55. Cucina, M.; De Nisi, P.; Sordi, S.; Adani, F. Sewage Sludge as N-Fertilizers for Crop Production Enabling the Circular Bioeconomy in Agriculture: A Challenge for the New EU Regulation 1009/2019. *Sustainability* **2021**, *13*, 13165. [CrossRef]

56. Lukas, V.; Neudert, L.; Širůček, P.; Novák, J.; Elbl, J. Effect of variable rate application of phosphorus and potassium fertilizers in sugar beet. *Listy Cukrov. A Řepařské* **2021**, *137*, 417–422.
57. de Souza Braz, A.M.; da Costa, M.L.; Ramos, S.J.; Dall'Agnol, R.; Fernandes, A.R. Long Term Application of Fertilizers in Eastern Amazon and Effect on Uranium and Thorium Levels in Soils. *Minerals* **2021**, *11*, 994. [CrossRef]
58. Issukindarsyah, I.; Sulistyaningsih, E.; Indradewa, D.; Putra, E.T.S. The Effect of Ammonium Nitrate Ratio and Support Types on the NPK Uptake and Growth of Black Pepper (*Piper nigrum* L.) in Field Conditions. *Poljoprivreda* **2021**, *27*, 25–33. [CrossRef]
59. Varga, I.; Kerovec, D.; Engler, M.; Popović, B.; Lončarić, Z.; Iljkić, D.; Antunović, M. Determination N-NO_3^- in sugar beet leaves). *Listy Cukrov. A Řepařské* **2022**, *138*, 69–72.
60. Trap, J.; Bonkowski, M.; Plassard, C.; Villenave, C.; Blanchart, E. Ecological importance of soil bacterivores for ecosystem functions. *Plant Soil* **2016**, *398*, 1–24. [CrossRef]
61. Harrison, R.B.; Footen, P.W.; Strahm, B.D. Deep soil horizons: Contribution and importance to soil carbon pools and in assessing whole-ecosystem response to management and global change. *For. Sci.* **2011**, *57*, 67–76.
62. Hoffmann, C.M.; Blomberg, M. Estimation of leaf area index of Beta vulgaris L. based on optical remote sensing data. *J. Agron. Crop Sci.* **2004**, *190*, 197–204. [CrossRef]
63. Tsialtas, J.T.; Maslaris, N. Leaf allometry and prediction of specific leaf area (SLA) in sugar beet (*Beta vulgaris* L.) cultivar. *Photosynthetica* **2008**, *46*, 351–355. [CrossRef]
64. Tsialtas, J.T.; Saulioti, E.; Maslaris, N.; Papakosta, D. Genotypic response to regrowth of defoliated sugar beets after re-watering in a water-limited environment: Effects on yield and quality. *Int. J. Plant Prod.* **2009**, *3*, 1–18.
65. Putnik-Delić, M. Fiziološki i Molekularni Aspekti Tolerantnosti Šećerne Repe Prema Suši. Doctoral Dissertation, University of Novi Sad, Novi Sad, Serbia, 2013.
66. Kosterj, A.; Repka, J. Quantitative indicators of growth, production process and yield-formation of sugar-beet. *Rostl. Vyrob.* **1993**, *39*, 1077–1086.
67. Tsialtas, J.T.; Maslaris, N. Leaf physiological traits and its relation with sugar beet cultivar success in two contrasting environments. *Int. J. Plant Prod.* **2012**, *6*, 15–36.
68. Kristek, A.; Kristek, S.; Antunović, M.; Varga, I.; Katušić, J.; Besek, Z. Utjecaj tipa tla i gnojidbe dušikom na prinos i kvalitetu korijena šećerne repe. *Poljoprivreda* **2011**, *17*, 16–22.
69. Brentrup, F.; Küsters, J.; Kuhlmann, H.; Lammel, J. Application of the Life Cycle Assessment methodology to agricultural production: An example of sugar beet production with different forms of nitrogen fertilizers. *Eur. J. Agron.* **2001**, *14*, 221–233. [CrossRef]
70. Draycott, A.P. *Sugar Beet*; Wiley-Blackwell: Hoboken, NJ, USA, 2008.
71. Last, P.J.; Draycott, A.P.; Messem, A.B.; Webb, D.J. Effects of nitrogen fertilizer and irrigation on sugar beet at Broom's Barn 1973–8. *J. Agric. Sci.* **1983**, *101*, 185–205. [CrossRef]
72. Starke, P.; Hoffmann, C.M. *Yield Parameters of Beta Beets as a Basis to Estimate the Biogas Yield*; Nutzung von Zuckerrüben für die Biogaserzeugung–Definition der Qualität sowie ertragsrelevante Parameter von Rübe, Blatt und Schossern; Institut für Zuckerrübenforschung: Gottingen, Germany, 2014; Volume 42.
73. Starke, P.; Hoffmann, C.M. Yield parameters of Beta beets as a basis to estimate the biogas yield. *Sugar Ind.* **2014**, *139*, 169–176. [CrossRef]
74. Monreal, J.A.; Jiménez, E.T.; Remesal, E.; Morillo-Velarde, R.; García-Mauriñoa, S.; Echevarría, C. Proline content of sugar beet storage roots: Response to water deficit and nitrogen fertilization at field conditions. *Environ. Exp. Bot.* **2007**, *60*, 257–267. [CrossRef]
75. Vielemeyer, H.-P.; Lux, H.; Weege, K.-H. Einfluß des zeitlichen N-Angebots auf den Ertragsbildungsprozeß der Zuckerrübe. Arch. *Acker-Pflanzenbau Bodenkd. Berl.* **1986**, *30*, 131–137.
76. Yadav, M.R.; Kumar, R.; Parihar, C.M.; Yadav, R.K.; Jat, S.L.; Ram, H.; Meena, R.K.; Singh, M.; Birbal; Verma, A.P.; et al. Strategies for improving nitrogen use efficiency: A review. *Agric. Rev.* **2017**, *38*, 29–40. [CrossRef]
77. Melino, V.J.; Tester, M.A.; Okamoto, M. Strategies for engineering improved nitrogen use efficiency in crop plants via redistribution and recycling of organic nitrogen. *Curr. Opin. Biotechnol.* **2022**, *73*, 263–269. [CrossRef] [PubMed]
78. Stevanato, P.; Chiodi, C.; Broccanello, C.; Concheri, G.; Biancardi, E.; Pavli, O.; Skaracis, G. Sustainability of the sugar beet crop. *Sugar Tech* **2019**, *21*, 703–716. [CrossRef]
79. Atlason, R.S.; Lehtinen, T.; Davíðsdóttir, B.; Gísladóttir, G.; Brocza, F.; Unnþórsson, R.; Ragnarsdóttir, K.V. Energy return on investment of Austrian sugar beet: A small-scale comparison between organic and conventional production. *Biomass Bioenergy* **2015**, *75*, 267–271. [CrossRef]
80. Hoffmann, C.M. Changes in N composition of sugar beet varieties in response to increasing N supply. *J. Agron. Crop Sci.* **2005**, *191*, 138–145. [CrossRef]
81. Laufer, D.; Nielsen, O.; Wilting, P.; Koch, H.J.; Märländer, B. Yield and nitrogen use efficiency of fodder and sugar beet (*Beta vulgaris* L.) in contrasting environments of northwestern Europe. *Eur. J. Agron.* **2016**, *73*, 124–132. [CrossRef]
82. Monteiro, F.; Frese, L.; Castro, S.; Duarte, M.C.; Paulo, O.S.; Loureiro, J.; Romeiras, M.M. Genetic and genomic tools to asssist sugar beet improvement: The value of the crop wild relatives. *Front. Plant Sci.* **2018**, *9*, 74. [CrossRef]

83. Barłóg, P.; Grzebisz, W.; Feć, M.; Łukowiak, R.; Szczepaniak, W. Row method of sugar beet (*Beta vulgaris* L.) fertilization with multicomponent fertilizer based on urea-ammonium nitrate solution as a way to increase nitrogen efficiency. *J. Cent. Eur. Agric.* **2010**, *11*, 225–234.
84. Hergert, G.W. Sugar beet fertilization. *Sugar Tech* **2010**, *12*, 256–266. [CrossRef]
85. Barłóg, P.; Nowacka, A.; Błaszyk, R. Effect of zinc band application on sugar beet yield, quality and nutrient uptake. *Plant Soil Environ.* **2016**, *62*, 30–35. [CrossRef]
86. Bronson, K.F.; Scharf, P.C.; Kitchen, N.R. *Use of GIS-Based Site-Specific Nitrogen Management for Improving Energy Efficiency*; USDA-ARS; UNL Faculty: Lincoln, NE, USA, 2011; Volume 677, pp. 359–384.
87. Clay, D.E.; Shanahan, J.F. *GIS Applications in Agriculture, Volume Two: Nutrient Management for Energy Efficiency*; CRC Press: Boca Raton, FL, USA, 2011.
88. Last, P.J.; Draycott, A.P. Growth and yield of sugar beet on contrasting soils in relation to nitrogen supply: II. Growth, uptake and leaching of nitrogen. *J. Agric. Sci.* **1975**, *85*, 27–37. [CrossRef]
89. Pospišil, M.; Pospišil, A.; Rastija, M. Effect of plant density and nitrogen rates upon the leaf area of seed sugar beet on seed yield and quality. *Eur. J. Agron.* **2000**, *12*, 69–78. [CrossRef]
90. Marinković, B.; Crnobarac, J.; Jaćimović, G.; Rajić, M.; Latković, D.; Aćin, V. Sugar yield and technological quality of sugar beet at different levels of nitrogen fertilization. *Res. J. Agric. Sci.* **2010**, *42*, 162–167.
91. Hoffmann, C.M.; Märländer, B. Composition of harmful nitrogen in sugar beet (*Beta vulgaris* L.)—Amino acids, betaine, nitrate—As affected by genotype and environment. *Eur. J. Agron.* **2005**, *22*, 255–265. [CrossRef]
92. Malnou, C.S.; Jaggard, K.W.; Sparkes, D.L. Nitrogen fertilizer and the efficiency of the sugar beet crop in late summer. *Eur. J. Agron.* **2008**, *28*, 47–56. [CrossRef]
93. Marinković, B.J.; Crnobarac, J.Ž. Dependence of sugarbeet quality and yield on the application of NPK nutrients. *Acta Period. Technol.* **2000**, *31*, 345–350.
94. Bolton, M.D.; Birla, K.; Rivera-Varas, V.; Rudolph, K.D.; Secor, G.A. Characterization of CbCyp51 from field isolates of *Cercospora beticola*. *Phytopathology* **2012**, *102*, 298–305. [CrossRef] [PubMed]
95. Kristek, S.; Jović, J.; Zmaić, K.; Kraljičak, Ž.; Kišpal, H.; Bešlo, D.; Horvat, D.; Stjepanović, B.; Rašić, B. Problem of Development of Resistance to Some Fungicide Active Sub stances Intended for Suppressing of *Cercospora beticola* Sacc. *Listy Cukrov. A Řepařské* **2017**, *133*, 222–226.
96. Vogel, J.; Kenter, C.; Holst, C.; Märländer, B. New generation of resistant sugar beet varieties for advanced integrated management of Cercospora leaf spot in central Europe. *Front. Plant Sci.* **2018**, *9*, 222. [CrossRef] [PubMed]
97. Ereš, H.; Dujković, A.; Vrandečić, K. Cercospora leaf spot. *Glas. Zaštite Bilja* **2021**, *44*, 52–54. [CrossRef]
98. Byford, W.J. *Ramularia beticola* in sugar-beet seed crops in England. *J. Agric. Sci.* **1975**, *85*, 369–375. [CrossRef]
99. Wieczorek, T.M.; Jørgensen, L.N.; Hansen, A.L.; Munk, L.; Justesen, A.F. Early detection of sugar beet pathogen Ramularia beticola in leaf and air samples using qPCR. *Eur. J. Plant Pathol.* **2014**, *138*, 775–785. [CrossRef]
100. Mcfarlane, J.S.; Bardin, R.; Snyder, W.C. An Alternaria leaf spot of the sugar beet. *Proc. Am. Soc. Sugarbeet Technol.* **1954**, *8*, 241–245.
101. Rosenzweig, N.; Hanson, L.E.; Mambetova, S.; Jiang, Q.W.; Guza, C.; Stewart, J.; Somohano, P. Fungicide sensitivity monitoring of Alternaria spp. causing leaf spot of sugarbeet (*Beta vulgaris*) in the Upper Great Lakes. *Plant Dis.* **2019**, *103*, 2263–2270. [CrossRef] [PubMed]
102. Khan, M.F.R.; Haque, M.E.; Bloomquist, M.; Bhuiyan, M.Z.R.; Brueggeman, R.; Zhong, S.; Sharma Poudel, R.; Gross, T.; Hakk, P.; Leng, Y.; et al. First Report of Alternaria Leaf Spot Caused by *Alternaria tenuissima* on Sugar Beet (*Beta vulgaris*) in Minnesota, USA. *Plant Dis.* **2020**, *104*, 580. [CrossRef]
103. Koenick, L.B.; Vaghefi, N.; Knight, N.L.; du Toit, L.J.; Pethybridge, S.J. Genetic diversity and differentiation in *Phoma betae* populations on table beet in New York and Washington States. *Plant Dis.* **2019**, *103*, 1487–1497. [CrossRef] [PubMed]
104. Chand, N.; Jones, E.E.; Casonato, S. Pathogenicity of *Phoma betae* isolates from red beet (*Beta vulgaris*) at seed farms in Canterbury, New Zealand. *Plant Prot.* **2019**, *72*, 21–26. [CrossRef]
105. Agarwal, P.C.; Dev, U.; Rani, I.; Khetarpal, R.K. Seed-borne fungi detected in sugar beet seeds imported into India during last three decades. *Plant Health Prog.* **2006**, *7*, 2. [CrossRef]
106. Veresoglou, S.D.; Barto, E.K.; Menexes, G.; Rillig, M.C. Fertilization affects severity of disease caused by fungal plant pathogens. *Plant Pathol.* **2013**, *62*, 961–969. [CrossRef]
107. Huber, D.M.; Haneklaus, S. Managing nutrition to control plant disease. *Landbauforsch. Volkenrode* **2007**, *57*, 313.
108. Long, D.H.; Lee, F.N.; TeBeest, D.O. Effect of nitrogen fertilization on disease progress of rice blast on susceptible and resistant cultivars. *Plant Dis.* **2000**, *84*, 403–409. [CrossRef]
109. Liu, X.; Lyu, S.; Sun, D.; Bradshaw, C.J.; Zhou, S. Species decline under nitrogen fertilization increases community-level competence of fungal diseases. *Proc. R. Soc. B Biol. Sci.* **2017**, *284*, 20162621. [CrossRef] [PubMed]
110. Skaracis, G.N.; Pavli, O.I.; Biancardi, E. Cercospora leaf spot disease of sugar beet. *Sugar Tech* **2010**, *12*, 220–228. [CrossRef]
111. Schmittgen, S. Effects of Cercospora Leaf Spot Disease on Sugar Beet Geno-Types with Contrasting Disease Susceptibility. Ph.D. Thesis, Universitäts-und Landesbibliothek der Heinrich-Heine-Universität Düsseldorf, Düsseldorf, Germany, 2015.
112. Vereijssen, J. Cercospora Leaf Spot in Sugar Beet. Epidemiology, Life Cycle Components and Disease Management. Ph.D. Thesis, Wageningen University, Wageningen, The Netherlands, 2004.

113. Vereijssen, J.; Schneider, J.H.; Termorshuizen, A.J. Root infection of sugar beet by *Cercospora beticola* in a climate chamber and in the field. *Eur. J. Plant Pathol.* **2005**, *112*, 201–210. [CrossRef]
114. Vukčević, R. Nova primena aviometode na zaštiti šećerne repe od cerkospore. *Agron. Glas.* **1964**, *14*, 245–256.
115. Westerveld, S.M.; McKeown, A.W.; McDonald, M.R. Relationship between nitrogen fertilization and Cercospora leaf spot and Alternaria leaf blight of carrot. *HortScience* **2008**, *43*, 1522–1527. [CrossRef]
116. Makheti Mutebi, C.; Atieno Ondede, D. Effect of nitrogen nutrition on the intensity of Cercospora leaf spot of Mulberry. *Int. J. Hortic. Sci. Technol.* **2021**, *8*, 335–342.
117. Bergmann, W. *Nutritional Disorders of Plants: Visual and Analytical Diagnosis*; Jena Gustav Fischer Verlag: New York, NY, USA, 1992.
118. Mäck, G.; Hoffmann, C.M.; Märländer, B. Nitrogen compounds in organs of two sugar beet genotypes (*Beta vulgaris* L.) during the season. *Field Crops Res.* **2007**, *102*, 210–218. [CrossRef]
119. Bilir, B.; Saltalı, K. The Effect of Nitrogen-Boron Application and Time on the Nitrate Content of Sugar Beet Leaves Used as Animal Feed. *Turk. J. Agric. Food Sci. Technol.* **2021**, *9*, 395–400.
120. Grzebisz, W.; Przygocka-Cyna, K.; Łukowiak, R.; Biber, M. An evaluation of macronutrient nutritional status of sugar beets in critical stages of growth in response to foliar application of multi-micronutrient fertilizers. *J. Elem.* **2010**, *15*, 493–507. [CrossRef]
121. Pi, Z.; Stevanato, P.; Sun, F.; Yang, Y.; Sun, X.; Zhao, H.; Geng, G.; Yu, L. Proteomic changes induced by potassium deficiency and potassium substitution by sodium in sugar beet. *J. Plant Res.* **2016**, *129*, 527–538. [CrossRef]
122. Hampe, T.; Marschner, H. Effect of sodium on morphology, water relations and net photosynthesis of sugar beet leaves. *Z. Pflanzenphysiol.* **1982**, *108*, 151–162. [CrossRef]
123. Marinković, B.; Crnobarac, J. Zavisnost kvaliteta i prinosa šećerne repe od primene NPK hraniva. *Acta Period. Technol.* **2000**, *31*, 345–350.
124. Milić, S.; Pejić, B.; Maksimović, L. Uticaj navodnjavanja i đubrenja na sadržaj šećera i prinos korena šećerne repe. *J. Sci. Agric. Res.* **2006**, *67*, 5–12.
125. Hergert, G.W.; Nielsen, R.A. Comparison of strip tillage versus broadcast N application for sugar beets. In Proceedings of the 2009 Biennial Meeting, Denver, CO, USA, 2 April 2009.
126. Barłóg, P.; Grzebisz, W.; Peplinski, K.; Szczepaniak, W. Sugar beet response to balanced nitrogen fertilization with phosphorus and potassium. Part II. Dynamics of beet quality. *Bulg. J. Agric. Sci* **2013**, *19*, 1311–1318.
127. Abdelaal, K.A.; Sahar, F.T. Response of sugar beet plant (*Beta vulgaris* L.) to mineral nitrogen fertilization and bio-fertilizers. *Int. J. Curr. Microbiol. Appl. Sci.* **2015**, *4*, 677–688.
128. Pulkrábek, J.; Brinar, J.; Javor, T.; Dvořák, P.; Bečková, L.; Kuchtová, P.; Hubáčková, J. Experience with variable fertilization of sugar beet. *Listy Cukrov. A Reparske* **2021**, *137*, 184–193.
129. Leilah, A.A.; Khan, N. Interactive Effects of Gibberellic Acid and Nitrogen Fertilization on the Growth, Yield, and Quality of Sugar Beet. *Agronomy* **2021**, *11*, 137. [CrossRef]
130. Pogłodziński, R.; Barłóg, P.; Grzbisz, W. Effect of nitrogen and magnesium sulfate application on sugar beet yield and quality. *Plant Soil Environ.* **2021**, *67*, 507–513. [CrossRef]
131. Varga, I.; Lončarić, Z.; Pospišil, M.; Rastija, M.; Antunović, M. Changes of Nitrate Nitrogen in Sugar Beet Petioles Fresh Tissue during Season with Regard to Nitrogen Fertilization and Plant Population. *Listy Cukrov. A Řepařské* **2020**, *136*, 198–204.
132. Last, P.J.; Tinker, P.B.H. Nitrate nitrogen in leaves and petioles of sugar beet in relation to yield of sugar and juice purity. *J. Agric. Sci.* **1968**, *71*, 383–392. [CrossRef]
133. Steinke, K.; Bauer, C. Enhanced Efficiency Fertilizer Effects in Michigan Sugarbeet Production. *J. Sugar Beet Res.* **2017**, *54*, 2–19. [CrossRef]
134. Varga, I.; Lončarić, Z.; Kristek, S.; Kulundžić, A.M.; Rebekić, A.; Antunović, M. Sugar Beet Root Yield and Quality with Leaf Seasonal Dynamics in Relation to Planting Densities and Nitrogen Fertilization. *Agriculture* **2021**, *11*, 407. [CrossRef]
135. Idris, M.; Baha, E.; Wael, A.M.; Abubaker Haroun, M.A. Effect of Nitrogen Fertilizer and Plant Spacing on Vegetative Growth of Sugar Beet (*Beta vulgaris*). *J. Agron. Res.* **2021**, *4*, 6–13. [CrossRef]
136. Barłóg, P.; Grzebisz, W.; Szczepaniak, W.; Peplinski, K. Sugar beet response to balanced nitrogen fertilization with phosphorus and potassium. Part II. Dynamics of beet quality. *Bulg. J. Agric. Sci.* **2014**, *20*, 1326–1333.
137. Lentz, R.D.; Lehrsch, G.A. Nitrogen availability and uptake by sugarbeet in years following a manure application. *Int. J. Agron.* **2012**, *2012*, 120429. [CrossRef]
138. Lauer, J.G. Plant density and nitrogen rate effects on sugar beet yield and quality early in harvest. *Agron. J.* **1995**, *87*, 586–591. [CrossRef]

Nitrogen MDPI

Review

Identifying Sustainable Nitrogen Management Practices for Tea Plantations

Rhys Rebello, Paul J. Burgess and Nicholas T. Girkin *

School of Water, Energy and Environment, Cranfield University, Bedfordshire MK43 0AL, UK; rhysnrebello@gmail.com (R.R.); p.burgess@cranfield.ac.uk (P.J.B.)
* Correspondence: n.t.girkin@cranfield.ac.uk

Abstract: Tea (*Camellia sinensis* L.) is the most widely consumed beverage in the world. It is mostly grown in the tropics with a heavy dependence on mineral nitrogen (N) fertilisers to maintain high yields while minimising the areas under cultivation. However, N is often applied in excess of crop requirements, resulting in substantial adverse environmental impacts. We conducted a systematic literature review, synthesising the findings from 48 studies to assess the impacts of excessive N application on soil health, and identify sustainable, alternative forms of N management. High N applications lead to soil acidification, N leaching to surface and groundwater, and the emission of greenhouse gases including nitrous oxide (N_2O). We identified a range of alternative N management practices, the use of organic fertilisers, a mixture of organic and inorganic fertilisers, controlled release fertilisers, nitrification inhibitors and soil amendments including biochar. While many practices result in reduced N loading or mitigate some adverse impacts, major trade-offs include lower yields, and in some instances increased N_2O emissions. Practices are also frequently trialled in isolation, meaning there may be a missed opportunity from assessing synergistic effects. Moreover, adoption rates of alternatives are low due to a lack of knowledge amongst farmers, and/or financial barriers. The use of site-specific management practices which incorporate local factors (for example climate, tea variety, irrigation requirements, site slope, and fertiliser type) are therefore recommended to improve sustainable N management practices in the long term.

Keywords: tea; fertiliser; soil amendments; biochar; nitrification inhibitor; application rate; eutrophication; yield; quality; nitrous oxide

1. Introduction

Tea is the most widely consumed beverage in the world [1] and is made from the dried new leaves of the *Camellia sinensis* (L.) plant, an evergreen shrub [2,3]. Cultivation is generally confined to the tropics, due to the specific climate and soil that is required. In 2019, the global tea production was approximately 6.49 Mt, with 42% being exported with a value of about 15 billion USD [4,5]. In 2019, the largest tea producers globally were China (2.77 Mt), followed by India (1.39 Mt), Kenya (0.46 Mt), and Sri Lanka (0.30 Mt) [4]. There has been significant growth in tea production over the past two decades attributable to the increasing acceptance of tea as a drink of choice, the increasing area of production, better cultivars, advanced technology, and the widespread use of nitrogen (N) fertilisers to increase yield [6]. The size of the tea industry has been predicted to grow at an annual growth rate of about 4% to 5.5% from 2017 to 2024 [7]. The entire tea sector is projected to reach a retail value of about 73 billion USD by 2024 [7]. In order to maintain high levels of production to meet demand there will be an increase in dependence on high N fertiliser inputs. This is likely to have significant adverse environmental consequences resulting in an urgent need to identify and implement alternative practices.

The tea plant grows as an evergreen bush and can attain a maximum height of 15 m in the wild [8]. However, for the ease of cultivation the shrub is kept at a maximum height of

0.6–1.0 m to allow for easier harvesting [9]. Tea is consumed mainly as 'black', 'green' or 'oolong' tea, with the type of tea mainly dependent on the country of origin. For example, Sri Lanka, India, Kenya, and Malawi produce mainly black tea while green tea is mainly produced in Japan and China [1]. Harvesting tea leads to the removal of macronutrients (N, P, and K) thereby making it necessary to replenish these nutrients to the crop [10]. The new shoots that are harvested to produce tea have a high proportion of nutrients, with N being the most abundant [11].

Application rates of N fertilisers in tea are high, with upper estimates reported of up to 544 kg N ha^{-1} $year^{-1}$ [12]. However, depending on the type of applied N, only 25–50% is taken up by the plants, meaning that the majority of the inorganic NPK added to the soil ends up lost through leaching, erosion, volatilisation, or is immobilised in soil organic matter [13]. The excessive dependence on and use of fertilisers negatively impacts the water quality in the regions where tea is grown. High NO_3^- concentrations are found in ponds near tea plantations with excessive leaching leading to the contamination of the underground water [14,15]. In addition to water contamination, excessive use of ammonium fertilisers also has the effect of lowering the soil pH [16]. Regular increases in soil acidity result in a reduction in the uptake efficiency of the new N applied, therefore requiring ever greater N additions, thereby creating a compounding effect. In some locations, high applications of N only increase the yield of the tea when there is adequate irrigation [17]. This becomes an issue in drought-prone tea-growing areas. High application rates of N fertiliser can increase the partitioning of assimilates to the canopy rather than the roots which can reduce drought resistance [18]. The application of synthetic N fertilisers also results in the emissions of N_2O. For example, Akiyama, Yan, and Yagi (2006) found that among the crops grown in agricultural fields in Japan, tea plantations had the highest levels of N_2O emissions [19]. Such findings have been replicated in other major tea growing countries [20]. This is problematic as N_2O is an extremely potent and long-lived greenhouse gas with a global warming potential approximately 300 times more than carbon dioxide (CO_2) for a 100-year time horizon [21].

In this study, we assess the impacts of synthetic N fertilisers in tea plantations and identify alternative sustainable N management strategies that can help to reduce the impact. In this paper, we conduct a systematic literature review to (i) assess the N cycle in context to the tea plantations, assess types of fertiliser inputs and common application rates for each, investigating differences between regions and systems, (ii) quantify N use efficiency rates and the consequences of excess application, including evidence of eutrophication, soil acidification, and greenhouse gas emissions, and (iii) identify and assess sustainable and alternative N management strategies, focusing on organic systems and alternative solutions (e.g., soil amendments to mitigate N_2O) which may already be applied in other plantation systems. On this basis, we identify areas of future research to better support the development and implementation of sustainable N management in tea plantations.

2. Methods

We used a systematic literature review to assess the impacts of synthetic N fertilisers in tea plantations and identify alternative practices. In order to identify and select the relevant papers, we applied the preferred reporting items for systematic reviews and meta-analyses (PRISMA) approach [22]. Applying a set of key search terms, the literature database on Scopus was searched in May–June 2021 to identify papers that were relevant to the study area. The search strings used were a combination of 'tea', 'nitrogen', 'fertiliser', 'fertilizer', 'application rate', 'eutrophication', 'greenhouse gas emission', 'yield', 'nitrogen management', 'sustainable' and 'plantations', selected as the most appropriate terms for identifying papers as in scope. These search strings were added to the Scopus search tool searching within the article title, abstract, and keywords database categories of original research papers published in peer-reviewed academic journals. A total of 590 papers were found through the initial search (Figure 1).

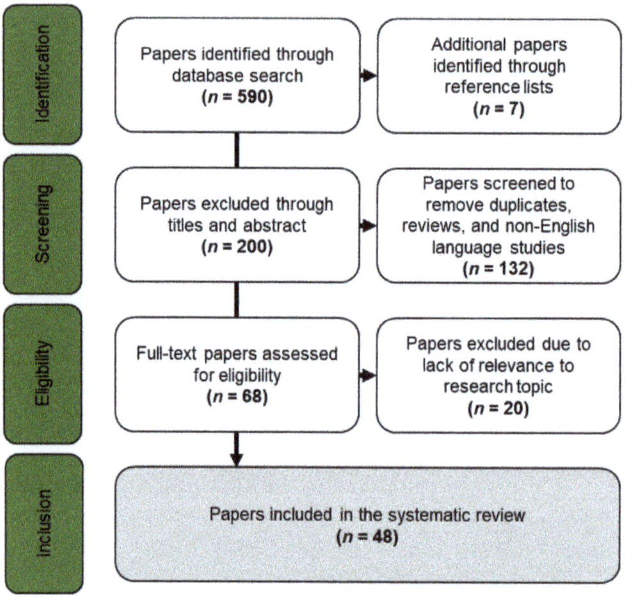

Figure 1. Selection process for the systematic literature review.

Of these, 394 papers were subsequently excluded based on the article title demonstrating that they were outside the scope of the study. A further 13 papers were excluded because they were reviews (book chapters and review articles), non-peer-reviewed conference papers and editorials, or commentary or perspective articles. Papers were subsequently excluded if they were not available either as open access publications, through institutional access, or online repositories (e.g., ResearchGate.net), reducing the number of studies to 144. Only English language studies were included, resulting in 132 papers. These were then shortlisted by reading the abstract and checking if the papers were within scope, reducing the total number of studies to 41. Seven additional relevant papers were found through the reference list of the 41 papers and were also included in the study, resulting in a total of 48 studies. A table of the final 48 papers is listed in the Supplementary Materials (Table S1). The selection criteria used may have excluded some relevant publications from other peer-reviewed literature which are published in non-English language journals and other "grey" literature. It was also noted that there was a large proportion of studies from China (n = 31) with comparatively few studies being from other tea growing regions including India (n = 2), Sri Lanka (n = 1), Kenya (n = 4), Japan (n = 4), Ethiopia (n = 1), Tanzania (n = 1), East Africa more widely (n = 1), and global trends (n = 3). Overall, there was also a lack of socioeconomic studies which looked into the impact of switching N fertilisers and utilising alternative methods and how this would impact the tea gardens and farmers. Such information is important to develop a better understanding of the impacts on the social as well as the economic aspects of tea growing, leading to the creation of more sustainable tea.

Data from each article was recorded including information on geographical distribution, tea variety, and its focus on either the need for nitrogen, the impacts of excessive nitrogen applications, or current or alternative nitrogen management. All units were converted into standard S.I. units with the exception of Xiao et al. (2018) for which insufficient data were available [23]. Overall, our approach allowed us to synthesise and identify the need for N, current practices adopted by tea plantations, their impacts, and consequences for the environment, as well as any practices that can help in mitigation highlighted in the literature. Effects and causes that were known to impact tea plantations but not found in the literature review were also noted and a separate search was conducted to find literature and practices that are used in other plantation types. This data was then used to identify sustainable and alternative N management strategies for tea plantations and identify gaps to inform future research recommendations [24].

This study did not investigate the availability of such approaches to various farming communities (particularly smallholders), instead focusing on synthesizing our understanding of the relevant biogeochemical pathways, and evidence for the benefits (or trade-offs of each technique). Making techniques rapidly available for smallholder farmers (for example through the outreach programs of non-governmental organisations) and making appropriate financial capital available to facilitate this is key to wider adoption, and important in terms of more widely improving the sustainability and adoption of principles of equity in the tea sector [24].

3. Results and Discussion

3.1. The Need for Nitrogen

3.1.1. Nitrogen and Crop Yields

N plays a vital role in plant growth and is one of the most deficient mineral elements in the soil [25]. Like all crops, tea bushes can suffer from N deficiency. The effects of N deficiency in tea include the leaves having a lighter green colour and shorter internodes [8]. In general, a tea plant is deficient when the mature leaf N concentration is below 3%, is mildly deficient between 3% and 3.5%, with an optimum concentration above 3.5% [26]. Harvesting, leaching, gaseous emissions, and immobilisation in the soil microbial communities all reduce available N for plant growth (Figure 2). As a consequence, maintaining yields requires soil N additions through fertiliser applications.

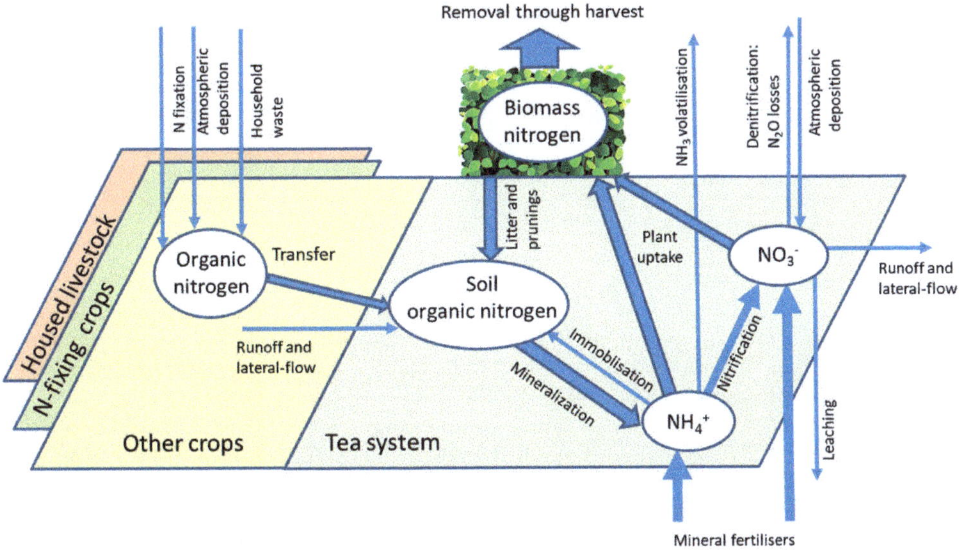

Figure 2. Nitrogen flows in tea plantations, focusing on organic management. Inorganic fertiliser additions could also be used to increase the size of the NO_3^- and NH_4^+ pools.

Five studies explicitly investigated the relationship between N application and yield (Table 1). Venkatesan et al. (2004) in South India found that the increase in the rate of N fertiliser application from no application to 150 kg N ha^{-1} year^{-1} resulted in an increase in the yield of 4.96 kg ha^{-1} year^{-1} per kg of applied N, and an increase of 2.38 kg ha^{-1} year^{-1} and 2.36 kg ha^{-1} year^{-1} per kg of applied N when the application rate was increased to

300 kg N ha^{-1} $year^{-1}$ and 450 kg N ha^{-1} $year^{-1}$, respectively [27]. Similarly, Stephens and Carr (1991) in Southern Tanzania noted that yield increases with higher NPK fertiliser application rates, but with a critical 'turn-over point' (375 kg N ha^{-1} $year^{-1}$ with well-irrigated tea and 300 kg N ha^{-1} $year^{-1}$ with unirrigated tea); there were reduced yield increases despite higher application rates [28]. Ma et al. (2021) in the Zhejiang province of China also reported a significant increase in the yield of harvested shoots with application of N fertilisers [29], with yields reaching a plateau at an application rate of 367 kg N ha^{-1} $year^{-1}$. In a short-term pot experiment, Xiao et al. (2018) reported a yield response of tea up to a level of 3.88 g N (kg $soil)^{-1}$ [23]. The potential tea yield in a given location is primarily dependent on the climate conditions [28].

Table 1. Comparative studies of application rates of fertilisers and impacts on yields in tea plantations. * Data from Stephens and Carr (1991) is reported for optimized irrigation management.

Location	Type of Fertiliser	Application Rate (kg N ha^{-1})	Yield (t ha^{-1})	Reference
Hangzhou, China	No fertiliser	0	1.25	Liu et al. (2012)
Hangzhou, China	Urea and $CaH_6O_9P_2$	548	2.97	Liu et al. (2012)
Hangzhou, China	Organic	548	2.60	Liu et al. (2012)
Hangzhou, China	Slow release	548	2.75	Liu et al. (2012)
Hangzhou, China	Urea	0	2.50	Ma et al. (2021)
Hangzhou, China	Urea	119	3.00	Ma et al. (2021)
Hangzhou, China	Urea	285	3.50	Ma et al. (2021)
Hangzhou, China	Urea	569	3.10	Ma et al. (2021)
South India	Urea	0	2.12	Venkatesan et al. (2004)
South India	Urea	150	2.37	Venkatesan et al. (2004)
South India	Urea	300	2.79	Venkatesan et al. (2004)
South India	Urea	450	3.11	Venkatesan et al. (2004)
Tanzania	NPK	0	1.87	Stephens and Carr (1991) *
Tanzania	NPK	110	2.80	Stephens and Carr (1991) *
Tanzania	NPK	170	2.91	Stephens and Carr (1991) *
Tanzania	NPK	230	3.40	Stephens and Carr (1991) *
Tanzania	NPK	280	3.70	Stephens and Carr (1991) *
Tanzania	NPK	350	4.50	Stephens and Carr (1991) *
Fujian, China	NPK	0	4.20	Ji et al. (2018)
Fujian, China	NPK	300	6.46	Ji et al. (2018)
Fujian, China	NPK (75%) + organic (25%)	300	6.65	Ji et al. (2018)
Fujian, China	NPK (50%) + organic (50%)	300	6.18	Ji et al. (2018)
Fujian, China	NPK (25%) + organic (75%)	300	6.23	Ji et al. (2018)
Fujian, China	Organic	300	5.36	Ji et al. (2018)

3.1.2. Crop Quality

A lack of N in tea plants affects the quality of both green and black tea [26,30–33]. Owuor (1997) demonstrated that the quality of green tea increases progressively with an increase in the N supply [26]. Tokuda and Hayatsu (2004) report that Japanese tea growers add as much as 1200 kg N ha^{-1} $year^{-1}$ to increase quality [32]. In contrast to green tea, the quality of black tea decreases with excessive N application [33]. The optimum N rate in black tea has been reported to be between 100–200 kg N ha^{-1} $year^{-1}$ [33,34]. Increases in the supply of N fertiliser were reported to increase amino acid concentrations and decrease polyphenol concentrations in leaves. These can be useful indicators for estimating tea quality [35] with amino acid concentration being a strong indicator for green tea [36]. Polyphenols are responsible for the bitter and acidic taste present in tea [37], so a higher concentration of polyphenols is less desirable for green tea.

3.1.3. Application Rates and Types of Fertilisers

Nitrogen and other nutrients can be applied in different inorganic forms. For example, urea and calcium superphosphate are reported in Liu et al. (2012) [30], urea in Ma et al.

(2021) [28] and Venkatesan et al. (2004) [27], and NPK fertilisers in Stephens and Carr (1991) [27] and Ji et al. (2018) [31]. Overall, the evidence suggests that organic fertilisers performed worse than traditional inorganic fertilisers in terms of benefits to yield (Table 1) but better than the application of no fertiliser. In some examples, a combination of organic and inorganic fertiliser for a given supply of N can provide a marginally higher yield than inorganic fertiliser alone [37]. However, our ability to make wider conclusions based on these results is limited given the sparsity of studies and limited information regarding soil type and other key soil parameters. Taken together, these results do, however, demonstrate that very high N application rates do not necessarily result in greater yields and highlight the need to manage N application rates depending on plant and soil requirements.

Ruan et al. (2007) reported that applications of ammonium (NH_4^+) rather than nitrate (NO_3^-) fertilisers improved nitrogen uptake due to the higher capacity of roots to take up NH_4^+ compared with NO_3^- [38]. This is one reason why tea plantation managers often select NH_4^+ or urea-based N fertilisers [38,39].

3.2. *Environmental Impacts of Excessive Nitrogen Applications*

3.2.1. Soil Health

Plant growth depends on a regular supply and balance of nutrients from the soil, which must also act as a suitable medium for root growth, maintaining moisture, and supporting the microbial community [40]. Soil acidification is a major problem and can be exacerbated by high N fertiliser use. Acidification results in the leaching of cationic nutrients including potassium (K), sodium (Na), calcium (Ca), and magnesium (Mg) leading to a reduction in plant productivity and potential increases in GHG fluxes [40–43].

In tea plantations, the heavy dependence on N fertilisers has led to significant pH reductions from 3.32 to 3.15 and 3.67 to 3.35 at depths of 0–0.4 m and 0.4–1.0 m, respectively [44]. Yan et al. (2018) also reported lower soil pHs in older tea gardens compared with younger tea gardens at depths of 0–2.0 m [43]. This was attributed to the continuous application of fertilisers throughout the life of the gardens. Cheng et al. (2015) found the soil pH of tea plantations declined from 5.3 to 4.9 after continuous application of fertilisers for about six years [44]. Willson (1975) noted that the increase in soil acidity leads to the accumulation of manganese (Mn) in the leaves which can be harmful at higher concentrations [45]. Wang et al. (2018) also noted that there was a decrease in the soil pH as the cultivation time increased but also noted that NO_3^- fertilisers had a more gradual decline in pH compared with NH_4^+ fertilisers [39]. Similarly, Wang et al. (2020) found that the effect of N addition on the soil pH was dependent on the type of N added, with ammonium bicarbonate and urea acidifying the soil while ammonium sulphate fertilisers had no effect on the soil pH [46].

Soil acidification impacts the soil microbial community, with consequences for fundamental ecosystem processes. Yang et al. (2019) noted that the long-term addition of N fertilisers led to the reduction in microbial biomass, inhibited respiration, and reduced the microbial diversity [47]. Similarly, Ma et al. (2021) found that there was microbial community destabilisation and a reduction in the complexity of bacterial networks [29]. Many microbes have evolved to produce enzymes that release N from recalcitrant organic matter. High levels of N or substrate N concentrations can inhibit the production of enzymes by microbes which in turn reduces C availability to microbes, and thereby reduces microbial biomass and activity.

3.2.2. Water Pollution

Excessive N fertilisers in tea plantations can result in N leaching into surface waters and groundwaters and can be aggravated through the improper application of fertilisers and irrigation [48]. N can be washed off from the tea plantations and seep into the groundwater or through surface waters (Figure 2). This leads to a risk of eutrophication and toxicity in water bodies and the water table [30].

Liu et al. (2012) investigated the application of different fertilisers in a tea plantation and studied the amount of N loss in the surface runoff [30]. They compared conventional fertilisers (urea), organic fertiliser (made from pig manure, chicken manure, oil tea seed meal, and humic acid), and slow-release fertiliser (polymer-coated fertiliser) and noticed that there was a higher nitrification rate in the chemical fertiliser treatment than in the slow release and organic fertilisers. The authors also noticed that there was an increase in nitrifying bacteria when ammoniacal fertilisers were added. Xu et al. (2017) also reported that high use of chemical fertilisers resulted in an annual N leaching loss of about 50 kg N ha^{-1} $year^{-1}$, equivalent to about 12% of the total N that was applied to the plots [49]. Maghanga et al. (2013) reported that application of NPK fertilisers can also increase surface water NO_3^- levels [50]. However, this increase was small and not to toxic levels, most likely due to the presence of well-draining soil. This suggests that site-specific soil properties [50], climate and fertiliser type [30], and the slope [51] play an important role in determining N loss through the surface water and groundwater leaching.

3.2.3. Gaseous Emissions

The extent of tea plantation greenhouse gas (GHG) emissions is dependent on the type of fertiliser applied and the application rate [52]. Globally, about 2–3% of the nitrogen applied as fertiliser is lost through N_2O emissions, with high rates requiring anaerobic conditions (water-saturated conditions) as well as a carbon source [53,54]. In China, where urea was applied to tea at a rate of 450 kg N ha^{-1}, emissions of N_2O (14–21 kg N ha^{-1} $year^{-1}$) and NO (18–19 kg N ha^{-1} $year^{-1}$) were greater than those for equivalent croplands [55]. Wang et al. (2020) reported mean N_2O emissions of 17.1 kg N ha^{-1} $year^{-1}$, equivalent to 8008 kg CO_2-eq ha^{-1}, approximately double the GHG footprint of rice [20]. In Japan, Tokuda and Hayatsu (2004) reported an annual N_2O emission of 25 kg N ha^{-1} after the application of 1200 kg N ha^{-1} $year^{-1}$, but only 0.6 kg N ha^{-1} with the application of 600 kg N ha^{-1} $year^{-1}$ [32]. Moreover, in Japan, Hirono and Nonaka (2012) reported mean annual N_2O emissions of 10.6–14.8 kg N ha^{-1} $year^{-1}$ from the application of 510 kg N ha^{-1} $year^{-1}$, with 65% of the emissions derived from an area of fertilizer application between the tea bushes and 35% from soils under the tea canopy [56]. The highest diurnal fluxes were associated with periods of high rainfall. The N_2O emissions from under the tree canopy were related to the falling leaves which were estimated to contribute 115–258 kg N ha^{-1} $year^{-1}$ [56]. Yao et al. (2015) also observed that rainfall events were a dominant driver of N_2O emissions [55].

Alongside N_2O, ammonia (NH_3) is also released from fertilisers (particularly urea) although the extent is dependent on soil conditions. NH_3 can have significant negative effects on human health as well as the natural environment [57]. In the atmosphere, NH_3 has the capacity to bind with other gases such as sulfur dioxide (SO_2) and nitrogen dioxide (NO_2) to form fine particulate matter less than 2.5 μm [58]. This can cause health issues when inhaled, including impacts on cardiovascular and respiratory health by contributing to chronic conditions such as heart attacks, cerebrovascular disease, chronic obstructive pulmonary disease (COPD), asthma, and lung cancer [59–62].

3.3. Alternative Nitrogen Management Strategies

3.3.1. Integrated Nutrient Management

As a result of the wide-ranging adverse environmental impacts, there is a growing need to identify and apply alternative N management practices, alongside altering the application rates of conventional mineral fertilisers. These range from using organic or control release fertilisers, using a mixture of organic and inorganic fertilisers, intercropping with legumes, adding nitrification inhibitors, biofertilisers, soil amendments, applying such approaches in combination, and changing farm-scale management practices (Figure 3). Alongside studies of individual techniques, no studies were identified that have specifically investigated potential interactive effects between different management practices (for example combined soil amendments (biochar) with organic fertilisers), despite evidence from other agroecosystems on the potential benefits of such approaches [63].

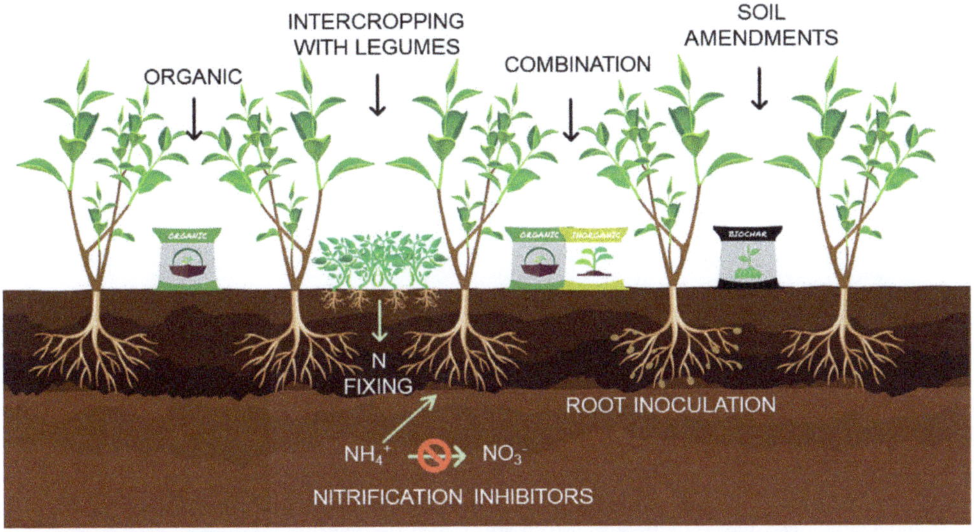

Figure 3. Integrating alternative nitrogen management practices for tea plantations.

There are a number of trade-offs associated with the use of these alternatives which range from lower yields, enhanced N_2O emissions, and the need for training or education to improve the adoption rates of the alternatives by farmers. This is highlighted by Hamid et al. (2021) who noted that when farmers are aware of potential benefits from reduced N fertiliser use, they are more likely to alter practices [64].

A limited number of studies assessed using a variety of integrated nutrient management practices and alternative application approaches to help address issues arising from the excessive use of N fertilisers. For example, Liu et al. (2012) [30] suggested the application of conventional fertilisers but at times when there was no rain anticipated to minimise surface runoff N losses. This was also suggested by Wang et al. (2018) but this approach depends on having accurate weather data and since tea plantations are located in regions that receive moderate to heavy rainfall, the reliance on there being a period of no rainfall for fertiliser application would be quite difficult [51]. Since many tea plantations are grown on hills with a slope, Wang et al. (2018) also suggested the use of contour planting to help in reducing the N losses due to surface runoff. However, such an approach is largely limited to the development of new plantations [51]. Liu et al. (2014) examined the effect of fertigation—applying nutrients in irrigation water. They showed that applying 83 kg N ha^{-1} through drip irrigation enabled similar tea yields as plots conventionally receiving 138 kg N ha^{-1}. Higher yields with fertigation were also matched by lower levels of nitrogen lost through runoff and leaching. Where irrigation is used, fertigation is one approach to maintain yield whilst minimising fertiliser inputs and negative environmental effects [65].

In China, Tang et al. (2021) investigated using the Nutrient Expert system, originally developed for optimizing fertiliser regimes in maize, rice, and wheat (i.e., non-perennial crops) [66]. The system relies on understanding the relationship between yield response (the difference between the yield in the treatment receiving NPK and that with no fertiliser) and agronomic efficiency (the yield increase per unit of N, P, and K applied). The use of the Nutrient Expert system contributed to a 14% increase in yields compared with conventional practices, as well as increased tea quality (increased total amino acid concentration).

Tang et al. estimated that the system was able to reduce N_2O emissions by approximately 28% compared with conventional practices [66].

3.3.2. Alternative Fertilisers

Organic fertilisers can be broadly classified into plant-based and animal-based fertilisers. Examples of plant-based fertilisers are compost, cottonseed meal, alfalfa and soyabean meal, and seaweed. Animal-based fertilisers include farmyard manure, blood meal, bone meal, and fish emulsion or fish meal [67]. Wang et al. (2014) found that when commercial organic fertilisers made from pig and chicken manure were used in tea plantations, the increase in soil bacteria and archaea communities was greater than with any of the other fertilisation methods [68]. Gu et al. (2019) observed that soils in tea plantations receiving long-term treatments with organic fertilisers showed an increase in the soil microbial diversity and thereby potential improvements in the ecosystem function when compared with synthetic fertilisers [69]. Xie et al. (2021) found that organic rapeseed cake fertilisers were useful in increasing the soil pH, total organic matter (from 1.6% to 2.0%), and total NH_4^+ and NO_3^- [70]. Xie et al. (2019) reported a reduction in NH_4^+ and NO_3^- concentrations in runoff compared with traditional fertilisers [71]. A decline in overall N losses and leaching was also reported by Liu et al. (2016) who reported a 59% reduction in total N losses and a 76% reduction in leaching into waters compared with traditional fertilisers [72].

However, organic fertilisers are also associated with several trade-offs. When compared with conventional fertilisers, Liu et al. (2012) found that organic fertilisers (made from pig manure, chicken manure, oil tea seed meal, and humid acid) resulted in a 21.5% and 12.2% lower yield compared with chemical fertiliser over two years [30]. Yao et al. (2015) and Han et al. (2021) also note that there was an increase in the N_2O emissions by 71% when organic fertilisers were used instead of chemical fertilisers, but it was suggested that this depended on the fertiliser type, climate, and geographical location of the tea plantation [55,73]. The dependence on fertiliser type was noted by He et al. (2019) who found that soya bean cake fertiliser significantly increased the N_2O emissions (5.4 kg N ha^{-1} $year^{-1}$) by 11% compared with chemical fertiliser (4.8 kg N ha^{-1} $year^{-1}$) [1], while livestock manures decreased the N_2O emission rates by 41–50% to 2.0–2.6 kg N ha^{-1} $year^{-1}$ [74].

To moderate the issues that arise with using only organic fertilisers, some studies have investigated the use of organic and inorganic fertilisers in combination. In China, Xie et al. (2019) found that combining 50% of organic fertilisers with 50% of conventional fertilisers and a 50% reduction in pesticide application resulted in yields of 5.4 t ha^{-1}, compared with 4.9 t ha^{-1} in a control treatment with only mineral fertilisers and full pesticide applications [71]. Xie et al. also found that organic fertilisers reduced the ratio of polyphenols to amino acids, broadly indicative of a superior aroma and taste in green tea. The authors also reported a 1.95% increase in caffeine levels. Debere et al. (2014) noted that combining fertilisers increased the height, the number of leaves per plant, the number of branches, and the root length of young plants [75]. Ji et al. (2018) found a significant increase in the soil bacterial community and diversity, the accumulation of soil carbon, and nutrients, with an application of 25% of organic fertiliser providing the best yield and quality [37].

Alongside organic fertilisers, biofertilisers (produced from substances that contain living microorganisms that colonise the plant rhizosphere to help in increasing the supply or availability of nutrients) may also provide benefits, although these are less well understood [76]. Xu et al. (2017) found that the application of a *Trichoderma viride* biofertiliser resulted in a 45% reduction in N leaching compared with the application of just conventional fertilisers [49]. This reduction was attributed to the ability of *T. viride* biofertilizers to promote root growth and nutrient uptake, and the slow release of NO_3^-.

3.3.3. Intercropping with Legumes

Leguminous plants can develop root nodules which enable them to fix N_2 in symbiosis with rhizobia bacteria and they may provide benefits through intercropping in tea plantations [77]. If the tea has an incomplete canopy cover, intercrops can help to reduce soil erosion and run-off [77–79]. Duan et al. (2019) investigated the impacts of intercropping young tea plants with soya bean and found that the soil NH_4^+ concentrations were higher in the intercropped tea than in the monoculture tea during the flowering period of the soybean. The variation in NH_4^+ and effective N remained the same, meaning that intercropping maintained soil N levels, findings that were also reported by Koenig and Cochran (1994) [79]. Duan et al. (2019) also reported increased free amino acids and caffeine concentrations with intercropping, suggesting that there was an improvement in the quality of the tea compared with monoculture tea plantations. The creation of tea varieties able to fix N through a symbiosis with rhizobia has also been proposed [80–83]. Rhizobia can fix N_2 [84] and increase N use efficiency [80].

3.3.4. Nitrification Inhibitors and Control Release Fertilisers

Nitrification inhibitors work by delaying the bacterial oxidation of NH_4^+ by depressing the activities of *Nitrosomonas* bacteria, which drive the transformation of NH_4^+ in the soil. The resultant NO_2^- is then reduced to NO_3^-, mediated by *Nitrobacter* and *Nitrosolobus* bacteria [85]. Considering the need to maintain the soil mineral N as NH_4^+ for longer and alleviate soil acidification, Wang et al. (2020) recommended that when fertilising with urea, urease and nitrification inhibitors are applied to limit N losses [46]. Qiao et al. (2021) found that with the application of DMPP (3,4–dimethylpyrazole phosphate, a nitrification inhibitor), soil pH increased at a urea application rate of 500 kg N ha^{-1} and above, due to reduced nitrification and increased plant uptake of NO_3^- [86]. The leaching losses of nitrates and total inorganic N were also noticed to decrease with the addition of DMPP.

A study by Han et al. (2008) found that control release fertilisers (CRFs) coated with nitrification inhibitors in a pot experiment resulted in an N use efficiency (46%) that was 17% higher than that (29%) with uncoated fertilisers [87]. CRFs which are sometimes referred to as slow-release fertilisers (SRFs) have been developed as an approach towards solving the issues that arise from the application of conventional fertilisers [85,88–90]. CRFs are designed with the aim to release the active ingredient at a controlled rate and to support by providing adequate nutrients for crop production [91,92]. Comparing three treatments each receiving 105 N ha^{-1} $year^{-1}$ from oilcake, Wu et al. (2018) found that the plots receiving 120 kg N CRF ha^{-1} $year^{-1}$ had 31% higher yields (4.1 t ha^{-1} $year^{-1}$) than those (3.1 t ha^{-1} $year^{-1}$) in plots receiving 345 kg N-urea ha^{-1} $year^{-1}$ [93]. However, the plots receiving 345 kg N CRF ha^{-1} $year^{-1}$ had the lowest yield (2.9 t ha^{-1} $year^{-1}$), which was associated with an inhibition of photosynthesis. In Sri Lanka, Raguraj et al. (2020) noted a 10–17% yield increase from the application of urea-coated "nanohybrid" CRF fertilisers compared with standard fertiliser treatments [92]. In China, Liu et al. (2012) [30] reported a 6–13% decrease in yields from CRFs-treated crops when compared with conventional fertilisers in a two-year study, but the difference was not statistically significant. Wu et al. (2018) reported that the N use efficiency was greater using CRFs compared with the conventional urea application when expressed as a function on tea yield [93]. However, in many instances, the benefits and trade-offs from CRFs and other fertiliser alternatives have only been investigated in the short term.

3.3.5. Soil Amendments

A range of soil amendments have been trialled in tea plantations with the view to improving soil health, including enhancing N availability to crops and mitigating GHG emissions. Pramanik, Safique, and Jahan (2018) found applying humic substances (HS) increased tea productivity and enhanced N availability, increasing the populations of nitrifying, ammonifying, and denitrifying bacteria [94]. An increase in yield was also reported. An HS application of 1.2 kg ha^{-1} delivered a yield of 1.8 t ha^{-1} for a period after application approximately 18% higher than non-HS treated fields. In China, Yang et al. (2021) reported that biochar increased annual tea yields from 2.6 t ha^{-1} to 2.8 t ha^{-1} for unfertilized tea and from 3.0 t ha^{-1} to 3.6 t ha^{-1} for tea receiving NPK fertiliser [95]. Yang et al. reported that the biochar-based fertiliser treatments had a higher diversity of fungal and soil bacteria, suggesting potential impacts on soil nutrient transformations. Wu et al. (2021) also reported an increase in mean annual tea yields by 17–26% from 3.1 t ha^{-1} to 3.6–3.9 t ha^{-1} when an addition of a wheat straw-derived biochar was introduced to the tea plantation [96].

Various studies also reported that additions of biochar decrease soil N_2O emissions in tea. In a field study, Han et al. (2021) used a biochar derived from herbaceous material in combination with different fertilizer types and reported a significant reduction in the soil N_2O emissions of 13% in the first year, irrespective of the type of fertiliser that was used [73]. Similarly, in a laboratory incubation, Oo et al. (2018) found that with the surface application of converted pruning residue biochar the soil N_2O emissions reduced by 36% compared with the control which had no addition of any biochar [97]. The field trial by Wu et al. (2021) found the N_2O emissions were dependent on the biochar application rate, with a low biochar amendment of 10 Mg ha^{-1} having mean annual N_2O emissions similar to conventional treatments [96]. A reduction in N leaching was also reported by Chen et al. (2021) in a laboratory leaching study. The authors reported that the addition of biochar significantly reduced the N leaching in the soils and enhanced soil N retention [98]. The authors also found that there was an inverse relationship between the application of biochar and N leaching, with an optimum application of 6% biochar (per unit mass of soil) found to have the highest mitigation of N leaching and retention of N, and also concurrently enhancing soil microbial biomass and enzyme activity. Lastly, in a field study, Yamamoto et al. (2014) assessed lime-N fertilisers as opposed to traditional inorganic fertilisers [99]. They found that the lime-N application significantly decreased soil N_2O emissions compared with conventional fertilisers.

4. Conclusions

Tea production requires the addition of N to replenish the N lost during the harvesting of tea leaves. Additional N can increase yields and the quality of green leaf tea, whereas the quality of black tea is reported to decline at high N applications. Higher yields are reported from the application of NH_4^+ rather than NO_3^- fertilisers. However, excessive and imprecise use of N fertilisers to increase tea yields has resulted in negative impacts on the environment such as soil acidification and the leaching of N into surface and groundwater, leading to eutrophication. The application of inorganic fertilisers also results in substantial emissions of N_2O, making tea plantations a net GHG source. A wide range of alternative N management techniques have been identified in the literature, and include the use of organic CRFs and biofertilisers, intercropping with legumes, root inoculation, use of nitrification inhibitors, and adding soil amendments. Collectively, these techniques may have an important role in reducing inorganic N inputs in tea plantations, but in general more evidence is needed before widespread adoption. Moreover, alternatives often have important trade-offs including lower yields and enhanced N_2O emissions.

Taken together, the available evidence suggests that the implementation of a site-specific management system which accounts for climate, soil nutrient levels, and field-specific fertiliser application data may help to combat the issues arising with the excess application of N fertilisers by monitoring the impacts and reducing them without negatively affecting the yield. The impacts of N fertilisers are site specific, with consequences that depend on the climate and the management practices used, therefore, requiring a tailored approach for each tea garden.

Supplementary Materials: The following supporting information can be downloaded at: https://www.mdpi.com/article/10.3390/nitrogen3010003/s1, Table S1: Studies identified on nitrogen management in tea plantations.

Author Contributions: Conceptualization, N.T.G. and P.J.B.; methodology, R.R.; formal analysis, R.R. and R.R.; writing—original draft preparation, N.T.G. and P.J.B.; writing—review and editing, supervision, N.T.G. and P.J.B. All authors have read and agreed to the published version of the manuscript.

Funding: Prof. Paul Burgess acknowledges funding from UKRI for the Innovate UK project: "Data-led Soil Management for Low Income Farmers" (Grant number 47864) and Dr. Nicholas Girkin acknowledges funding from HEIF for the project "Sustainable Management for Climate Resilient Plantation Agriculture".

Institutional Review Board Statement: Not applicable.

Informed Consent Statement: Not applicable.

Data Availability Statement: Not applicable.

Acknowledgments: The authors thank Ruben Sakrabani and Daniel Evans for feedback on earlier drafts of the manuscript, as well as the comments and feedback of the three anonymous reviewers.

Conflicts of Interest: The authors declare no conflict of interest.

References

1. Kaison, C. *World Tea Production and Trade: Current and Future Development*; Food and Agriculture Organization: Rome, Italy, 2015.
2. Hilal, Y.; Engelhardt, U. Characterisation of white tea—Comparison to green and black tea. *J. Verbrauch. Leb.* **2007**, *2*, 414–421. [CrossRef]
3. Khan, N.; Mukhtar, H. Tea and health: Studies in humans. *Curr. Pharm. Des.* **2013**, *19*, 6141–6147. [CrossRef] [PubMed]
4. FAO. FAOSTAT. 2021. Available online: http://www.fao.org/faostat/en/#data/QC (accessed on 27 June 2021).
5. United Nations. UN Comtrade Database. 2021. Available online: https://comtrade.un.org/data (accessed on 24 June 2021).
6. Majumder, A.B.; Bera, B.; Rajan, A. Tea statistics: Global scenario. *Int. J. Tea Sci.* **2010**, *8*, 121–124.
7. Vivek, V.; Steffany, B.; Cristina, L. Global Market Report: Tea. The International Institute for Sustainable Development, Sustainable Commodities Marketplace Series. 2019. Available online: https://www.iisd.org/system/files/publications/ssi-global-market-report-tea.pdf (accessed on 20 June 2021).
8. Hajiboland, R. Environmental and nutritional requirements for tea cultivation. *Folia Hortic.* **2017**, *29*, 199–220. [CrossRef]
9. de Costa, W.A.J.M.; Mohotti, A.J.; Wijeratne, M.A. Ecophysiology of tea. *Braz. J. Plant Physiol.* **2007**, *19*, 299–332. [CrossRef]
10. Tabu, I.M.; Kekana, V.M.; Kamau, D.M. Effect of Varying Ratios and Rates of Enriched Cattle Manure on Leaf Nitrogen Content, Yield and Quality of Tea (*Camellia sinensis*). *J. Agric. Sci.* **2015**, *7*, 175–181. [CrossRef]
11. Sultana, J.; Siddique, J.M.N.; Kamaruzzaman, M.; Halim, M.A. Conventional to ecological: Tea plantation soil management in the Panchagarh District of Bangladesh. *J. Sci. Technol. Environ. Inform.* **2014**, *1*, 27–37.
12. Hirono, Y.; Nonaka, K. Effects of application of lime nitrogen and dicyandiamide on nitrous oxide emissions from green tea fields. *Soil Sci. Plant Nutr.* **2014**, *60*, 276–285. [CrossRef]
13. Roberts, T.L. Improving nutrient use efficiency. *Turk. J. Agric. For.* **2008**, *32*, 177–182.
14. Hirono, Y.; Watanabe, I.; Nonaka, K. Trends in water quality around an intensive tea-growing area in Shizuoka, Japan. *Soil Sci. Plant Nutr.* **2009**, *55*, 783–792. [CrossRef]
15. Tanaka, T.; Ii, H.; Hirata, T.; Nishikawa, M.; Umehara, K.; Ogawa, Y. Groundwater metal and nitrogen contaminations caused by nitrogen fertilizer in tea plantation catchment, the center of Shizuoka, Japan. *J. Hydrosci. Hydraul. Eng.* **2002**, *20*, 37–47.
16. Thenmozhi, K.; Manian, S.; Paulsamy, S. Influence of long term nitrogen and potassium fertilization on the biochemistry of tea soil. *J. Res. Agric.* **2012**, *1*, 124–135.
17. Carr, M.K.V. The Climatic Requirements of the Tea Plant: A Review. *Exp. Agric.* **1972**, *8*, 1–14. [CrossRef]
18. Cheruiyot, E.K.; Mumera, L.M.; Ng'etich, W.K.; Hassanali, A.; Wachira, F.N. High fertilizer rates increase susceptibility of tea to water stress. *J. Plant Nutr.* **2010**, *33*, 115–129. [CrossRef]
19. Akiyama, H.; Yan, X.; Yagi, K. Estimations of emission factors for fertilizer-induced direct N_2O emissions from agricultural soils in Japan: Summary of available data. *Soil Sci. Plant Nutr.* **2006**, *52*, 774–787. [CrossRef]
20. Wang, Y.; Yao, Z.; Pan, Z.; Wang, R.; Yan, G.; Liu, C.; Su, Y.; Zheng, X.; Butterbach-Bahl, K. Tea-planted soils as global hotspots for N_2O emissions from croplands. *Environ. Res. Lett.* **2020**, *15*, 104018. [CrossRef]
21. IPCC (Intergovernmental Panel on Climate Change). *Climate Change 2007: The Physical Science Basis. Contribution of Working Group I to the Fourth Assessment Report of the IPCC*; Solomon, S., Qin, D., Manning, M., Eds.; Cambridge University Press: Cambridge, UK, 2007; p. 996.
22. Moher, D.; Liberati, A.; Tetzlaff, J.; Altman, D.G.; PRISMA Group. Preferred reporting items for systematic reviews and meta-analyses: The PRISMA Statement. *PLoS Med.* **2009**, *6*, e1000097. [CrossRef] [PubMed]
23. Xiao, F.; Yang, Z.; Huang, H.; Yang, F.; Zhu, L.; Han, D. Nitrogen fertilization in soil affects physiological characteristics and quality of green tea leaves. *HortScience* **2018**, *53*, 715–722. [CrossRef]
24. Onduru, D.D.; de Jager, A.; Hiller, S.; van den Bosch, R. Sustainability of smallholder tea production in developing countries: Learning experiences from farmer field schools in Kenya. *Int. J. Dev. Sustain.* **2012**, *1*, 714–742.
25. Hawkesford, M. Functions of Macronutrients. In *Marschner's Mineral Nutrition of Higher Plants*, 3rd ed.; Academic Press: Cambridge, MA, USA, 2011.
26. Owuor, P.O. Fertilizer use in tea: The case of nitrogen. *Tea* **1997**, *18*, 132–143.

27. Venkatesan, S.; Murugesan, S.; Ganapathy, M.N.K.; Verma, D.P. Long-term impact of nitrogen and potassium fertilizers on yield, soil nutrients and biochemical parameters of tea. *J. Sci. Food Agric.* **2004**, *84*, 1939–1944. [CrossRef]
28. Stephens, W.; Carr, M.K.V. Responses of tea (*Camellia sinensis*) to irrigation and fertilizer. i. yield. *Exp. Agric.* **1991**, *27*, 75–85. [CrossRef]
29. Ma, L.; Yang, X.; Shi, Y.; Yi, X.; Ji, L.; Cheng, Y.; Ni, K.; Ruan, J. Response of tea yield, quality and soil bacterial characteristics to long-term nitrogen fertilization in an eleven-year field experiment. *Appl. Soil Ecol.* **2021**, *166*, 103976. [CrossRef]
30. Liu, Z.; Yang, J.; Yang, Z.; Zou, J. Effects of rainfall and fertilizer types on nitrogen and phosphorus concentrations in surface runoff from subtropical tea fields in Zhejiang, China. *Nutr. Cycl. Agroecosystems* **2012**, *93*, 297–307. [CrossRef]
31. Ji, L.; Wu, Z.; You, Z.; Yi, X.; Ni, K.; Guo, S.; Ruan, J. Effects of organic substitution for synthetic N fertilizer on soil bacterial diversity and community composition: A 10-year field trial in a tea plantation. *Agric. Ecosyst. Environ.* **2018**, *268*, 124–132. [CrossRef]
32. Tokuda, S.-I.; Hayatsu, M. Nitrous oxide flux from a tea field amended with a large amount of nitrogen fertilizer and soil environmental factors controlling the flux. *Soil Sci. Plant Nutr.* **2004**, *50*, 365–374. [CrossRef]
33. Owuor, P.O.; Ng'etich, W.K.; Obanda, M. Quality response of clonal black tea to nitrogen fertilizer, plucking interval and plucking standard. *J. Sci. Food Agric.* **2000**, *80*, 439–446. [CrossRef]
34. Owuor, P.O.; Obanda, M.; Nyirenda, H.E.; Mphangwe, N.I.; Wright, L.P.; Apostolides, Z. The relationship between some chemical parameters and sensory evaluations for plain black tea (*Camellia sinensis*) produced in Kenya and comparison with similar teas from Malawi and South Africa. *Food Chem.* **2006**, *97*, 644–653. [CrossRef]
35. Yao, L.; Liu, X.; Jiang, Y.; Caffin, N.; D'Arcy, B.; Singanusong, R.; Datta, N.; Xu, Y. Compositional analysis of teas from Australian supermarkets. *Food Chem.* **2006**, *94*, 115–122. [CrossRef]
36. Hara, Y.; Luo, S.J.; Wickremasinghe, R.L.; Yamanishi, T. Special issue on tea. *Food Rev. Int.* **1995**, *11*, 371–545.
37. Ding, Z.; Kuhr, S.; Engelhardt, U.H. Influence of catechins and theaflavins on the astringent taste of black tea brews. *Z. Lebensm. Unters. Forsch.* **1992**, *195*, 108–111. [CrossRef]
38. Ruan, J.; Gerendás, J.; Härdter, R.; Sattelmacher, B. Effect of nitrogen form and root-zone pH on growth and nitrogen uptake of tea (*Camellia sinensis*) plants. *Ann. Bot.* **2007**, *99*, 301–310. [CrossRef]
39. Wang, H.; Zhang, H.; Zhao, Y.; Xu, R. Soil acidification of alfisols influenced by nitrate and ammonium nitrogen level in tea plantation. *Int. J. Agric. Biol. Eng.* **2018**, *11*, 168–172. [CrossRef]
40. Dai, Z.; Zhang, X.; Tang, C.; Muhammad, N.; Wu, J.; Brookes, P.C.; Xu, J. Potential role of biochars in decreasing soil acidification—A critical review. *Sci. Total Environ.* **2017**, *581*, 601–611. [CrossRef]
41. Zhang, Y.; He, X.; Liang, H.; Zhao, J.; Zhang, Y.; Xu, C.; Shi, X. Long-term tobacco plantation induces soil acidification and soil base cation loss. *Environ. Sci. Pollut. Res.* **2016**, *23*, 5442–5450. [CrossRef]
42. Bouwman, A.F.; Boumans, L.J.M.; Batjes, N.H. Modeling global annual N_2O and NO emissions from fertilized fields. *Glob. Biogeochem. Cycles* **2002**, *16*, 1080. [CrossRef]
43. Yan, P.; Shen, C.; Fan, L.; Li, X.; Zhang, L.; Zhang, L.; Han, W. Tea planting affects soil acidification and nitrogen and phosphorus distribution in soil. *Agric. Ecosyst. Environ.* **2018**, *254*, 20–25. [CrossRef]
44. Cheng, Y.; Wang, J.; Zhang, J.-B.; Müller, C.; Wang, S.-Q. Mechanistic insights into the effects of N fertilizer application on N_2O-emission pathways in acidic soil of a tea plantation. *Plant Soil* **2015**, *389*, 45–57. [CrossRef]
45. Willson, K.C. Studies on the mineral nutrition of tea II—Nitrogen. *Plant Soil* **1975**, *42*, 501–516. [CrossRef]
46. Wang, J.; Tu, X.; Zhang, H.; Cui, J.; Ni, K.; Chen, J.; Cheng, Y.; Zhang, J.; Chang, S.X. Effects of ammonium-based nitrogen addition on soil nitrification and nitrogen gas emissions depend on fertilizer-induced changes in pH in a tea plantation soil. *Sci. Total. Environ.* **2020**, *747*, 141340. [CrossRef]
47. Yang, X.; Ma, L.; Ji, L.; Shi, Y.; Yi, X.; Yang, Q.; Ni, K.; Ruan, J. Long-term nitrogen fertilization indirectly affects soil fungi community structure by changing soil and pruned litter in a subtropical tea (*Camellia sinensis* L.) plantation in China. *Plant Soil* **2019**, *444*, 409–426. [CrossRef]
48. Li, X.; Hu, C.; Delgado, J.A.; Zhang, Y.; Ouyang, Z. Increased nitrogen use efficiencies as a key mitigation alternative to reduce nitrate leaching in north china plain. *Agric. Water Manag.* **2007**, *89*, 137–147. [CrossRef]
49. Xu, S.; Zhou, S.; Ma, S.; Jiang, C.; Wu, S.; Bai, Z.; Zhuang, G.; Zhuang, X. Manipulation of nitrogen leaching from tea field soil using a *Trichoderma viride* biofertilizer. *Environ. Sci. Pollut. Res.* **2017**, *24*, 27833–27842. [CrossRef] [PubMed]
50. Maghanga, J.K.; Kituyi, J.L.; Kisinyo, P.O.; Ng'Etich, W.K. Impact of nitrogen fertilizer applications on surface water nitrate levels within a Kenyan tea plantation. *J. Chem.* **2012**, *2013*, 1–4. [CrossRef]
51. Wang, W.; Xie, Y.; Bi, M.; Wang, X.; Lu, Y.; Fan, Z. Effects of best management practices on nitrogen load reduction in tea fields with different slope gradients using the SWAT model. *Appl. Geogr.* **2018**, *90*, 200–213. [CrossRef]
52. Yang, X.-D.; Ni, K.; Shi, Y.-Z.; Yi, X.-Y.; Zhang, Q.-F.; Fang, L.; Ma, L.-F.; Ruan, J. Effects of long-term nitrogen application on soil acidification and solution chemistry of a tea plantation in China. *Agric. Ecosyst. Environ.* **2018**, *252*, 74–82. [CrossRef]
53. Reay, D.S.; Smith, K.A.; Hewitt, C.N. Methane: Importance, sources and sinks. In *Greenhouse Gas Sinks*; Reay, D.S., Hewitt, C.N., Smith, K.A., Grace, J., Eds.; CAB International: Oxfordshire, UK, 2007.
54. International Atomic Energy Agency. *Manual on Measurement of Methane and Nitrous Oxide Emissions from Agriculture*; International Atomic Energy Agency: Vienna, Austria, 1992.

55. Yao, Z.; Wei, Y.; Liu, C.; Zheng, X.; Xie, B. Organically fertilized tea plantation stimulates N_2O emissions and lowers NO fluxes in subtropical China. *Biogeosciences* **2015**, *12*, 5915–5928. [CrossRef]
56. Hirono, Y.; Nonaka, K. Nitrous oxide emissions from green tea fields in Japan: Contribution of emissions from soil between rows and soil under the canopy of tea plants. *Soil Sci. Plant Nutr.* **2012**, *58*, 384–392. [CrossRef]
57. Guthrie, S.; Dunkerley, F.; Tabaqchali, H.; Harshfield, A.; Ioppolo, B.; Manville, C. *Impact of Ammonia Emissions from Agriculture on Biodiversity: An Evidence Synthesis*; RAND Corporation: Santa Monica, CA, USA, 2018. [CrossRef]
58. Stokstad, E. Ammonia Pollution From Farming May Exact Hefty Health Costs. *Science* **2014**, *343*, 238. [CrossRef]
59. Han, Y.; Zhu, T. Health effects of fine particles (PM2.5) in ambient air. *Sci. China Life Sci.* **2015**, *58*, 624–626. [CrossRef] [PubMed]
60. World Health Organization. Health Effects of Particulate Matter: Policy implications for countries in eastern Europe, Caucasus and central Asia. *J. Korean Med. Assoc.* **2013**, *50*.
61. Xing, Y.F.; Xu, Y.H.; Shi, M.H.; Lian, Y.X. The impact of PM2.5 on the human respiratory system. *J. Thorac. Dis.* **2016**, *8*, 69–74.
62. Kloog, I.; Ridgway, B.; Koutrakis, P.; Coull, B.A.; Schwartz, J.D. Long-and short-term exposure to PM2.5 and mortality: Using novel exposure models. *Epidemiology* **2013**, *24*, 555–561. [CrossRef] [PubMed]
63. Liu, L.; Chen, T.; Wang, Z.; Zhang, H.; Yang, J.; Zhang, J. Combination of site-specific nitrogen management and alternate wetting and drying irrigation increases grain yield and nitrogen and water use efficiency in super rice. *Field Crop. Res.* **2013**, *154*, 226–235. [CrossRef]
64. Hamid, F.; Yazdanpanah, M.; Baradaran, M.; Khalilimoghadam, B.; Azadi, H. Factors affecting farmers' behavior in using nitrogen fertilizers: Society vs. farmers' valuation in southwest Iran. *J. Environ. Plan. Manag.* **2021**, *64*, 1886–1908. [CrossRef]
65. Liu, R.; Kang, Y.; Zhang, C.; Pei, L.; Wan, S.; Jiang, S.; Liu, S.; Ren, Z.; Yang, Y. Chemical fertilizer pollution control using drip fertigation for conservation of water quality in Danjiangkou Reservoir. *Nutr. Cycl. Agroecosystems* **2014**, *98*, 295–307. [CrossRef]
66. Tang, S.; Zheng, N.; Ma, Q.; Zhou, J.; Sun, T.; Zhang, X.; Wu, L. Applying Nutrient Expert system for rational fertilisation to tea (*Camellia sinensis*) reduces environmental risks and increases economic benefits. *J. Clean. Prod.* **2021**, *305*, 127197. [CrossRef]
67. Singh, R.P. *Organic Fertilizers: Types, Production and Environmental Impact. Organic Fertilizers: Types, Production and Environmental Impact*; Nova Science Publishers: Hauppauge, NY, USA, 2012.
68. Wang, H.; Yang, S.-H.; Yang, J.-P.; Lv, Y.-M.; Zhao, X.; Pang, J.-L. Temporal changes in soil bacterial and archaeal communities with different fertilizers in tea orchards. *J. Zhejiang Univ. Sci. B* **2014**, *15*, 953–965. [CrossRef]
69. Gu, S.; Hu, Q.; Cheng, Y.; Bai, L.; Liu, Z.; Xiao, W.; Gong, Z.; Wu, Y.; Feng, K.; Deng, Y.; et al. Application of organic fertilizer improves microbial community diversity and alters microbial network structure in tea (*Camellia sinensis*) plantation soils. *Soil Tillage Res.* **2019**, *195*, 104356. [CrossRef]
70. Xie, S.; Yang, F.; Feng, H.; Yu, Z.; Liu, C.; Wei, C.; Liang, T. Organic fertilizer reduced carbon and nitrogen in runoff and buffered soil acidification in tea plantations: Evidence in nutrient contents and isotope fractionations. *Sci. Total. Environ.* **2020**, *762*, 143059. [CrossRef] [PubMed]
71. Xie, S.; Feng, H.; Yang, F.; Zhao, Z.; Hu, X.; Wei, C.; Liang, T.; Li, H.; Geng, Y. Does dual reduction in chemical fertilizer and pesticides improve nutrient loss and tea yield and quality? A pilot study in a green tea garden in Shaoxing, Zhejiang Province, China. *Environ. Sci. Pollut. Res.* **2019**, *26*, 2464–2476. [CrossRef] [PubMed]
72. Liu, R.; Kang, Y.; Wan, S.; Pei, L. Evaluation of methods of nutrient and water management on tea performance and nutrient loss in the Danjiangkou Reservoir area, China. *Arch. Agron. Soil Sci.* **2016**, *62*, 1123–1135. [CrossRef]
73. Han, Z.; Wang, J.; Xu, P.; Li, Z.; Liu, S.; Zou, J. Differential responses of soil nitrogen-oxide emissions to organic substitution for synthetic fertilizer and biochar amendment in a subtropical tea plantation. *GCB Bioenergy* **2021**, *13*, 1260–1274. [CrossRef]
74. He, T.; Yuan, J.; Luo, J.; Wang, W.; Fan, J.; Liu, D.; Ding, W. Organic fertilizers have divergent effects on soil N_2O emissions. *Biol. Fertil. Soils* **2019**, *55*, 685–699. [CrossRef]
75. Debere, N.; Lemessa, F.; Urgessa, K.; Berecha, G. Influence of Combined Application of Inorganic-N and Organic-P Fertilizers on Growth of Young Tea Plant (*Camellia sinensis* var. assamica) in Humid Growing Area of SW Ethiopia. *J. Agron.* **2014**, *13*, 179–186. [CrossRef]
76. Bhattacharjee, R.; Dey, U. Biofertilizer, a way towards organic agriculture: A review. *Afr. J. Microbiol. Res.* **2014**, *8*, 2332–2343.
77. Graham, P.H.; Vance, C.P. Legumes: Importance and Constraints to Greater Use. *Plant Physiol.* **2003**, *131*, 872–877. [CrossRef] [PubMed]
78. Duan, Y.; Shen, J.; Zhang, X.; Wen, B.; Ma, Y.; Wang, Y.; Fang, W.; Zhu, X. Effects of soybean–tea intercropping on soil-available nutrients and tea quality. *Acta Physiol. Plant.* **2019**, *41*, 140. [CrossRef]
79. Koenig, R.T.; Cochran, V.L. Decomposition and nitrogen mineralization from legume and non-legume crop residues in a subarctic agricultural soil. *Biol. Fertil. Soils* **1994**, *17*, 269–275. [CrossRef]
80. Yanni, Y.G.; Rizk, R.Y.; Corich, V.; Squartini, A.; Ninke, K.; Philip-Hollingsworth, S.; Orgambide, G.; de Bruijn, F.; Stoltzfus, J.; Buckley, D.; et al. Natural endophytic association between Rhizobium leguminosarum bv. trifolii and rice roots and assessment of its potential to promote rice growth. *Plant Soil* **1997**, *194*, 99–114. [CrossRef]
81. Noel, T.C.; Sheng, C.; Yost, C.; Pharis, R.P.; Hynes, M.F. Rhizobium leguminosarum as a plant growth-promoting rhizobacterium: Direct growth promotion of canola and lettuce. *Can. J. Microbiol.* **1996**, *42*, 279–283. [CrossRef]
82. Höflich, G.; Wiehe, W.; Hecht-Buchholz, C. Rhizosphere colonization of different crops with growth promoting Pseudomonas and Rhizobium bacteria. *Microbiol. Res.* **1995**, *150*, 139–147. [CrossRef]

83. Biswas, J.; Ladha, J.; Dazzo, F. Rhizobia Inoculation Improves Nutrient Uptake and Growth of Lowland Rice. *Soil Sci. Soc. Am. J.* **2000**, *64*, 1644–1650. [CrossRef]
84. Urquiaga, S.; Cruz, K.H.S.; Boddey, R.M. Contribution of nitrogen fixation to sugar cane: Nitrogen-15 and nitrogen-balance estimates. *Soil Sci. Soc. Am. J.* **1992**, *56*, 105–114. [CrossRef]
85. Trenkel, M.E. *Controlled-Release and Stabilized Fertilizers in Agriculture*; International Fertilizer Industry Association: Paris, France, 1997.
86. Qiao, C.; Mia, S.; Wang, Y.; Hou, J.; Xu, B. Assessing the effects of nitrification inhibitor dmpp on acidification and inorganic n leaching loss from tea (*Camellia sinensis* L.) cultivated soils with increasing urea–n rates. *Sustainability* **2021**, *13*, 994. [CrossRef]
87. Han, W.-Y.; Ma, L.-F.; Shi, Y.-Z.; Ruan, J.-Y.; Kemmitt, S.J. Nitrogen release dynamics and transformation of slow release fertiliser products and their effects on tea yield and quality. *J. Sci. Food Agric.* **2008**, *88*, 839–846. [CrossRef]
88. Du, C.W.; Zhou, J.M.; Shaviv, A. Release characteristics of nutrients from polymer-coated compound controlled release fertilizers. *J. Polym. Environ.* **2006**, *14*, 223–230. [CrossRef]
89. Ibrahim, A.A.; Jibril, B.Y. Controlled release of paraffin wax/rosin-coated fertilisers. *Ind. Eng. Chem. Res.* **2005**, *44*, 2288–2291. [CrossRef]
90. Ni, B.; Liu, M.; Lü, S.; Xie, L.; Wang, Y. Multifunctional Slow-Release Organic−Inorganic Compound Fertilizer. *J. Agric. Food Chem.* **2010**, *58*, 12373–12378. [CrossRef]
91. Ni, B.; Liu, M.; Lü, S.; Xie, L.; Wang, Y. Environmentally Friendly Slow-Release Nitrogen Fertilizer. *J. Agric. Food Chem.* **2011**, *59*, 10169–10175. [CrossRef]
92. Raguraj, S.; Wijayathunga, W.M.S.; Gunaratne, G.P.; Amali, R.K.A.; Priyadarshana, G.; Sandaruwan, C.; Karunaratne, V.; Hettiarachchi, L.S.K.; Kottegoda, N. Urea–hydroxyapatite nanohybrid as an efficient nutrient source in *Camellia sinensis* (L.) Kuntze (tea). *J. Plant Nutr.* **2020**, *43*, 2383–2394. [CrossRef]
93. Wu, Y.; Li, Y.; Fu, X.; Shen, J.; Chen, D.; Wang, Y.; Liu, X.; Xiao, R.; Wei, W.; Wu, J. Effect of controlled-release fertilizer on N_2O emissions and tea yield from a tea field in subtropical central China. *Environ. Sci. Pollut. Res.* **2018**, *25*, 25580–25590. [CrossRef] [PubMed]
94. Pramanik, P.; Safique, S.; Jahan, A. Humic substrates facilitated nitrogen uptake in tea bushes by influencing biochemical and microbiological properties in soil of Northeast India. *J. Plant Nutr.* **2018**, *41*, 2376–2385. [CrossRef]
95. Yang, W.; Li, C.; Wang, S.; Zhou, B.; Mao, Y.; Rensing, C.; Xing, S. Influence of biochar and biochar-based fertilizer on yield, quality of tea and microbial community in an acid tea orchard soil. *Appl. Soil Ecol.* **2021**, *166*, 104005. [CrossRef]
96. Wu, Y.; Li, Y.; Wang, H.; Wang, Z.; Fu, X.; Shen, J.; Wang, Y.; Liu, X.; Meng, L.; Wu, J. Response of N_2O emissions to biochar amendment on a tea field soil in subtropical central China: A three-year field experiment. *Agric. Ecosyst. Environ.* **2021**, *318*, 107473. [CrossRef]
97. Oo, A.; Sudo, S.; Win, K.; Shibata, A.; Sano, T.; Hirono, Y. Returning Tea Pruning Residue and Its Biochar Had a Contrasting Effect on Soil N_2O and CO_2 Emissions from Tea Plantation Soil. *Atmosphere* **2018**, *9*, 109. [CrossRef]
98. Chen, P.; Liu, Y.; Mo, C.; Jiang, Z.; Yang, J.; Lin, J. Microbial mechanism of biochar addition on nitrogen leaching and retention in tea soils from different plantation ages. *Sci. Total. Environ.* **2020**, *757*, 143817. [CrossRef]
99. Yamamoto, A.; Akiyama, H.; Naokawa, T.; Miyazaki, Y.; Honda, Y.; Sano, Y.; Nakajima, Y.; Yagi, K. Lime-nitrogen application affects nitrification, denitrification, and N_2O emission in an acidic tea soil. *Biol. Fertil. Soils* **2014**, *50*, 53–62. [CrossRef]

MDPI AG
Grosspeteranlage 5
4052 Basel
Switzerland
Tel.: +41 61 683 77 34

Nitrogen Editorial Office
E-mail: nitrogen@mdpi.com
www.mdpi.com/journal/nitrogen

www.ingramcontent.com/pod-product-compliance
Lightning Source LLC
LaVergne TN
LVHW070447061126
851858LV00011B/179
* 9 7 8 3 7 2 5 8 8 0 7 9 9 *